WHY IN THE WORLD

ADVENTURES in GEOGRAPHY

George J. Demko

with Jerome Agel and
Eugene Boe

Produced by Jerome Agel

⚓

Anchor Books
DOUBLEDAY
NEW YORK LONDON SYDNEY AUCKLAND TORONTO

AN ANCHOR BOOK
PUBLISHED BY DOUBLEDAY
a division of Bantam Doubleday Dell Publishing Group, Inc.
666 Fifth Avenue, New York, New York 10103

ANCHOR BOOKS, DOUBLEDAY, and the portrayal of an anchor
are trademarks of Doubleday, a division of Bantam Doubleday
Dell Publishing Group, Inc.

BOOK DESIGN BY CLAIRE VACCARO

Library of Congress Cataloging-in-Publication Data
Demko, George J., 1933–
Why in the world: adventures in geography / George J. Demko with
Jerome Agel and Eugene Boe : produced by Jerome Agel.
— 1st Anchor Books ed.
p.   cm.
1. Geography.   I. Agel, Jerome. II. Boe, Eugene. III. Title.
G116.D46 1992
910—dc20   91-39397
CIP

ISBN 0-385-26629-4
Copyright © 1992 by Jerome Agel
ALL RIGHTS RESERVED
PRINTED IN THE UNITED STATES OF AMERICA
FIRST ANCHOR BOOKS EDITION: MAY 1992
10   9   8   7   6   5   4   3

Frontispiece: Earth, whirling 1,120 miles a minute around our nearest star, the Sun, is more romantic, more exotic, more misunderstood than ever. To discern where and how all 5.3 billion of us live, geographers with locational perspectives explore the planet's continuously-changing spatial relationships. The highest form of geographical analysis is predicting how those flows and processes will alter places. (A steel simulacrum of the globe—the 12-story-tall unisphere—has been showered with huzzahs since its unveiling at the New York World's Fair in 1964.) THE BETTMANN ARCHIVE

WHY IN THE
WORLD

CONTENTS

CONTENTS

WHY IN THE
WORLD

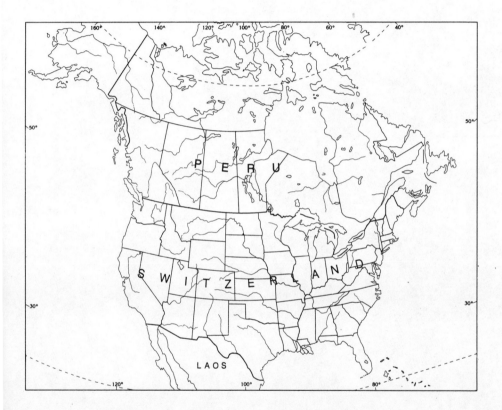

Confused?

Don't be. Geography is never the same from one day to the next. "Only flux and becoming are real," the Greek philosopher Heraclitus of Ephesus observed 25 centuries ago. "Permanence and constancy are merely apparent."

WHAT
IN THE WORLD
IS GEOGRAPHY?

Geography—everybody's favorite subject—turns out to be much more, and much more significant, than many of us realized.

It is much more than knowing where countries like Niger and Sri Lanka and cities like Montevideo and Ulan Bator are located, or knowing the height of the tallest mountains in Britain and Indonesia.

It is much more than knowing what natural resources and crops are indigenous to a region, such as the oil that makes Russia the world's number-one producer and the cocaine that makes Colombia the number-one trafficker, or that Bolivia has the most tin and Canada the most wheat.

It also is much more than knowing where rivers and deserts and tundra are on a map of the world—a map that distorts the configuration of the seven continents and the four oceans and is outdated in many of its other details.

Geography, in other words, is far, far more than an inventory of burdensome information that benumbs the memory without conferring an understanding of the never-ending drama of the interaction of man and planet Earth—the world of incessant change. The year 1991 saw nothing less than an epochal upheaval that will rank as one of the greatest events of the twentieth century: the world's largest country and its Communist party disintegrated. No one is prescient enough to predict what will happen to the late Soviet empire and its beleaguered 280 million people.

Geography—*real*-world geography—is the art and science of location, or place. It is about spatial patterns and spatial processes. It is about which way the wind blows from Chernobyl, the Pacific "ring of fire," AIDS, terrorists, and refugees. It is about acid rain, El Niño, ocean dumping, cultural censorship, droughts and famines, and it is about MiTTs. (MiTTs are pure geography, the measured *mi*nutes of *t*elecommunication *t*raffic—voice, facsimile, and data transmission on public circuits—information flows between places internationally. MiTTs compute what places are connected geographically in what proportion to other places, creating a critically important map of economic and social interdependence. In terms of MiTTs per 1,000 people, Hong Kong leads the growing spatial process of globalism with 56,296, one third of them to the People's Republic of China, which takes over the British crown colony-island in 1997.)

Real-world geography also explores things in locations: why something is where it is and what processes change its distribution. Geography is the why of where of an ever-changing universe. Its surpassing objective is to discover the processes that move over space and connect places and continually transform the location and character of everything.

As a matter of fact, confining knowledge of geography to certain circumscribed areas of factual knowledge is like restricting mathematics to adding, multiplying, subtracting, and dividing. It is like trying to gain a facility with language by memorizing the dictionary. A person confined to the ABCs of geography—K2 is the world's second-tallest mountain, Greenland is the world's second-largest island (surpassed in size only by Australia)—can never progress to knowledge of global phenomena. The world is just too competitive, just too dangerous to be a blur of memorized places, names, and stats. Loading up the brainpan with locational trivia and other random data is as futile as Sisyphus rolling the heavy stone up the hill. Frankly, it is also impossible, given the changes occurring on a globe in ceaseless flux. "My head is a Department of Terrific Trivia," F. Scott Fitzgerald once lamented. "If I just knew anything, I'd be sensational."

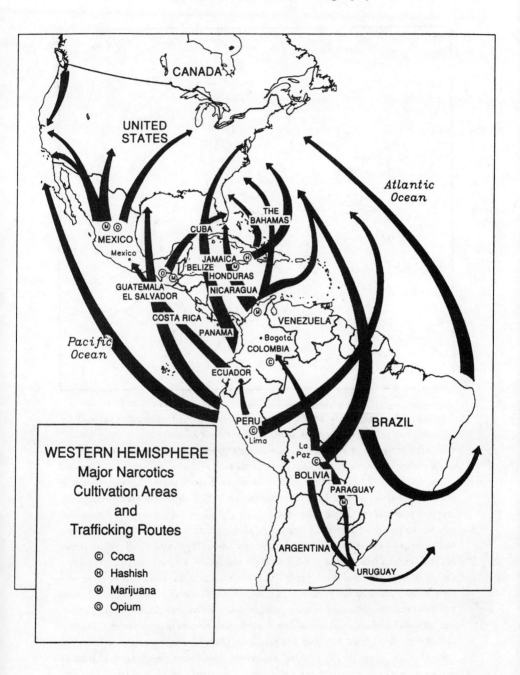

WESTERN HEMISPHERE
Major Narcotics
Cultivation Areas
and
Trafficking Routes

Ⓒ Coca
Ⓗ Hashish
Ⓜ Marijuana
Ⓞ Opium

This is geography: a set of processes uniquely spatial and continually changing—narcotics' places of origin, places of processing, places and routes of shipment, places of destination.

Chernobyl

The spatial flows of air, in both the upper and lower levels, unexpectedly diffused the radioactive cloud west and north from the nuclear-reactor explosion at the Soviet Union's Chernobyl power plant in Ukraine, immediately tipping off European monitors to the world's worst nuclear accident. Air flows in northern Ukraine are invariably eastward across Asia. (It had taken more than two decades for Western scientists to learn details of the explosion at the Soviets' Kyshtym nuclear-waste facility, which created a virtual desert over a wide area of the remote Chelyabinsk region of the Urals; it is today the most polluted spot on the planet.) Chernobyl's radioactive plume spread over the northern hemisphere and as far away as the western United States and Japan. It polluted a half-million people in the Chernobyl region alone and left 300,000 needing annual medical care. Researchers dispute Moscow's official death figure of 31, asserting that the actual toll of those who perished as a result of the "cleanup" is 5,000 to 7,000 and that many thousands more will die of radiation poisoning or related cancer. Just five days after the April 1986 disaster, tens of thousands of unsuspecting children in nearby Kiev paraded through unseen radioactive dust and ash to celebrate May Day and the glories of the Communist reign—a "march of contamination" along the city's central boulevard. (One Soviet expert claims at least 35 million people were contaminated worldwide.) Chernobyl, which is west of the Kiev Reservoir and near Belarus, is today an irradiated ghost town. Some former residents have crept back, preferring to chance it in familiar, though deadly, surrounds.

Human pressures are shaping the planet as never before. Borders and boundaries are vanishing. Millions of refugees are seeking sanctuary in neighboring countries and elsewhere. Forests and farmlands are being laid waste and paved over. AIDS is increasingly rampant throughout the world and diffusing over space to new victims. Homelessness, hunger, and overpopulation are savaging parts of the planet. Everywhere, our food, water, and the very air we breathe are being endangered. These are all geographic processes.

There is a close kinship between geography and history. The historian describes, analyzes, and explains events over *time*—events as well as sequences that are science history, political history, intellectual history, and so on. Historical analyses help us avoid the errors of the past and understand historical processes well enough to predict, even if crudely, the future. The geographer describes, analyzes, and explains anything and everything over *space*—events as well as processes that affect and effect other sets of events and actions. Even an ordinary room is crammed with action and change: movements of dust particles, radio waves, air currents. Every piece of the planet, in fact, is densely crisscrossed by multiple spatial processes. "For as Geography without History seemeth a carcass without motion," wrote the English colonist in America John Smith in 1624, "so History without Geography wandreth as a Vagrant without a certain habitation." The two disciplines are indeed often spoken of in connubial terms, like husband and wife. Geography provides the space within which historical events take place.

I use the word "spatial" frequently. It is a key term in the lexicon of geography. By spatial, I mean related or connected to space. Everything is spatial. Everything has location in space; therefore, everything has spatial meaning. Location becomes the crucial element when places interact with other places.

Space is part of everyday life. Geographers examine what part it plays in how and why we do what we do, where we do it, how we view other places, and how we might better plan and organize our activities over space.

There is no absolutely unchanging space. Nor is there absolute time. Space and time are dynamic quantities in a universe in perpetual motion. They are the fundamental shores on which human activity occurs.

Canals are examples of pieces of space through which spatial processes connecting places occur. The Suez Canal, which was constructed in the third quarter of the 19th century, is a 105-mile-long waterway that brought Europe and the Far East a month closer together. In this century, the 40-mile-long Panama Canal—"the path between the seas"—began saving marine transport up to 8,000 miles.

Geographers, with their expertise on spatial patterns, help plan the layout of new suburbs and the revitalization of cities. They advise on the management and conservation of natural resources. They study voting patterns (over political units) and the migration of "killer" bees (over space). With the analysis of spatial flows and changes, geography provides an understanding of the consequences of events at particular places. It can map the pattern of nuclear and toxic wastes, the spreading population epidemic, the depletion of natural resources—problems global in nature.

Geographers in the good old days were not surprised that the invention of the silk hat led to the formation of many small lakes and bogs in the United States, because there was now a reduced demand for beaver furs and the reprieved beavers set about their normal business of reshaping geography. Earlier, they were not surprised that Manhattan was valuable to the British during the American Revolution. Britain was a naval power, the United States had almost no navy. Britain's command of Manhattan Harbor and the strategic waterways leading from it all but cut the fledgling United States in half. (King George's Redcoats were never driven out of New York City; they marched out in formation two years after Yorktown.)

Ignorance of dynamic geography has placed us at a disadvantage with other countries when it comes to international relations, business, and technology. What happens in that colossus of the marketplace, Japan, indeed affects American jobs, consumer prices, planning—and a multitude of other things. A sound geographic education would also improve knowledge of the billions with whom we share Mother Earth. If there is one thing we know for sure, it is that we are going to become increasingly interdependent and interconnected. Marshall McLuhan described Earth as a "global village." I happen to prefer Wendell Willkie's designation, "one world."

Time and space, especially the flows and links between places, were dramatically altered by construction of the 105-mile-long, sea-level Suez Canal, which cleft east Africa from Asia in 1869, and the 40-mile-long, lock-type Panama Canal, whose opening during the first month of the First World War marked the passage of the first ship through the American land mass. For mariners, the desert swath brought Europe and the Asian subcontinent 5,800 miles closer; the giant cut through Central America—"the greatest liberty man has ever taken with Nature," one observer declared—eliminated an 8,000-mile voyage around turbulent Cape Horn. Mines sown and ships sunk during the Six Day War closed the Suez between 1967 and 1975; dredging now allows passage by aircraft carriers and supertankers. The United States–directed herculean enterprise at the Isthmus of Panama was built as a lock type at President Theodore Roosevelt's command to save the additional five years it would have taken to construct a sea-level canal. Each transit of a vessel through the Panama sends 52 million gallons of water into the ocean. The path between the Atlantic and the Pacific oceans will be turned over by treaty to the Republic of Panama on January 1, 2000. President Jimmy Carter has said that the 1977 agreement "will always be" one of his "proudest moments." (The Senate barely ratified it.) Japan has on the drawing board a sea-level canal 10 miles west of the outdated Panama which could be built in six years.

THE BETTMANN ARCHIVE

"The father of geography" was the blind poet Homer. His ninth century B.C. epic *The Odyssey* is an account of a journey to the edges of the known world. It traces the efforts of Odysseus to return home to Ithaca after the fall of Troy; the adventure covered 10 years and much wandering, because Odysseus kept being blown off course.

The ancient Greeks blessed geography with balance. The philosopher and mathematician Pythagoras (582–500 B.C.) gave it a mathematical tradition; the philosopher Anaximander (611–547 B.C.), a cartographic tradition; Hecataeus (d. 436 B.C.), a literary tradition; the historian Herodotus (484–425 B.C.) considered geography to be the "handmaiden of history."

There are two divisions of geography. One is the physical realm. The shell of Earth consists of four closely interrelated elements—the atmosphere (the air that surrounds us—an invisible ocean—is a mixture of gases, which includes water vapor and dust); the hydrosphere (all the water on the surface of the globe, in the ground, and in the surrounding air); the lithosphere (the solid outer crust of the planet extending to a depth of approximately 60 miles); and the bio-

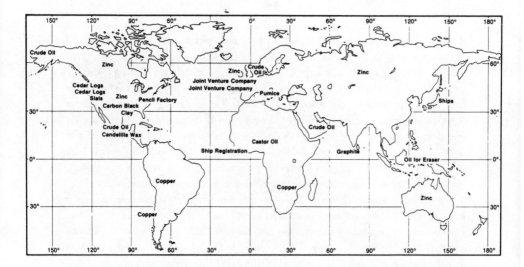

The manufacture of wooden pencils brings together resources from all over the globe. Not to put too fine a point on it, this is another example of geographical connections. (Map by Lawrence C. Wolken, who teaches in the department of finance, Texas A&M University, and is director, summer assembly for global education. His publications include the high school text Invitation to Economics.*)*

sphere (an envelope made up of Earth's water, land, and atmosphere, a 12-mile region top to bottom, where all organisms can exist). The four involve all phenomena of land, sea, and air. The second division is the human realm. Its moods and mysteries and its intercourse with the natural environment present the social sciences and humanities with the challenge of discovery. Today, there is almost a universally accepted view that humans and environment are equal partners in an interacting unit and that *Homo sapiens* is a species that methodically "adjusts" the physical landscape.

The combination of the physical realm and the human realm is regarded by most geographers as providing the core, the very reason for being, of geography. There can be no geography without both. Without geography, knowledge of Earth remains fragmented and partial. Geography, in short, gives us a broad case of earth knowledge. The word "geography," in the literal Greek, means "to describe earth."

The Second World War jolted geography out of its ivory tower more than any other single event in recorded history. The war was an enormous geographic shock for many people who for the first time learned about the great big world out there. The world became smaller as one learned about other places and how those places were connected. Geographers began to participate in the profound changes sweeping the globe and to expand the meaning of the discipline. A series of global processes were initiated: foreign aid and the spread of ideologies, to name two. The end of colonialism began. Cries for self-determination reverberated through the eastern hemisphere. The sun set on the British Empire and the "natives" took back their lands from the imperialists. In Africa, in particular, their sense of identity moved indigenous peoples to return to the earlier names of their countries and cities. Middle Eastern countries took control of the oil under their feet, altering the flows and distributions of wealth over the globe and becoming awesomely rich.

Central America was once "seen" only as an isthmus in a tropical region connecting North America and South America, a series of "banana republics" dominated by American corporations. Modern geographers recognize it as an area beset by endless political and social turmoil, a fascinating blend of Hispanic and Indian cultures, seven small, independent Third World countries seeking development in the western hemisphere.

The new geography opens minds to fresh areas of discovery. Tracing the routes of explorers and immigrants to the "New World," we can fathom the role geography played in the choice of destination. Civil War buffs can gauge how much the roads, rivers, terrain, weapons, social habits, slavery, and communications of mid-19th-century America affected the conduct of the war: *did* field commanders make best use of geographically available resources? The British and now the Japanese developed world power far out of proportion to their available resources. Napoleon and Hitler failed to grasp geographic factors in their attempts to conquer Russia—the uniqueness of patterns of climate, vast breadth of space, history, and patriotism.

We are curious about where places are, and now we are adding a more fruitful compound question, What is there—and why? Why is Silicon Valley in northern California and Hollywood in southern California? ("Hollywood" was once in Fort Lee, New Jersey, and Astoria, New York, in the sense that the first movies were produced there; "the Coast" in those days was the East Coast.) Why are so many countries vying for control of the reefs known as the Spratly Islands in the central part of the South China Sea? Why did the physiography of Vietnam and Afghanistan negate the technologically advanced armies of the United States and the Soviet Union? Knowing where something is located is only the precursor question to in-depth knowledge and new insights.

Geography, in contradistinction to the trivia of names and locations, has nearly universal applications. Agronomists, wildlife biologists, urban planners, and foresters use geographical knowledge to evaluate resources and plan for the future. To function effectively, other professional groups—lawyers, economists, physicists, developers, and anthropologists, to name a few—need basic geographical knowledge in terms of the distribution of natural resources, transport systems, and boundaries. And all of us in general need geography to be informed citizens, to be able to read the front page of a newspaper with intelligence and comprehension—to know, in short, what is happening where and to know why there.

"How can there be anything new in geography?" I am often asked. "Everything has been discovered—hasn't it?"

The "geography" in that inquiry obviously refers to physiography, the physical landscape. At least 70 percent of that world has never been seen. It consists of mountains, valleys, plains, thriving forms of life, and natural resources—all under thousands of feet of water, at the bottom of oceans. The 70 percent is the 98 percent of the still-unexplored ocean bottoms. To those interested in the physical aspects of geography, there is nothing to match the lure of the unknown.

More to the point is the fact that *every* place is always changing and requiring new exploration. Cities increase or decrease in population. Natural catastrophes occur, eviscerating cities, towns, and villages, swallowing buildings, propelling emigration. Virulent diseases and pestilences germinate and take their peculiar paths of destruction. Religions and political parties flourish, spread, and wane, sometimes exerting enormous influence on human events. Crime and terrorism become epidemic. Governments and ideologies are overthrown with breathtaking speed. All of the processes that affect places and people in them must be understood, explained, and coped with—and it's all geography.

There is good news. The United States Congress and geographic organizations promote an annual Geography Awareness Week. From the outset, in 1987, the sponsors have emphasized that a "sound geographic education provides perspectives, information, concepts, and skills to understand ourselves, our relationship to Earth, and our interdependence with other people over the many places in the world. It reinforces and extends the process of critical thinking and problem solving. And it becomes critical when we formulate national policies that rely on imprecise information and on unclear interpretations about our own geography and that of other nations."

No day passes that we don't use geographic principles or feel the effects of geographic influences. We learn about topography and drainage, and such knowledge enables us to build safe, secure, and desirable housing. We learn the whereabouts of good schools and medical facilities and well-laid-out transportation systems that significantly inform the decisions affecting where we should live. We develop progressive public policies. Society requires the participation of a populace reasonably informed about its own constitution and how it relates both to the adjacent communities and to the entire world. Issues such as where to store toxic wastes, where to develop low-income housing, where to build freeways, where to locate huge government-funded supercolliders—they are only a few of the issues that require geographic analysis.

Most of us tend to exaggerate the size of our own country in relation to the size of other countries. Americans are genuinely surprised to learn that one South American country, Brazil, is larger than the continental United States; that Turkey is twice the size of the U.S.'s most populous state, California; and that Zaire, one of Africa's 52 states, is about the size of the U.S. east of the Mississippi River. Twenty states west of the Mississippi are larger than the largest state east of the Mississippi, Georgia.

The well-informed student of geography begins to see a world never seen before. The sense of where things are and how they interact broadens. The planet, by and large, is a rational system, and there are deep, ongoing connections running between all places on the globe.

The boundaries of geography are boundless. They lie beyond infinity. Not even the sky is the limit.

Welcome to the world of geography, the map of human existence.

A
SENSE
OF PLACE

■ ■ ■ ■ ■ ■ ■ ■ ■ ■ ■ ■ ■ ■ ■

The truth in fiction depends for its very life on place. Location is the crossroad of circumstance, a proving ground of what happened, who's here, who's coming . . .

—EUDORA WELTY

E very place on Earth is unique. Each has characteristics distinguishing it from all other places. Geographers usually describe places by their characteristics, both human and physical. Sometimes they group places that are somewhat similar, thereby "creating" regions, such as "the Corn Belt," or the "American South," which is a cultural region that is changing rapidly. Regions can be created from any set of places with something in common.

It is hard to think of any place that is untouched by human contact, even the cruelest, most hostile environments, such as the glaciers of Antarctica and the broiling, waterless Sahara. All places bear the imprimatur of human visits and habitation, or the vestiges of such connections.

Few of us in the West envy the peoples of Third World countries, their hardships and often-hostile environments. But these peoples are usually in touch with their indigenous places in a very real, a very physical sense. Changes in locales, whether in vegetation and agricultural methods or available building materials, come slowly. These peoples rarely leave their corners of the world. Neither are they uprooted by choice or necessity, as are so many Westerners. (The average American moves his residence 18 times in his lifetime.) They remain relatively unaware of the variety of places that lie outside their borders. Billions of people have never heard of the United States.

People often say of their childhood, "We were poor, but we didn't know it." They may not be exaggerating. But what they are really saying, I think, is that they knew who they were, they had the security of belonging somewhere. Why else do so many of us want to go home again? When we give ourselves to a place, we put it on, the surroundings included, as if it were our very own clothing. We are truly "in place."

Too many Westerners, in their getting and spending and laying waste, have lost their understanding of place in any meaningful sense. Oh, we may remember our hometowns, the house we grew up in, the schools we went to, where we went to college, where we got inducted into the armed services, where we proposed marriage and spent our honeymoon. We may be swept away on a tide of memory if we revisit these scenes, recollecting beloved faces and good times long ago. But we do not have that sense of the totality of place that is the essence of geography. We have static images of places frozen in time.

"One place comprehended," Eudora Welty remarks, "can make us understand other places better. A sense of space gives equilibrium, a sense of direction."

All places change. They change in themselves and they change relative to other places, and they may cause change in other places. We may imagine there are certain places magically untouched by time or change. But we have to turn to literature to find Shangri-la and Brigadoon.

Maps can conjure up images of strange and enchanting landscapes, unintelligible languages, and rainbow-hued peoples. Tweakings of the imagination are the magic of geography, a magic that stays with some of us forever, enriching our lives as we travel, observe, work, and dream.

Political leaders can also be guilty of primitive locational ignorance. President William McKinley, as commander in chief, ordered a United States fleet of six warships into Manila Bay during the "splendid little" Spanish-American War, in 1898. "God told me to take the Philippines," he later said to news reporters, but confessed he had had no idea where the Philippines were. Only the incredible ineptitude of the Spaniards and the phenomenal luck of the Americans kept the war from stretching into a struggle as long and as full of disasters as the Boer War became for the British. With the acquisition of the 7,100-island archipelago approximately 500 miles off the southeast coast of Asia, the United States became a world colonial power. LIBRARY OF CONGRESS

Returning to the old homestead rouses mixed emotions and nostalgia. So much has changed, even street names. The vicissitudes of the elements and time itself tend to alter familiar characteristics. Sensitivity to place is one of the hallmarks of a great civilization, but as Thomas Wolfe observed, you can't go home again. By home, one assumes he meant more than a structure or a street. JESSICA JULIAN

Places and their contents, and the processes that continually change places and their contents, are the wondrous ingredients of geography and the seductive potion that excites minds and imaginations. All places, even after they have been found, described, and added to maps, can still thrill us in their rediscovery and reexploration.

We can search for an understanding of why places are continually changing and try to predict how they will change. There can be no doubt that the Germany east of the former Berlin Wall and barbed-wire boundary will be a different place in 1999 than it was in 1989. The spatial processes bombarding the former East Germany—migration, the flow of capital, the flow of ideas—will transform its towns, villages, and cities, farms, landscapes, population, and industry. They will require new explorations by all of us to know it again.

I don't believe anyone has written more sensitively about the sense of place and the inevitability of change than the novelist Willa Cather. In *My Antonia,* two characters reminisce about their childhood in a town in Nebraska. They recall the burning summers with everything green and billowy under a brilliant sky and the smell of heavy harvests and the ferocious winters when the whole country was bare and gray as sheet iron. "We agreed that no one who had not grown up in a little prairie town could know anything about it," one of the characters remarks. "It was a kind of freemasonry, we said."

Civilization, as we know it, is the poorer when we lose the sense of place. Every piece of space is unique. Processes over space and time keep them so. A sense of place is central to our very comprehension of the world, "this spherical universe wrapped layer around layer with the cunning of nesting dolls."

"YOU MEAN, BOSTON'S *NOT* IN TIBET?"

■ ■ ■ ■ ■ ■ ■ ■ ■ ■ ■ ■

We have all seen the alarming statistics:

—Forty-five percent of Baltimore's junior and senior high-school students could not locate the United States on a map of the world.

—Thirty-nine percent of Boston's high-school graduates could not name the six New England states, and on a map of the world some of the youngsters placed Boston in Tibet.

—Twenty-five percent of high-school students in Dallas could not name the country directly south of Texas.

—Forty percent of high-school seniors in Kansas City, Missouri, could not name even three of South America's dozen countries.

—Sixty-nine percent of 2,200 students in North Carolina could not name a single African country south of the Sahara.

In a geography quiz administered to students in many countries, Americans ranked at the bottom.

"Do you realize," one teacher is said to have put to her class, "that we Americans rank 19th worldwide in our knowledge of geography?"

"That's okay," one of her students responded. "As long as we're still in the top ten."

Much of our place-name "geographical illiteracy" is attributable to the fact that for some time geography had all but disappeared from the curriculum of our secondary and high schools. In a survey of 3,000 private and public high schools conducted by the United States Department of Education, geographic skills were described as feeble and disturbing. There was little ability, for instance, to interpret and analyze geographical facts as they relate to environment.

On balance, I would have to say I am in favor of the bees sponsored by the National Geographic Society every spring. The event is high-powered and spreads the word about basic geography to just about every school in the country, and it really is something to hear youngsters respond correctly to such questions as where Mt. Erebus is, which country has the most islands, and what is the most populous English-speaking country in the world after the United States. The winners and runners-up have at the tip of their tongues a ton of information classifiable as "geography." But such data should be only the prelude to real geographical discovery. One wishes that the winners would use their generous prizes to pursue an education in true geography. They could help to predict terrorist attacks or to map the origin and likely spread of the next plague or to learn what can be done to alleviate famine in Africa or to redirect water supplies. At the local level, attention might be paid to trying to predict school-age population growth for a school district or the future distribution of the aging population.

It is indeed disturbing when a student, or any of us for that matter, cannot locate the major countries and cities on a map of the world. But to the geographer, it is more vexing that most Americans, even our well-educated citizens, do not know the purpose and function of geography. "Where is this place?" "Where is that place?" the questions usually go. A good atlas can answer such questions readily. Rarely are the questions the essential "Why is it there?" or even "How is it connected physically, socially, and intellectually with its neighbors?" Does relative ignorance stem from our physical isolation from Europe and Asia and cultural isolation from even our immediate neighbors, Canada and Latin America?

Time was when geographers mostly explored, and people were excited to hear for the first time what lay over the horizon or deep in the jungle and at the bottom of the ocean. Today, the excitement is discovery of new knowledge about familiar places that are constantly changing.

PEANUTS

Harvard University eliminated its geography department shortly after the Second World War. More recently, the universities of Pittsburgh, Michigan, Northwestern, Chicago, and Columbia have eliminated theirs. None has provided a reasonable explanation as to why, except to save money! Geography, when it is taught at all, has become the neglected stepchild of social sciences. Those who are ignorant of history, we have heard a thousand times, are doomed to repeat its mistakes. I know that those who are ignorant of geography are doomed to make costly errors in judgment about space and location.

A sparkling exception to the waning interest in geography at the college level is Clark University, in Worcester, Massachusetts. One of its early presidents was a geographer. Geography today, notes my friend B. L. Turner II, head of Clark's graduate school of geography, has no subject matter that is special unto itself. Geographers focus on anything and everything but relate their investigations to place and space. A linkage to other disciplines is basic to addressing and answering the why of where. Concentration on why things occur where they do is pivotal to understanding most phenomena.

If colleges required geography for admission, more high schools

---

Trashing the geography department and its curriculum has gone on at major universities in the United States, including Harvard, Northwestern, Columbia, Pittsburgh, and Michigan, resulting in educational privation. The study of geography makes a unique contribution to the understanding of humanity, space, and place, as well as to humankind's relationship to the environment. JANE REED/HARVARD MAGAZINE

would start teaching it, and that would help to close the "geography gap." Inasmuch as American students notoriously place near or even at the bottom in geography test scores, they might even learn where things are simply by learning where on a map are those countries with students who outscored them, and then posing the question "Why there?" The study of geography should be reintroduced into the curriculum before the ninth grade as a separate subject, not as a part of social studies or any other grouping.

Knowing countries, states, capitals, and rivers will always have a place in geography—and in the repertoire of well-informed people. But if geographers can convince the rest of us of the depth and breadth of the subject, geography will assume a greater importance than ever in academic education, and the public will insist that geography be brought to their children in a meaningful and exciting way to serve their needs as fully educated human beings. Why be stunned to learn that Russia's evergreen forests are so vast that they cover an area the size of the continental United States? Or that if the world were miniaturized to a town of 1,000 people, there would be 564 Asians, 210 Europeans, 86 Africans, 80 South Americans, and 60 North Americans; and that 700 of the 1,000 people would be illiterate and 500 would be hungry? Perception tends to be reality.

ONCE

UPON

A

TIME

Earth has been compared to a soft-boiled egg: a shell-like crust, a thick middle layer of semiplastic basalt representing the albumen, and a fiery, partly liquid core equivalent to the yolk.

A glowing fireball at birth about 4.5 billion years ago, Earth's form congealed into a Stygian world from spinning, contracting clouds of dusts and gases. At first the planet was entirely molten rock. Gradually, the heavier elements, like nickel and iron, settled into a dense inner core. Some of the lighter rocky materials—basalt and granite among them—melted, floated upward, and cooled into a thin crust, the mantle around the core.

It rained. Not for just an hour, not for just a day, but for millennia, perpetually, filling sculptured basins, which became oceans.

The surface of the whole world was water. Mount Everest and Mauna Kea were mere rocks under boiling, billowing waters. We were the planet Water, not Earth.

Bit by bit, during the primordial, cooling process, water evaporated, and land emerged and floated around.

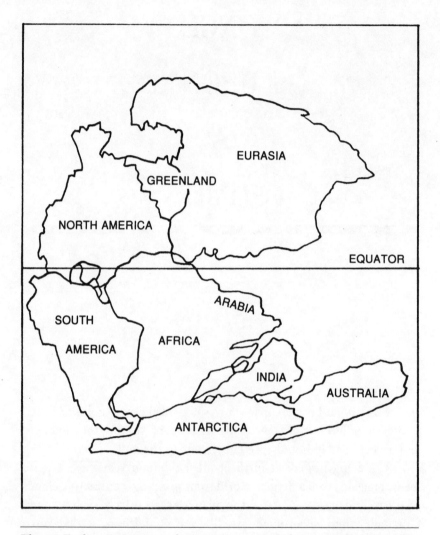

This was Earth once upon a time, the way we were, near the beginning. About 500 million years ago, Antarctica was probably connected to the west coast of North America when both were a single continent near the equator; North America looked like a slice of pizza between Antarctica and South America. About 240 million years ago, today's landmasses were altogether, forming the supercontinent Pangaea—masses of light granite rock floating on denser basalt; temperatures in the interior bounced up and down wildly. As recently as 120 million years ago, South America and Africa were still breaking apart and North America was moving away from Europe; California and much of the Pacific Northwest, Canada, and Mexico did not exist. The seven continents are still moving around—the process is called plate tectonics—setting off earthquakes in relatively predictable places and zones. The oldest rock yet found on the ocean floor is practically a baby; it is around 3.85 billion years younger than the oldest rock found on land, chunks of 4-billion-plus-years-old granite in the Northwest Territories of Canada.

In the fullness of time, the scattered chunks of land in the southern seas came together along the equator, forming one single, desertlike supercontinent with wildly fluctuating temperatures: Pangaea, which is "all lands" in Greek. Antarctica and Australia were linked, and Greenland was caught between North America and Eurasia.

A hundred million years later, the protocontinent began to break up, fractured by shifts in the molten core of the planet. In the process of plate tectonics—the interaction of the constantly moving thick slabs of rigid rock called plates, which make up Earth's hard shell— the fragmented lands inched apart. The slabs, each about 60 miles thick, were like rafts on a sea.

What we now know as South America separated from Africa and floated thousands of miles to the west. India, which had become an island, plowed into Asia and drove the Himalayan range out of the ocean and into the heavens. North America and Europe split up and went their separate ways. Gibraltar was part of a slab that wandered from somewhere north of Algiers and ended up wedged between the colliding continental masses of Africa and Spain. The Mediterranean was once waterless. Antarctica, which, incredibly, once basked in a temperate climate, and Australia broke apart, Australia sailing into the Indian Ocean, Antarctica moving south to the antipode of the globe. Only Africa stayed more or less in place.

Today, the planet, still restless, consists of seven continents, or land masses, of light granite rock "floating" on a denser basalt layer and sustained largely by a sheet of soil derived from, and covering, the rock. Still, water makes up 70 percent of the total surface of the planet.

Tomorrow? Crete will probably knock into Libya. The Farallon Islands off San Francisco will continue to move up the West Coast of the United States. The Indian Ocean could rise a few inches and drown the 1,200 Maldives. The Indo-Australian plate, or slab of rock, which is still colliding with the Eurasian plate, could lift the Himalayas even higher than their present 22,000-plus feet. Africa's Great Rift Valley could sink enough to let seawater pour in, flooding the valley and splitting the already-splintering continent in two. One of the planet's annual million earthquakes and its attendant shudders could send California, which already is breaking up, shuddering toward the Pacific Ocean. The "ring of fire," the zone with most of

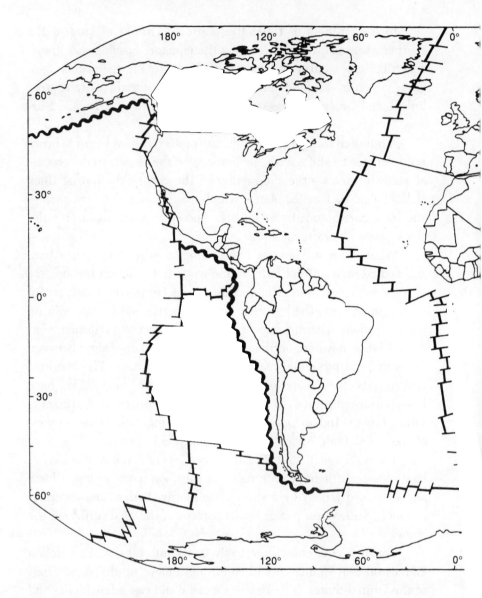

Earth's rocky outermost crust is a series of gigantic, rigid slabs, or plates, which hold both the ocean floor and the continents. They are constantly in motion, creeping and shuffling around, separating, colliding—they recreate physiography and space. The interactions of the slabs, called plate tectonics, unleash earthquakes (the most lethal natural disasters) and volcanoes, and form new lands. Such turmoil led the planet's prehistoric supercontinent Pangaea to come apart and eventually assume the present alignment of continents. In the 1906 San Francisco earthquake, the Pacific Ocean's tectonic plate, which usually is inching

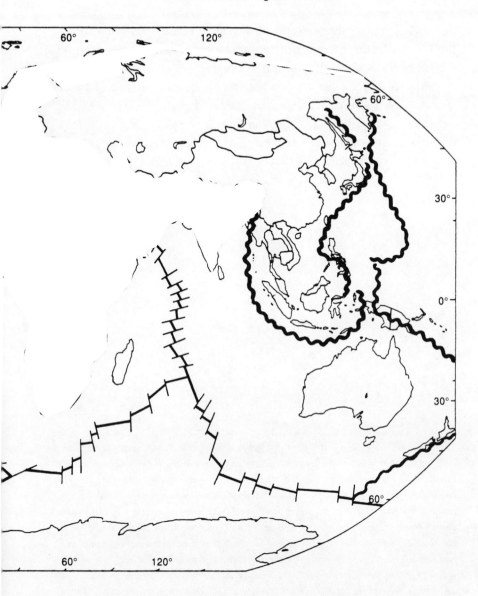

northward, lurched 16.5 feet in moments. In 1970, a 7.7-Richter-scale quake killed
around 75,000 Peruvians and destroyed 186,000 buildings—the worst natural disaster on
record in the western hemisphere. The collision between India and Asia that raised the
Himalayas is still reverberating, its shock waves fracturing the eastern section of the largest
continent and triggering many devastating earthquakes in the People's Republic of China.
About 800,000 died there in a quake in 1976.

the world's active volcanoes and earthquakes, could ignite and reshape the entire Pacific basin, threatening Japan. The plates are always moving, always impacting on Earth's places.

Alaska didn't exist 150 million years ago; it was formed by 50 terranes, groups of rock formations plastered one upon another. The Atlantic Ocean won't exist in 200 million years; you'll be able to drive from the East Coast of the United States (wherever it happens to be at the time) to Africa (if Africa is still where it is today).

Volcanic eruptions dramatically alter topography, landscape, and the flora and fauna cover, unforgettable reminders of the awesome power in motion in the belly of our hot-hearted planet. Reacting to events below the Pacific Ocean—the grinding between water and continental plates—the cataclysmic 1980 Mount Saint Helens explosion in the Cascade range in the state of Washington packed a force 2,500 times greater than the Hiroshima atomic bomb. Debris erupted at speeds up to 400 miles an hour, triggering the largest landslide in recorded history. One side of the 40,000-year-old volcano, which had been silent for 123 years, was blown away and the summit decapitated. Fir trees in a 15-mile radius (4 billion board feet in all) were denuded and devastated. In three minutes, some 232 square miles were totally wasted. (Nearly a century earlier, Indonesia's Krakatau, lying in the collision zone of two tectonic plates, unleashed a series of titanic detonations heard 2,900 miles across the Indian Ocean.) Other Cascade mountains could blow. An eruption far exceeding St. Helens's may occur in Yellowstone National Park in northwest Wyoming and southern and eastern Montana; subterranean hot spots continuously pump heat, keeping the park on a low simmer. Eruptions expel particles into the atmosphere that increase precipitation levels and also create wonderful sunsets. Active volcanoes are not confined to Earth. There are eight continually erupting volcanoes on the Jovian moon Io. The physiography of all nine planets is always changing! THE BETTMANN ARCHIVE

HOME
SWEET HOME

EARTH:	*the third planet from the Sun and bound by gravity to the Sun; the fifth-largest planet in the solar system of nine known planets*
AGE:	*4.5 billion years*
SHAPE:	*oblate spheroid*
TOTAL AREA:	*196,940,400 square miles*
MASS:	*6,585,000,000,000,000,000,000 tons*
VOLUME:	*259,875,300,000 cubic miles clasped by intense belts of trapped radiation*

INTERIOR COMPOSITION:	*principally hot, liquid rock and iron*
SURFACE COMPOSITION:	*70.92 percent water (about 139,628,046 square miles); 29.08 percent dry land (about 57,308,437 square miles)*
THICKNESS OF THE CRUST:	*6 miles*
TILLABLE SOIL:	*6 percent*
ATMOSPHERE:	*gases, principally nitrogen (80 percent) and oxygen (the weight of the atmosphere is equal to the weight of a layer of water 34 feet deep covering the entire planet)*
POLAR CIRCUMFERENCE:	*24,859.73 miles*
EQUATORIAL CIRCUMFERENCE:	*24,901.46 miles*
EQUATORIAL DIAMETER:	*7926.677 miles*
POLAR DIAMETER:	*7899.99 miles*
TILT OF AXIS:	*approximately 66.5 degrees*
REVOLUTIONS AND ORBIT:	*1,120 miles per minute, 67,000 miles per hour, 590 million miles a year around the Sun*

MEAN DISTANCE TO THE SUN:	93,020,000 miles (if the Sun were the size of an orange, the Earth at the same scale would be the size of a pinhead and lie about one foot away)
MEAN DISTANCE TO THE Moon:	238,857 miles
POPULATION:	5,300,000,000—and multiplying
MOST-POPULATED COUNTRY:	The People's Republic of China (1.13 billion)
SOVEREIGN STATES:	173
LARGEST COUNTRY:	The former Soviet Union (8,582,330 square miles)
DESERTS OR SEMIDESERTS:	about a third of the land surface (not including the polar and subpolar "cold deserts")
LARGEST HOT DESERT:	Sahara (3,500,000 square miles)
LARGEST COLD DESERT:	Antarctica (about 5,000,000 square miles)
ACTIVE VOLCANOES:	600
LARGEST ISLAND:	The continent of Australia (2,967,909 square miles)
CONTINENTS:	7 (Asia is the largest, 16,988,000 square miles)

LONGEST RIVER: *Nile (4,145 miles)*

LARGEST LAKE: *Caspian Sea (143,550 square miles; the second-largest, Lake Superior, is only 31,800 square miles)*

DEEPEST LAKE: *Baykal, in southern Siberia (5,715 feet)*

TERRAIN: *forest, steppe, marsh, permafrost, heath, moor, desert, bog, tundra, swamp, plain, prairie, grassland, mountain, canyon, plateau, cliff, glacier, volcano, rock, terrace, savanna, valley, sinkhole, loess, mesa, jungle, hill—all varying over the space of Earth*

MEDIAN TEMPERATURE: *approximately 60° F.*

GLACIATION: *10.5 percent of the land surface*

THE HOTTEST PLACE EVER RECORDED: *Azizia, Libya (136.4° F., in 1922)*

THE COLDEST PLACE EVER RECORDED: *Vostok Station, Antarctica (minus 128.6° F., in 1983)*

THE WETTEST PLACE: *Tutunendo, Colombia (463.4 inches of precipitation annually)*

THE DRIEST PLACE:	*Atacama, on the Pacific coast of northcentral Chile, between Arica and Antofagasta (less than ¹⁄₁₆ of an inch of precipitation annually)*
THE HIGHEST POINT:	*Mount Everest (29,078 feet, in the eastern Himalayas, between Tibet and Nepal)*
THE LOWEST POINT ON LAND:	*the shores of the Dead Sea (1,312 feet below sea level in Israel and Jordan)*
OCEANS:	*4 (the Pacific Ocean alone is larger in area than all the land in the world combined: 64,186,300 square miles and 3,496,000,000,000,000,000, 000 gallons*
SEAS:	32
THE DEEPEST POINT IN THE OCEANS:	*the Mariana Trench (35,810 feet in the western Pacific)*
MOST POPULATED CITY:	*Mexico City (15,000,000)*
MOST PLENTIFUL METALLIC ELEMENT:	*aluminum*
LARGEST CROP:	*rice*

LIVING SPECIES
(APPROXIMATE FIGURES):

BIRDS:	*8,600*
PLANTS:	*300,000*
FISH:	*25,000*
INSECTS:	*2,000,000*
AMPHIBIANS:	*2,500*
REPTILES:	*6,000*
MAMMALS:	*5,000*
PROTOZOA (AMOEBAS, ETC.):	*30,000*
FUNGI:	*80,000*
PARAZOA (SPONGES):	*5,000*
MOLLUSKS:	*100,000*
CRUSTACEANS:	*40,000*
MYRIAPODS (CENTIPEDES, MILLIPEDES):	*11,000*
ARACHNIDS (SPIDERS, SCORPIONS, TICKS):	*65,000*

OLDEST LIVING THING: *a 4,700-year-old gnarled bristlecone pine, "Methuselah," in California's White Mountains*

The matrix of time and space. "All who live under the sun are plaited together like one big mat."—Malagasy proverb. "And we are here as on a darkling plain / Swept with confused alarms of struggle and flight / Where ignorant armies clash by night."—Matthew Arnold NASA

PEOPLE WITHOUT SPACE:
TERRITORY AND CONFLICT

■ ■ ■ ■ ■ ■ ■ ■ ■ ■ ■ ■ ■ ■ ■

All I want is my own country where I can live
freely. You can't imagine what it is like not to have
a state, to be an eternal guest, threatened and hu-
miliated.
—PALESTINIAN WOMAN AFTER THE GULF WAR, *1991*

Territory, or space, has always attracted humankind's attention. It has been divided and bounded and owned from the earliest
times, a fact well demonstrated in history, mythology, and religion.
The Garden of Eden, the promised land of Moses, and Shangri-la
were regions with implied boundaries of significance to specific groups
of people.

There has always been a nearly perfect correlation between
political organizations and geographically defined areas. Even in pre-
historical eras, humans defined their space. Nomadic herders carved
out routes of movement with the seasons. Aboriginal peoples utilized
well-defined areas, separate from other groups. If one group "tres-
passed" into another's, conflict erupted. With the evolution of political / economic systems, individuals and groups shifted dependence
and / or loyalty from a tribal leader or sovereign to a set of institutions
that exercised authority over legally defined, geographical nation-

states bounded by space. The partitioning of the planet acquired an increasingly significant role of rule over space, coinciding with the effective jurisdiction of nation-states.

Space, or place, is one of the most important and pervasive attachments that people and their governments develop. It is this obvious, strong relationship that is at the root of most conflicts among and between groups and states. (Groups can be defined as any collection of people with a strong common identification, which may include religion, language, cultural tradition, and residential patterns. The notion is similar to "nation," but more inclusive.)

The origin of conflict is complex, and often difficult to determine in explicit, conclusive terms. Most intergroup or interstate conflicts have multiple, interrelated causes. Groups most often in conflict are those that differ over race, ethnicity, social class, religion, gender, language, and, of course, territory, or "turf." As a result of perceived or even real inequities in power or wealth, one group may attempt to redress imbalances or seize power under given conditions (e.g., when coercion is removed, the pent-up hostility of one such group is released). The Armenians and the Azerbaijani in the former Soviet Union have different languages and religions, and claim pieces of the same territory. Conflict was not possible under earlier Soviet regimes, but broke out violently under glasnost and Mikhail Gorbachev.

Territory itself can be a causal factor in conflicts. A group may identify with a specific territory and seek to redress wrongs or usurp power. There have been many region-based conflicts in which the population is a mix of ethnic, racial, language, and class groups; for example, the Confederacy in the United States Civil War and South Korea and North Korea today. Incipient regionalism in which the population identifies with one another and perceives a persistent bias from the political center can be found today in northeastern Brazil, in Catalonia in Spain, in the Mezzogiorno in Italy, in Siberia, and especially in certain of the Middle Eastern countries.

Clearly, space can be claimed by two groups or states by virtue of historical attachment. Nagorno-Karabakh has been claimed by Armenians and Azerbaijanis, Kosovo in Yugoslavia has been claimed by Serbs and Albanians. The Malvinas (or Falklands) in the South Atlantic are claimed by the British and the Argentines, and have been a battleground. Root causes can be historical, cultural, economic, elitist, or ideological, and can combine with a desire for long-

The rush for resources by European countries influenced the configuration of resource-rich Africa deeply, and usually negatively. The colonial powers capriciously drew boundaries to suit their own needs while ignoring tribal and economic structures. One result was raging conflict, generally bloody, among many of the renamed African states. Nevertheless, the Organization of African Unity accepted the inequitable colonial demarcations. Individual states, however, often do not accept such historical legacies. Complicating the situation has been the hundreds of tribes throughout the continent; an African's allegiance to his or her tribe in the pockets outlined here is often far stronger than allegiance to country.

term effective control. Given the long and raucous history of political groups, migrations, population displacements, the drawing and re-drawing of boundaries, and the creation and re-creation of empires, countries, and nation-states, there have been, and are, an extraor-dinary number of potential and real territorial-space conflicts around the globe.

The entire territory of the planet is essentially occupied, claimed by political entities with no space for expansion to relieve economic or demographic pressures. This situation, combined with long his-torical periods of occupation of particular territories by one group or another, has created a zealousness on the part of governments, groups, and / or individuals to preserve and protect what they see as *their* space. Some governments can verge on the paranoid when seeking to guard their territory. They create a tense zone around their space, which may include the ocean and certainly includes the space to the top of the atmosphere. As we all know, many, many conflicts have been sparked over the issue of boundaries. The pretext of boundary violation has been used to ignite war. Hitler's Germany violated Po-land's sphere of influence; Hirohito's Japan did the same to China. Individuals and groups evolve as a society with strong identification / ties with *their* territory.

Actual analysis in the search for causal factors of conflict has not been a well-developed area of geographic research. The principal reason is the complexity of most conflicts and the poor data bases available. A study of 310 cases of interstate conflict from 1945 to 1974 revealed that more than half of them took place between adjacent states and that more than 20 percent involved conflict over the location of land boundaries. Border disputes seem to be a distinctive and homogeneous type of conflict; 63 percent of all state conflict between 1945 and 1968 occurred between states that shared boundaries. In all wars between 1816 and 1980, factors relating to regional-territorial systems, that is, local causes, were more important in causation than were international factors or reasons. Obviously, spatial-diffusion processes, that is, the spread over space and places, are significant in border / territorial wars. A study of countries and environments concluded that the history of war and peace is largely identical with the history of territorial changes as results of war. Territorialization of political problems makes for near-irreconcilable conflicts. The evi-dence for the significant role of space or territory in interstate conflict is indeed compelling. Humans have the habit of attaching themselves to space, be it a backyard fence or a miserable, barren, mountainous area between India and the People's Republic of China.

The role of territory in conflicts *within* states is not at all well-documented, primarily because of the vast numbers of such conflicts and disputes, and the difficulty of receiving news about such clashes in remote places with strange names. In 1990 and early 1991, 65 territorial disputes, claims, and demands could be identified *within* the Soviet empire alone.

However, even with limited data on wars and conflicts, a strong case can be made for the significance of territory, or some aspect of territory, as a provoking factor. Just ponder the following types of spatial conflicts seen around the world today:

1. Conflict over sacred, historical, originating territory. The disputed space has psychological significance for two or more groups, or for one group threatened by another group. Examples include the Temple Mount in Jerusalem, a mosque in northern India, mountainous Nagorno-Karabakh in the former Soviet Union, and a golf course in Canada on land sacred to Native Americans.

2. Conflict caused by space or territorial claims by groups, majority or minority, as a result of long-term occupation or settlement. Examples abound: the Tamils in Sri Lanka, the Russians in northern Estonia, the indigenous of New Caledonia, the British in the Falklands.

3. Conflict related to space or territory belonging to one group and infringed upon by another's when, for example, a new state or province is created or a state's internal boundaries are reorganized. The Sikhs in the Punjab and many republics in the former Soviet Union are examples of this type of conflict.

4. Conflict or potential conflict that is caused by a group's loss of territory is exemplified by South Africa's former homeland policies and the deportation of nationalities in the Soviet Union during Stalin's dictatorship.

5. Spatial conflict caused by groups that are ignored or passed over in the process of a country's development. The Kurds stretch across five countries but have no sovereign space of their own. Ab-

original populations in Australia and North America have shared similar fates.

6. Conflicts that arise from the arbitrary drawing of boundaries by external political powers. Such processes were common during the colonial period. Most of Africa and most of the Middle East are deeply afflicted by such spatial colonialism; many current and future conflicts have resulted and will result. The Iraqi claim of Kuwait that started the Gulf War in 1990 can be traced to a colonial boundary drawing.

7. Conflicts that result from the forceful incorporation of a group's sovereign territory by an external, usually adjacent political power. Such actions were common in the period of empires. There are similar events even today: China's incorporation of Tibet, Iraq's attempt to "engulf" Kuwait.

We have entered a period in which political and economic alignments and systems are being restructured rapidly and significantly. Processes have set loose a number of latent disputes; some have already broken out into civil strife between groups that had lived quiescently in the same region for decades. Examples abound from outright war in Yugoslavia, bloody clashes in the former Soviet Union, and severe tensions in Czechoslovakia and Romania, to identify only a few. It is crucial that geographers and others accelerate efforts to understand the origins of these spatial, ethnic, and psychological conflicts if we expect to manage or resolve them. Perhaps we can even learn enough to predict and prevent them.

WHEN IS A NATION
NOT A COUNTRY?

■■■■■■■■■■■■■■■■■■■■■■■■

The Gulf War in 1990–1991 introduced many of us to the Kurds, a people, a nation without sovereign space, who have been long-time residents of Iraq and also occupy land in Iran, Syria, Turkey, and in Armenia and Azerbaijan. They are Muslims but not Arabs. Whoever they are, they seem to be *personae non gratae*. Without provocation, Iraq killed thousands of Iraqi Kurds—with chemical weapons during its war with neighboring Iran in the 1980s—deliberate mass murder of Saddam Hussein's own people. Kurds have been a "people" for centuries, but they are, quite simply, people without a country.

There are generally recognized criteria for defining "country":
1. It must have space, or territory, which must be useful.
2. It must have a more or less stationary population with common cultural, lingual, and historical traits.
3. It must have a government, an administrative system, and political organization.
4. It must have an organized economy, much of which is supervised or policed by the government.
5. It must have a distribution system, that is, a system of spatial interaction in which communications, money, credit, and transportation can move within and without the space, back and forth.

In addition, there are two political prerequisites for "statehood":
1. It must exert sovereignty over the space and people—that is, it must be able to protect and defend that space and system.
2. It must be recognized by other sovereign states.

The word "country" is frequently used interchangeably with the more correct term "sovereign state." The common usage of "nation" confuses the issue of appellation. Most people think of "nation" as synonymous with "country" or "sovereign state." In the strictest political and geographical sense, "nation" refers to a group of people who identify with one another because of a compelling historical, cultural, and lingual similarity. A "nation" may or may not have its own sovereign space. The Kurds qualify as a "nation" because of their commonly shared language and historical experience. They aspire to their own state: Kurdistan.

There are literally thousands of such nations. There are Armenians, Sikhs, Basques, Singhalese, Inuits, Tamils, Palestinians, and Biafrans, to name a few. Some have states but many do not. Many—such as the nations within Russia, the United States, and the People's Republic of China—share sovereign states. Almost every sovereign state, or country, is multinational, in that it includes more than one nation. Sweden has Lapps, Japan has Ainus. The only state that comes close in a true sense to being a single nation-state is Iceland.

Many stateless peoples tend to live in or even create unstable areas. Peoples without sovereignty and who are alienated from a space of their own may invariably breed terrorism, crime, political instability—and refugees.

The Sikhs are sparing little bloodshed to carve out a place of their own in India. Extremist Sikhs dream of an independent state called Khalistan. Delhi has long been apprehensive about the Sikhs, who dominate the Punjab. Many also live in Pakistan. If all the Sikhs united and created their own nation within India, or even within Pakistan, both countries would feel they had a dangerous situation on their doorstep. (India's first woman prime minister, Indira Gandhi, was murdered in 1984 by two Sikh members of her own bodyguard— more than 20 bullets—another killing motivated by religious hatred.)

Native Americans have enriched our heritage by being the source of, among many things, the names of 26 states, 18 of our greatest cities, most of our larger lakes and longest rivers, some of our highest mountains, thousands of smaller towns and natural features—and our respect for the earth. But these earliest settlers, whose lands were wrested from them, have yet to win their space and control of that space that would make them a country. They call themselves a nation (the Sioux Nation, the Navajo Nation, the Chippewa Nation, et al.).

The Tamils have been fighting a long war to secure their own space in Sri Lanka, the majority of whose 18 million population is Singhalese. The Tamils claim they are discriminated against by the Singhalese, who control the top ranks of the government. The Tamils claim a special right to the northern and eastern parts of Sri Lanka, where *they* are the majority population. The Basques in Spain, the Armenians in Azerbaijan, and the Palestinians in the Middle East persist decade after decade without much encouragement in their claims for independent statehood or at least for their own space.

Stateless persons often become a conspicuous majority in someone else's space. Native Catalans, who are descendants of a succession of invading Greeks, Carthaginians, Romans, and Visigoths, declare, "Catalonia is a nation, we have a language and a culture different from Castile, Andalusia, and the other parts of the Iberian Peninsula." The Corsican nationalists on that famous Mediterranean island want independence from France. The Eritrean partisans demand separation from Ethiopia. Turks are still fighting for the division of Cyprus and separation from the Greeks. (Gypsies, by the way, seek no country, though they do hold an international conference every four years to decry the discrimination against them in nearly every one of the many countries in which they are located.) Intense feelings of nationalism—the strong identification of individuals with a group—quite naturally leads to demands for some type of political autonomy and often for space.

Irredentism, a concept related to nationalism, defines a state's demand to recover space with strong linguistic or historical connections. The reuniting of the two Germanys in 1990 is an instance of successful irredentism. Claims by extremist Germans for the return of the former German-occupied areas in Czechoslovakia (Sudetenland) and Poland are dangerous signs of irredentism reflecting latent conflicts.

Chauvinism is nationalism run amok, a blind patriotism characterized by a delusionary belief in the superiority of one's nation or group. In times of crisis or deep division, governments wrap themselves in the flag and play on long-entrenched feelings of patriotism. Such chauvinism, as we have observed in this century, can lead to paranoid biases, creating insatiable hungers for more territory and authority. The Nazis were an excellent example of chauvinism at its worst.

Independent statehood for peoples who feel themselves a country apart can be an almost hopeless uphill struggle, but the effort persists. How do the stateless get to be countries or sovereign states? There are several ways. One is by occupying a given space over a long historical span. As peoples "settle in," boundaries become fixed around the space, and eventually negotiations are entered into with neighbors. States may acquire territory by annexation and conquest. The United States has proved a master of this, aggrandizing the land of Native Americans and former colonial powers. Concession of space can be voluntary, and accretion of space can occur by natural or political processes.

The newest members of the United Nations, such as Brunei Darussalam, Liechtenstein, and the Baltic states are uniformly small. But being small need not be a handicap. As J. B. S. Haldane, the prominent British biologist who lived in India, observed, every organism has its own size. Being without natural resources need not be a negative factor in this age of complex technology. Tiny Singapore proves that economic development to a high level is possible by exploiting location and an educated population. The Netherlands and Switzerland, both small countries, are highly developed and wealthy by world standards. (Switzerland, with its rigid neutrality, is the world's only prominent country that is not a member of the United Nations.)

Is there an optimum-size state? Not really. Switzerland, as noted, is a small state that works superbly. The United States is a large state that, on balance, works. The behemoth former Soviet Union vividly proved that vastness does not always work; once an inefficient authoritarian regime was removed, the empire crumbled. A country's history, demography, population, homogeneity, and location, more so than its size, can be the major reasons for success or failure. India, a very large country, is not really a success in either economic or political terms.

One characteristic shared by sovereign states is ecumene. Ecumene is defined as effective economic-political territory—that part of a state's sovereign space fully integrated into the economic, political, and cultural life of the state. Just about every bit of space in the United States today is effectively in the ecumene. In frontier days, the United States west and northwest were "outside." Today, only bits and pieces, like Appalachia, are still "outside." Some countries, like Brazil, are trying to tie in remote regions to the rest of the state; the effort seeks to integrate the resource-rich Amazon into the ecumene. Siberia on the eastern frontier was integrating into the ecumene of the U.S.S.R.

Some geographers consider the literal shape of the state to be a key factor in its being. Elongated, attenuated countries like Chile, The Gambia, and Vietnam have problems of communication and development because of their configuration. Many countries like Belgium, the Netherlands, and England are compact entities. There are prorupt countries—countries with an extension, a corridor, or a panhandle leading away from the main body of the territory; such a configuration is a detriment to development, and usually results in a piece of space left vulnerable to outside forces. Thailand and Namibia are prorupt countries. Oklahoma is a panhandle state.

Politically, irregular pieces of space come about through unusual, if not unique, arrangements. There are militarily occupied areas, such as Israel's West Bank. Before the reunification of Germany, West Berlin was an enclave of the Federal Republic of West Germany, lying inside East Germany.

Arbitrary partitions of states exist, such as those that separate South Korea from North Korea and those divisions that earlier separated the two Vietnams and the two Germanys. There are quasistates like Andorra, Vatican City, San Marino, and Monaco. There are protectorates from colonial leftovers, like Macao and Puerto Rico. By treaty, Hong Kong will continue to be a British protectorate until 1997, when it comes under the sovereignty of the People's Republic of China.

In some areas of the world, where tension exists between major powers, a common practice is to create buffer zones, or strips of space, that act as warning areas or "breathing space." These are

In 1913, seven European colonial powers—France, Britain, Germany, Portugal, Spain, Italy, and Belgium—had sovereign control of almost all of Africa, the world's second largest land mass. The only independent states were Liberia, the black patch on the Atlantic coast, and Ethiopia, the black mass on the Horn. Today, there are 52 sovereign countries.

spatial situations that relate to contemporary political activity, geo-concepts in the contemporary sense. For years, eastern and central Europe was the buffer zone between East and West in the Cold War. Demilitarized zones separate the Koreas.

Boycotts, blockades, and embargoes are barriers developed by antagonists to limit an enemy's freedoms and thereby weaken him. They are also erected to isolate pieces of space and states from other spaces and states.

The long-popular "domino theory" is a spatial concept that lacks validity. It became an obsession with United States policymakers justifying American involvement in the Vietnam civil war. If American troops were withdrawn, the theory went, South Vietnam would be captured, and the remaining countries of Southeast Asia would fall one by one, like dominoes, to Communist forces. In the early 1960s, President John F. Kennedy's chairman of the Joint Chiefs of Staff, General Maxwell Taylor, told a House subcommittee that if the United States pulled out its forces from South Vietnam, the U.S. would soon be pushed out of the western Pacific and back to Honolulu.

The United States has also withdrawn from its suspect policy of encirclement or containment, the strategy of employing spatial barriers, controlling space around an enemy through military bases and aid to friendly countries. This strategy was applied particularly by the United States to prevent Soviet expansion in the post–Second World War era. Regional integration is a more recent concept, with promising potential. The EC (European Community) brings 12 countries together into one regional, economic, spatial, integrated powerhouse, but it must still confront the issue of political integration. The EC has joined with the 7-member European Free Trade Association to form the European Economic Area, the world's largest trading block, embracing 380 million Western European consumers from the Arctic Circle to the shores of the Mediterranean. Now that's a *powerhouse!*

Colonialism was an economic-political-geographic process that was nothing more than the establishment of sovereignty over areas not part of the colonizing country. The lust for gold, glory, and God was the driving force for European expansion and colonialism. Western European countries became the first political economic powers to go far afield in search of resource-rich and easily grabbed pieces of space. South America and, later, Africa proved to be magnets for the raiders, Africa becoming "the white man's burden."

In an earlier era, the Romans were colonial powers, as were the Persians. The religious polity of Islam spread itself around the world in the 11th, 12th, and 13th centuries, though it was not until Spain had defeated and routed the Moors in January 1492, that Columbus was able to convince the monarchs that Spain should pursue its destiny as a mighty colonial power.

Colonialism of adjacent polities is not uncommon when one state simply crosses a border and gobbles up a neighbor. The United States, emboldened by dreams of manifest destiny, helped itself to a huge chunk of Mexico. Russia and its successor, the Soviet Union, the People's Republic of China, Austro-Hungary, and Nazi Germany have also been space-grabbers. Overseas expeditions to occupied or unoccupied space, often associated with exploration and the search for new lands, led to extended colonialism over large spatial divides. England, Portugal, Spain, France, Belgium, Germany, and the Netherlands were indefatigable explorers and colonizers, propelled by the lure of overseas resources. Imperialistic colonialism is based on nationalism, geopolitics, and missionary zeal, but it almost amounts to the same thing: the establishment of political and economic rule over another's space.

Decolonization came about through a number of processes. Presumably, enlightenment was one of them. The spread of democracy after the devastation wrought by Hitler was another. Generally, wars, with their costly sacrifices, inhibit further colonial exploits and create a resistance to them in hitherto subdued peoples. (Over the sweep of history of the past couple of centuries, I believe that the creation of the United States probably has had one of the most profound influences on prompting have-not peoples to seek the accreditation of real statehood and have their voices heard in world affairs.)

Classic colonialism, to say the least, was an interesting phenomenon. It made monarchies wealthy and powerful beyond their wildest fantasies. A dismantled British Empire to this day benefits from the legacy of economic exploitation, even though the freed peoples may still be impoverished. Colonial powers geared their activities toward selfish development rather than toward the well-being of the colonial spaces as a whole. Today, most colonies are gone or are on the way toward dissolution. But there is still "economic" colonialism. Major economic powers can influence—sometimes actually control—the economies of poorer states. Multinational corporations have enormous influence over some states. It is one of the more interesting and complex issues of the day—indirect economic colonialism.

Many of the world's bitterest disputes have occurred when colonial powers drew boundaries dividing nations, groups, tribes, cultures, resources, and communication systems. The folly of that arrogant practice has come home to roost, particularly among the newest African countries. To understand Third World hostility toward the First World, one has only to consider the rapacious thrust and brutal insensitivities of colonialism.

The world's states fall into four groupings. The First World, the one the United States inhabits, consists of the developed Western democracies. The Second World is the designation for states with developed socialist economies, such as North Korea and the People's Republic of China. This group's membership is rapidly changing as a result of the new directions taken by Mikhail Gorbachev in 1985. The Third World signifies those countries that are in varying stages of development, such as India, many of the newly independent African countries, and several Latin American ones. The Fourth World is the status conferred upon peoples like the Kurds and the Sikhs— stateless peoples in search of statehood.

The World Bank classifies countries by economic development, especially gross national product. It places the oil-exporting countries into a category of their own because of their high gross national product. If they did not have oil-derived revenues, these states would have little influence, because they are otherwise underdeveloped. Some Third World countries are classified as NICs—Newly Industrializing Countries; they include the Asian "Four Tigers," or "Little Dragons": Singapore, Hong Kong, Taiwan, and South Korea, all with impressive, rising economies.

Characteristics of Third World countries can change. They employ an equity policy to stimulate the growth of lagging regions so that large chunks of territory don't languish in economic penury or inequality. Capital grows and multiplies. The core expands. The ecumene expands and grows. Frontiers disappear. The undeveloped edges of the ecumene develop and are absorbed. But when the state expands to the margins of its sovereign space, boundary disputes can arise.

Toward the goal of a people becoming a sovereign state, economic development must keep pace with cultural and political development. As a state evolves from a core area with a capital city and expands its economic, cultural, and political processes into the undeveloped space of the state, the ecumene expands to the edges, to the boundaries. The United States, for example, started as a strip along the eastern seaboard, then gradually diffused its hegemony over space to beyond the shores of the Pacific Ocean. Economic growth dominates in the beginning; economic development follows. There is an analogy with human development. Children start out life uncoordinated and clumsy. As they play and grow, they become coordinated, they make connections, they flex their muscles, they become fluid and athletic. Countries do much the same thing.

POLITICAL SPACE
AND POWER

■■■■■■■■■■■■■■■■■■■■■

Of all the subfields of geography, modern political geography has become the most ideological. There are geographers whose approaches to research are radical, capitalist, communist, nihilist, determinist, and even fascist.

Soviet-style Communism was an attempt at global domination. Party-controlled states expanded in numbers during the decades after the Second World War. But the first and most powerful of these, the Soviet Union, was not able to impose its "system" irrevocably on the countries of Eastern Europe—military occupations, at times brutal, notwithstanding. With vanishing support from Moscow, a doddering Cuba may expire in its own rhetoric and the civil-warring sub-Saharan countries may suffer veritable genocide.

In this century, chauvinism, which evolved into determinism, has proved to be one of the most pernicious of the geopolitical ideologies. During the flowering of Nazism—the 1,000-year Reich that lasted 12 years—there were German geopoliticians who argued that the state was analogous to a biological organism. Like any other living organism, it must have nourishment and living space in order to bloom and grow. Determinism became the goad to Hitler's crazed ambition, the need for more space, *Lebensraum*—Germany couldn't breathe

without more territory. He "needed" nothing less than all of Europe—
and then the world. The rationale provided by Germany's geopoliticians in the Führer's lawless territorial expansion cast disrepute on
the field of geopolitics. For years, the very identity of geopolitics had
to be changed to geostrategy within the disciplines of geography.

An American admiral and historian, Alfred Thayer Mahan
(1840–1914), was convinced that sea power was the key to world
domination. The great waterways would determine the wealth and
status of countries. The skipper pointed to sea power as the linchpin
in the supremacy of the British. The impotence of the Russians, he
believed, was due to the lack of a port on an ice-free navigable sea.
Faith in Mahan's geopolitical notions gave the United States the

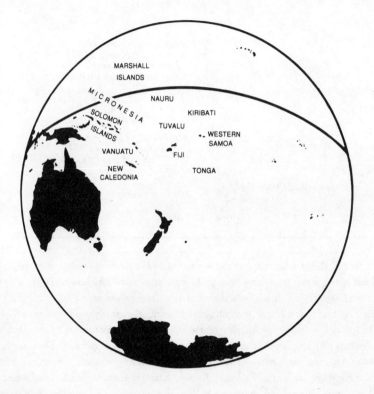

*Our planet could just as well have been named Water. Water, after all, comprises 70
percent of the surface of Earth. One of the four oceans—the Pacific—is 12 million square
miles larger than all the land area in the world put together. Space is the most important
resource of our oceans, the only saltwater oceans in the solar system, as far as we know.
The oceans, like the land, vary over their space remarkably, differing in depth, salinity,
fish, and resources.*

impetus to seize the Philippines in 1898 and to increase its coastal
and maritime presence wherever it could.

Around the turn of the century, the English geographer Sir
Halford John Mackinder (1861–1947) wrote an influential essay con-
tending that world power was characterized by a recurring conflict
between landsmen and seamen, a land-water antagonism. He saw

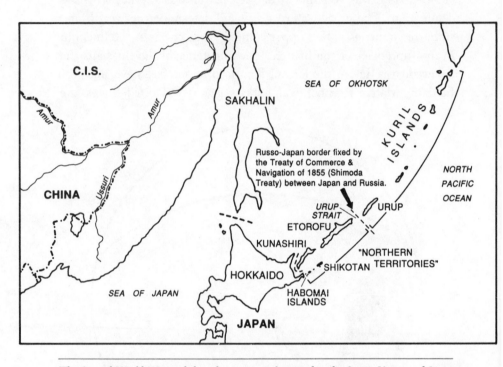

*The Second World War ended in the summer of 1945, but the Soviet Union and Japan
did not sign a peace treaty and technically are still at war. The* raison d'être: *a bitter
territorial dispute, which has been described as a disagreement over a microscopic bit of
territory of negligible strategic value to either country. Moscow refused to return to Tokyo
several islands and a group of islets in the volcanic Kuril chain north of the Japanese
archipelago which were seized at gunpoint in the closing days of the war. There are now
military bases and nuclear-armed submarine ports on these "northern territories," which
form a buffer for the Sea of Okhotsk. Relinquishing the territories could limit access to
the open seas and anger the military and Russian nationalists who consider it "Russian"
space. (Passages through the chain are considered to be international straits.) In 1991,
Moscow agreed that resolving the technically obscure topic was a proper subject of talks,
which are arduous, and said that its armed forces on the islands would be reduced. Return
of the territories would be conducive to desirous Japanese economic aid and investment in
Russia. Every February, the Japanese celebrate Northern Territories Day—it is a very
emotional piece of territory for them.*

the ascendancy of land control—now that railways had connected the land and the water—as the key to dominance. He placed the center of land power in the middle of Eurasia, between the Caspian Sea and Afghanistan. After the First World War, Mackinder saw the geographical pivot of history as shifting eastward from Eurasia to the heartland of Central Asia. Before he died, after the Second World War, he was having second thoughts about his land-sea theory.

To contemporary geopoliticians, the Arctic presents a vista of considerable complexity. Lying to the north of the 65th latitude, the Arctic is rich in resources: fish, fur, timber, petrochemicals, oil, diamonds, coal, nickel, and gas. It occupies 15 percent of the land area of the planet, but only three tenths of 1 percent of the world's population lives there. In spite of its sparseness of population, this remote space could play an increasingly vital role in our lives by the early part of the next century—politically, economically, geographically. Crossing this territory provides the shortest air route among the countries that border its rim: Canada, the United States (Alaska), Russia, Finland, Norway, Sweden, Denmark, and Iceland, as well as Greenland. But developing the "top of the world" is not going to be an easy feat. Conferences and negotiations are being held to try to enhance cooperation in the use of Arctic resources while safeguarding the environment and banning nuclear activities.

Political geographers are interested in the power arrangements around the world. Why do some countries come to the fore while others recede to secondary roles globally? At one time, Venice was the most powerful city-state on earth; today, it can hardly keep its head above water. The Spain of Columbus and the conquistadors had its moment in history, then faded. In our time, we have seen the diminution of the colonial powers. And now we are hearing on all sides, "The United States is no longer Number One." Who, then, is Number One? Japan? Germany? The People's Republic of China? Or is there no longer a Numero Uno? Could it be that the coming coalition of European sovereign states will be Number One? The "new world order" is unclear but it will be determined by political, geographical, and economic processes: geopolinomics!

At the beginning of this century, eight great powers dominated the world: Austria-Hungary, Germany, Great Britain, Japan, France,

Gerrymander: to create an electoral district in such a form as to give unfair advantage to the political party manipulating space. New districts may be created with bizarre loops, legs, and zigzags. The first gerrymander: the fancied resemblance to a salamander (made famous by caricature) of the irregularly shaped outline of a state senatorial election district in northeastern Massachusetts that was formed for partisan purposes, in 1812, with Governor Elbridge Gerry's complicity; thus, a gerrymander. (A historical aside: Gerry, a Founding Father at the Constitutional Convention in Philadelphia in 1787, was the second United States Vice President to die in office.) To politicians, mapmaking is lifeblood. In 1964, the Supreme Court of the United States ruled that both congressional and state legislative voting districts must be of approximately equal population.

Italy, Russia, and the United States. Clearly, the power arrangements have altered significantly since then. Power is fleeting. The question arises, how is power acquired and retained?

If every country had similar political and socioeconomic systems and the same advantages in resources, location, and sensibility, there still would be significant differences in power among the states—a pecking order. We may not know for certain in which direction power among the countries will flow, but we can predict that geographical variations in power will occur.

There are specific measures that geographers have used to help abort the arms race. Spatial simulations projecting climatic and other zonal impacts have measured the probable extent and intensity of destruction of nuclear war, worldwide and country by country. Such forecasts have influenced public opinion and military expenditures.

Geographers provide facts on the impact of military operations on the nature, population, and economies of many countries. While some geographers have been studying nuclear winter, others have been applying themselves to political and economic issues to help ease global tension and to promote peace. They monitor such situations as maritime disputes, terrorism, and international tension that could lead to political and economic coercion and the force of arms.

Modern political geography runs the gamut from the international level to neighborhood concerns. It is quantitatively analyzed by mathematical models and computer technology. Of late, it has concentrated on local and national issues. A hypothetical study, but all too rooted in reality, of vested interests in an American city vying for power over space might go like this: an interstate highway has to be routed across space within a city, and obviously no one in its proposed path finds this a desirable prospect. The citizens with political clout—the well-organized and well-heeled (and statistically white)—apply pressure in the appropriate places to reroute the interstate far away from them. The NIMFY (Not In My Front Yard) Syndrome, as some call it (or NIMS—Not In My Space!).

The political map of the United States is complex and naturally in flux. There are ever-changing variations in electoral power and party strength nationally, statewide, locally. More than one recent presidential election has turned up seemingly curious voting patterns. Governor George Wallace, whose appeal was principally to southern racists, showed surprising strength in certain precincts in Ohio and Indiana when he was a candidate for the Democratic nomination in 1972. The populous, ethnically heterogeneous northeastern states, once the bastion of the Democratic Party, have in recent presidential elections, with few exceptions, given a plurality of their votes to Republicans.

It is fascinating to trace the politicizing of a particular issue from the local level to the regional level to the national level and finally to the international level. Many of the world's concerns, among them environmental matters, are politically significant and have significant spatial implications. Most environmental problems respect no boundaries; thus, international treaties are required to address them.

The ravaging of a rain forest in Madagascar robs the globe of a unique source of a component essential for a certain medicine. An accident in a nuclear-power facility discharges radioactivity into the winds, which carry it across continents. Some parts of the world have ignored expediency to declare themselves nuclear-free zones; New Zealand, for one. Its relations with the United States have become strained, because American naval skippers neither confirm nor deny that their vessels are carrying nuclear weapons. Proclaiming one's space or place a nuclear-free zone becomes a political process by which a sovereign state like New Zealand can limit nuclear activity in designated places in space.

The countries of the world, as we know, start out unequal in resources. But do resources determine whether a country is going to be rich or poor? Resources go through a spatial process in which they are exchanged, traded from one place to another. If an enterprising region or country lacks a needed resource, it tries to get it elsewhere. More often than not, one region or country shrewdly bests the other in the exchange. Being in the right place can be advantageous. Singapore has promoted its port and location on the trade routes of the world to enrich its people.

A country's wealth can also be measured by its moral leadership, its ideas, its science, its technological expertise. For example, states with natural resources such as oil can control political and economic events for a time. It was not long ago that the OPEC countries had

---

Oil is the fuel used for nearly 50 percent of the world's daily energy needs. It is both cleaner and cheaper than coal and easier to transport than gas. Unfortunately, it is geographically maldistributed, a lot of it being in the Middle East and the former Soviet Union. Conflicts involving oil-producing countries provide valuable geography lessons. In 1951, Premier Muhammad Mussadegh nationalized BP's Iranian holdings. In 1956, Egypt blocked the Suez Canal; turmoil piled up. In 1967, the Six-Day War. In 1973, another Israeli-Arab war was followed by quadrupled oil prices. In 1979, the Iranian revolution. In 1990–1991, invasion, sanctions, and war in the Persian Gulf. Yet, developed countries continue to rely on the goodwill of Arab countries rather than decrease their dependence on volatile sources and resources. Liberating Kuwait from Iraqi domination in February 1991 was an attempt to maintain a tenuous political stability in the Middle East. New pipelines promise an easing of dependency on tanker traffic through the Strait of Hormuz. The movement of the world's capital—petro dollars—to the oildoms has been an extraordinary process. At present consumption rates, known oil reserves will be exhausted by the middle of the 21st century. No new Saudi Arabia exists; major new deposits, if there are any, will be found only under water.

just such a stranglehold on industrial powers. But thanks to the self-restraint practiced by some clients and the cost-cutting competitiveness within the OPEC community itself, oil soon was flowing at more reasonable rates. The fact that 732 oil wells in Kuwait were set afire by the retreating Iraqi army—"the bonfire of the insanities"—made no appreciable difference in the price of oil. With conservation, alternative fuels, and related policies, spatial-resource blackmail can be rendered neutral.

England now has no measurable coal reserves. But the fact that

Boundaries in the Arabian Peninsula as the Gulf War Loomed.

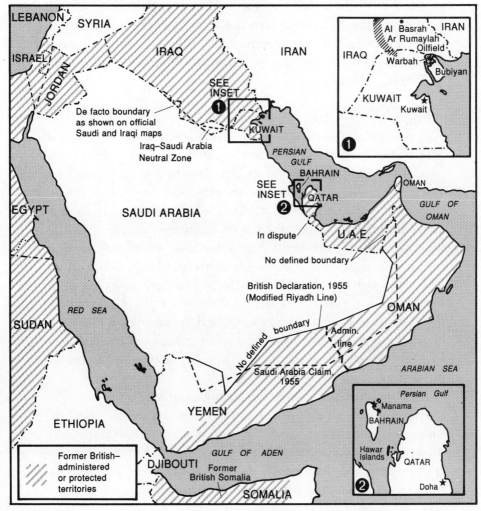

coal was once abundantly available in the North and in Wales enabled Britain to stoke its satanic mills and build a far-flung colonial empire and exploit other lands. The wheel has come full circle. The once-rich industrial North is now as poor and helpless as any area of a First World country, a reflection both of England's lack of natural resources and the unprofitability of serving diminishing markets.

The new focus on international political geography received a big boost from the demise of the Cold War and the increasing importance on the world scene of players other than countries. The growth of international, intergovernmental organizations (for example, the UN, the OAS), nongovernmental organizations such as Greenpeace and Amnesty International, and newly emerging super-states like the European Community are challenging both the balance of power and the sovereignty of states in matters ranging from environment to human rights.

The art and science of understanding and predicting shifts in political power among groups, nations, states, and other units began in a formal sense about a century ago. At that time, as empires and states vied for shares of power or security from the power of others, theories to explain them were put forth by geopoliticians. These early geopoliticians laid heavy emphasis on resources in space, especially in terms of the military technology of the time. Power was weighted according to manpower and now-obsolete capabilities of destruction. The technological revolutions over the past century have changed all that. Economic variables, nearly instantaneous global communications, weapons of mass annihilation, and threatening political ideologies have robbed the world of its innocence, and made it more volatile. The key players in the global political arena have changed markedly.

The gradual evolution of human organizational units occupying terrestrial space—from small groups to tribes, nations, empires, and states—continues unchecked. The preeminence of the state (multinational and nation-state) is of recent origin. The breakup of the great empires and their gradual replacement by states is usually associated with European history and the gradual formation of states such as modern France and Germany.

The variables and processes that define the dimensions of political power and vulnerability have increased and become more interrelated. The acquisition of territory and its natural resources remains a major international issue. Old land boundaries continue to be a source of conflict and tension. This process has been extended and complicated as states have incorporated the seas and the sky above in the passion for aggrandizement. Recent claims to Antarctica, the deep-sea bed, and outer space threaten to prolong the process.

The maldistribution of food and other basic human needs remains a major source of international tension and strife. The rapid growth of population and its concentration in huge cities, especially in the Third World, has created massive national and international pressures. Environmental degradations that ignore boundaries have led to a flurry of accusations, disputes, and negotiations, aggravating political and economic relations and endangering the safety and well-being of all humankind. Deadly plagues (terrorism, drugs, AIDS) are forcing global leaders to formulate governance policies and modes of attack that must be coordinated regionally and sometimes globally in order to be effective.

The level of a state's technological development is increasingly a key force in a state's ability to compete in the global markets and circles of power. The major technologies include ingenious advances in communications. The control and transfer of these technologies are in large measure directed and influenced by the fairly recent global pacesetters of transnational corporations.

The new geopolinomics is a discipline that attempts to understand and explain the prevailing distribution of power and the processes that alter this distribution.

DON'T
FENCE ME
IN

■■■■■■■ ▪ ▪ ▪ ▪ ▪ ▪ ▪ ▪ ■■■

"**S**omething there is that doesn't love a wall."

True, certainly if the wall separates friendly and trustworthy neighbors. Robert Frost also wrote, "Good fences make good neighbors."

Walls and fences are still much in evidence around the globe. The most fabled wall of them all, the Great Wall built in China by the Ming dynasty in the 1500s to keep out enemy forces, is the country's foremost tourist attraction and generator of hard currency today. It may be the best role model for all boundaries.

The United States has been flirting for years with the notion of excavating wide trenches along the entire 2,200-mile border with Mexico. Its purpose would be to thwart the influx of illegal aliens. Morocco actually built a 1,550-mile barrier of sand and stone—a berm—along the desert frontier to defend its claim to Western Sahara (formerly Spanish Sahara); the northwest African country claimed its sand-wall strategy succeeded in deflecting guerrillas who fought for years to make Western Sahara an independent sovereign state.

In the 1980s, South Africa erected an electrified fence, topped with coils of razor-sharp wire to put a fine point on it, along strategic sections of its border with Zimbabwe to keep men from crossing over in search of work or trouble. On the Malay Peninsula, a fence with dozens of watchtowers stretches across 35 miles of Malaysia's narrow northern border.

One of the most high-tech secured borders in the world lies along the Golan Heights between Israel and Syria. An authority on international-border regions has speculated that the barbed-wire fences there are so sensitized that a single wandering sheep could trip a device and bring half the Israeli army to the site. The Soviet Union's borders with Finland and Turkey were once ominously and closely guarded with a series of heavily fortified fences.

Boundaries are three-dimensional. They extend to the top of the atmosphere and to the center of the globe—formidable indeed! World-wide, there are at least 100 officially recognized boundary disputes at any given time, and nasty new ones arise regularly.

A map of political boundaries is a veritable treasure of geographic inquiry. It reveals the recent and ongoing disputes between El Salvador and Honduras, Ecuador and Peru, Guyana and Venezuela, Brazil and both Paraguay and Uruguay, Argentina and Uruguay, Senegal and The Gambia, Chad and Libya, the Arab world and Israel, Ethiopia and Somalia, India and Pakistan, China and India, China and Russia, Bangladesh and India, Laos and Thailand, and Cambodia and Vietnam, to name only a handful.

No recent event prior to the havoc in the Soviet Union in August 1991 so stunned the world as the sudden collapse of the Berlin Wall and the quick unification of the Federal Republic of West Germany and the German Democratic Republic of East Germany. To be sure, the new Germany is a fragment of its former geographic self. There was a time before the First World War when the boundaries of Germany took in Alsace and Lorraine in the west, the territories of East Prussia and West Prussia, which are now part of Russia or Poland, and Austria and Czechoslovakia (through annexation).

For years, the United States' friendly relationship with Canada has been strained by a disagreement over the sovereignty of the Northwest Passage and five other maritime boundary disputes. Canada claims that the waters of the Northwest Passage are internal Canadian waters and subject to its exclusive control. The United States does not argue Canada's claim to the islands bordering the Passage, but insists that the Passage itself is international water open to all countries.

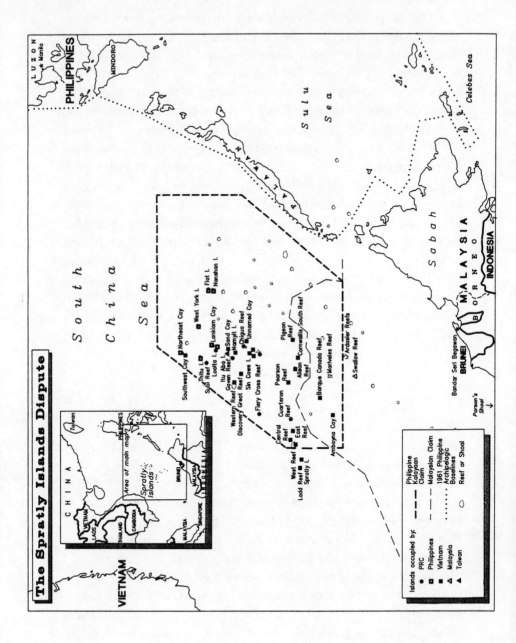

Coastal countries have pushed their territorial waters far beyond those that were acceptable in the 1920s. There was a dispute for decades between the former Soviet Union and Norway (and its allied signatories to the Treaty of Spitsbergen) over the Svalbard Islands in the Arctic Ocean. The islands remain under Norwegian authority, but 41 signatory states now share equal rights to mine coal there. The Soviets mined coal on the islands until 1991.

Historically, undeveloped frontiers were the fuzzy edges of a country's sovereign space. But frontiers are transitory, the last space between places. Civilization eventually dissolves all frontiers as they become part of a state's ecumene. Even frontiers of outer space are being pushed back.

The states of Ohio and Kentucky have been in contention for decades over a stretch of the Ohio River, much of it involving fishermen, the licensing of boats, and taxes from the sale of the vessels. In a generous gesture described as the biggest land transaction since the U.S. purchases of Alaska and the Louisiana Territory in the 19th century, Canada agreed to cede land rights to an area the size of Texas, along the boundaries of the Arctic Ocean, to Inuits, Indians, and peoples of mixed ancestry living there.

The constant flow of illegal aliens trying to escape poverty or hunger or political threats south of the United States border is an

---

The Spratlys, 400 or so desolate and otherwise forgettable island-reefs and shoals in the central South China Sea about 775 miles northeast of Singapore, are semi-enclosed, disputed places that could become the site of a power play for their underwater petroleum and natural-gas resources, possibly East Asia's most lucrative. Four countries—the Philippines, Malaysia, Vietnam, and the People's Republic of China—and one non-nation— Taiwan—claim sovereignty. By international law, a 200-mile exclusive economic zone surrounds every island for the party that establishes sovereignty over the island. China's interest also includes having access from its southern development zone to the main sealanes through the busy Strait of Malacca chokepoint, or potential stranglehold, which could be pinched off. Geopolinomics—connectivity between places, claims over space—are manifest here. Unencumbered navigation across the oceans and through narrow international straits is essential to geostrategic planning and global shipment of key commodities, a cardinal doctrine of international law for more than three centuries. The 1982 Law of the Sea Conventions recognized the right of all countries to sail through straits used for international navigation, including narrower ones that lie totally within the territorial waters of bordering countries. The convention also established the right to fly over straits. The paired straits of Malacca and Singapore form a 1,200-nautical-mile shortcut between the Indian Ocean and the South China Sea and the Pacific Ocean.

aggravation but does not dispute boundaries or borders. The cause of the flow will not be altered by better walls. The solution lies in creating a stronger economy in Mexico that would provide more jobs for Mexican citizens. The 60-year United States protest concerning the millions of gallons of raw sewage that originate in Tijuana, Mexico, and empty into the ocean off San Diego will also not be solved by a better boundary but rather by cooperation with Mexico to solve the problem at its *place* of origin.

Boundaries are not just drawn, they undergo a process of development. Initially, two countries decide on a political division of territory, a kind of general allocation. "Delimination" is a step in boundary formation, encompassing the selection of the boundary site and its definition. "Demarcating" is the pinning down of boundary space. Administration is established for the development of procedures for maintaining the boundary.

Boundary disputes are common and continuous, though they seldom become a threat to international equilibrium. Boundaries were drawn haphazardly, before states had developed their salient characteristics and before precise surveying tools were available. Today, there are boundaries and there are boundaries. Some coincide with physical features (rivers and mountains, for example) or cultural divides (such as language). There are ethnic boundaries and religious boundaries. Rivers make undesirable boundaries because they change course, meander, overspill their banks, and are generally unreliable.

The function of boundaries is obvious. International boundaries define the legal sovereign space of a nation-state and are continually monitored and maintained. There is a United States–Mexico Boundary Commission and a United States–Canada Boundary Commission. Most countries have similar commissions with countries with which they share borders.

The Koreas have been exceptionally rigid in safeguarding their separateness since their civil war (1950–1953). Ten million families have been segregated by the jointly fortified border at the 38th parallel. They were not allowed to make telephone calls or send letters to relatives in the "other" Korea. Not until 1990 was free travel between the two countries permitted, and then it was only for a few days. But in December 1991, the leaders of the two Koreas signed a

Germans went to great lengths to see what was on the other side of the Berlin Wall, for 28 years an icon to futile attempts to stop change by building walls and brick-and-barbed-wire barriers and drawing deadlines in the sand. Eighty-one fleeing East Europeans were killed trying to scale the partition and seek asylum in the West. All arbitrary boundaries are artificial and porous: the Maginot Line, the Iron Curtain, the 38th Parallel, the Saddam Line. Political boundaries invariably have a pragmatic function, dividers helping sovereign states identify spatial responsibilities. Before being knocked down in late 1989 by both East and West Berliners, the Wall had further isolated the former German capital as a West German island in Soviet-dominated East Germany. OUTLET BOOKS

treaty of nonaggression renouncing armed force against each other and agreeing to end formally their war of forty years earlier.

The three Baltic republics of the U.S.S.R. became independent of the Soviet Union in 1991. There is a fourth Baltic state. It consists of the northern half of the old German province of East Prussia. The Soviets annexed this province in 1945 and named it the Kaliningrad Oblast of the Russian Republic. Occupying a small area between Lithuania and Poland, it was a forbidden-area zone from 1945 to 1989, closed to everyone, even most Soviet citizens. The southern half of East Prussia was held by Poland. East Prussia's German population was expelled in 1945, though a significant minority of Germans went on living there. The Kaliningrad Oblast is a troublesome piece of space, totally cut off from Russia by Lithuania. Some solution will have to be found to resolve the problem of this cutoff piece of space.

Sea boundaries are awash in controversy. Most coastal countries used to consider their offshore boundary to be 3 miles from shore. In 1927, the Soviet Union claimed its coastal boundary to be 12 miles offshore. After the Second World War, the United States reconsidered its ocean boundaries. It declared that it would regulate fisheries contiguous to its shores and would have jurisdiction over resources on the continental shelf and the subsoil of the shelf. The Truman administration defined "contiguous" to be the 200-meter isobath, the designation, like a line of longitude, that connects places under the sea. The Law of the Sea now allows any country to claim sovereignty of waters up to 12 miles off its coast and ownership of resources in the waters and sea bottom up to 200 miles offshore.

The continental shelf is the underwater land that borders a continent. It extends from the coastline to a point in the ocean called the shelf break. These shelves are extensions of the continents. Their width averages 40 miles. Along the Pacific coast of Siberia the shelf is 800 miles. Along the state of California, it is no more than a half mile. The shelves lie under less than 10 percent of the total area of the oceans, but they are fertile, productive areas of ocean, and most of the plants and animals in the ocean are found there.

Who owns the resources at the bottom of the high seas—those areas beyond the 200-mile offshore line? What about the goal of arriving at agreements that redound to the common good? Third World countries protest that the "far out" resources, like manganese nodules, should be shared by all countries. Such claims have been rejected by the United States and some other First World countries.

(Arguments over the heritage of resources are not limited to the sea bottom or the high seas. The resources of Antarctica and outer space, intellectual resources, and a host of other resources fall into the category of continuing inquiry and dispute.)

An island by international definition is a detached or isolated piece of land in the water that is above water at high tide and can sustain life. The Japanese have been known to build up an island with a concrete tower to make sure that the island doesn't disappear underwater at high tide; if the island went out of sight, Japan's claim to any resources that may be within the 200-mile limit would also disappear.

The war over the Falklands between Great Britain and Argentina in 1982 was not over a few forbidding islands in the South Atlantic that Argentina wanted back, but for the resources, oil among them, to be found within 200 miles of the islands. For the same reason, the Spratlys, comprising 400 islands and reefs in the South China Sea, are being claimed by China, Vietnam, the Philippines, Malaysia, and Taiwan.

"Uncommon space" is the last frontier. Here, the boundaries are not drawn and the space is unique. Uncommon space includes the high seas, Antarctica, the polar ice caps, and outer space, and they all generate controversy.

Humankind's passion for claiming all kinds of space goes on despite harsh conditions. The most uncommon of uncommon space is the 5.5 million square miles of Antarctica. About 40 percent larger in area than the United States, the southernmost continent consists of landmass, shelf ice, and pack ice off its shores. The landmass is terra firma, the shelf ice and the pack ice are transitory. Antarctica, known as "the Ice" to visitors, remains unclaimable space through international agreements.

Outer space is clearly of enormous significance with commercial, scientific, and military applications. And where exactly is outer space? It is the universe beyond Earth's atmosphere. The United Nations as long ago as 1958 established a committee to promote exploration and exploitation of outer space, in accordance with international law, which made outer space a common resource. By international law, a state cannot claim sovereignty in outer space. No one can claim celestial bodies, our moon, or the moons of other planets, the other planets themselves, or the stars. No one can plant a flag out there and proclaim, "This place is my space!"

By land or sea or air, boundaries and borders and the disputes over them are likely to continue until we live in a world almost beyond envisaging. Humans covet space; it's just the way we are.

THE KINGDOMS
OF APHRODITE
AND HEAVEN

■ ■ ■ ■ ■ ■ ■ ■ ■ ■ ■ ■ ■ ■ ■

When I was The Geographer of the United States, in the Department of State, I maintained a file I labeled Mythical Kingdoms, Kooky Kings, and Pretending Princes—claims, credentials, declarations, affidavits, and maps from eccentrics, pranksters, and con artists manqué who attempted to found their own countries.

The file also included memos from a variety of United States government agencies inquiring about the legitimacy of such "countries." Immigration officers would ask about some unfamiliar place. "We've got this couple from the Republic of Six Happinesses. Where on earth is it?"

Where indeed.

There are five basic requisites for being a country: space, population, economic activity, government structure, and recognition by other countries. In recent years, many aspirants have met the criteria. Among them: Zimbabwe, which was Southern Rhodesia, in south-central Africa; Vanuatu, formerly called the New Hebrides, a group of islands in the southwest Pacific Ocean; Antigua and Barbuda, an independent state in the eastern West Indies; Belize, formerly British Honduras, in Central America; St. Christopher Nevis, an independent state in the West Indies; and Brunei Darussalam, an independent sultanate in the northeast part of Borneo.

In most cases, claimants actually have *bone fide* dominion over land that exists, though the land is usually a tiny, isolated speck in the sea or an area that is underwater. Some of the self-styled kings and princes proclaim their independence, adopt constitutions, seek diplomatic recognition, send out special envoys, display coats of arms, mint coins, and issue stamps, passports, and visas.

The file of pretenders showed claims for sovereignty from the Kingdom of Aphrodite, an 8-foot by 10-foot islandlike platform west of Tijuana, Mexico; the Realm of Redonda, a pile of rocks off Bermuda, which flew a pair of pajama bottoms as its flag; the United Kingdom of Arya, a 500,000-square-mile whites-only sanctuary in Antarctica; the Republic of Minerva, a spit of land near Fiji, whose prospective rulers declared it would become an eternal haven free of smog and income taxes; and the Federal District of El-Eastmoor, a landlocked "country" off the coast of Florida, whose "ruler" was seeking verification that it was not under the jurisdiction of the state of Florida.

Petitions for sovereignty go back decades. In 1948, a number of drunken fishermen proclaimed a small island off Nova Scotia to be the Principality of Outer Baldonia; its independence would give citizens the right to drink, cuss, gamble, and, of course, lie about the fish they had caught. Outer Baldonia disbanded in the 1970s after squabbling with Russians over fishing rights. The "principality" is now a bird sanctuary.

One of the wackiest claims for recognition came from Ernest Hemingway's brother. In 1964, Leicester Hemingway founded the "country" of New Atlantis. It turned out to be a bamboo platform anchored by a railroad axle and an old Ford engine block six miles off the coast of Jamaica.

Morton F. Meads, an American businessman, based his claim to the island-republic of Morac-Songhrati-Meads, which was formerly the Kingdom of Heaven, on his being a descendant of the sea captain who discovered the island in the 1870s while sailing under the British flag. The "republic," with its oil rights, is in the Spratly Islands chain in the South China Sea. Meads began his reign there more than 30 years ago. He designated Meads Island as his capital and chose a bar or two of Beethoven's Fifth Symphony as the national anthem.

Some years ago, Australia imposed a wheat quota that was anathema to one of its citizens, Leonard George Casley. He turned his 18,500-acre Western Australia farm into the Hutt River Province (he later called it a kingdom) and appointed himself Prince. Hutt River formally seceded from Australia on April 21, 1970, and the date became Mr. Casley's national holiday. The "kingdom" has a population of around 35. Australia refuses to recognize the enclave.

King and Absolute Ruler Michara Heatra signed a declaration of independence in 1985 creating the Maori Kingdom of Tahibiti Islands in the South Pacific. It was to reassure the Maori people ("a race that is speeding to oblivion") they would have a home to "call their own, their rightful place in the sun." Heatra's envoy even offered the United States a long-term lease on one island about 500 miles off the New Zealand coast.

Among claimants in the South Pacific was the Principality of Castellania, which was founded in 1974 by a group of disgruntled Austrians. Castellania turned out to be more a state of mind than an actual place.

Thinking globally, two would-be rulers served notice on Secretary of State Henry Kissinger of their rights as "sole owners of all the ocean floor and sealand." A British couple created their own half-acre "nation" in the North Sea, a Second World War radar-and-gun platform. They christened their lilliputian site Sealand.

Mythical Kingdoms was my favorite file. The examples illustrate one of the basic primal needs of humans: a place, a piece of space, a where! These people wanted recognition for their private Shangri-las. Cynics might charge they were seeking an impregnable tax-free haven.

KNOWING WHERE

THINGS ARE:

THE ART OF THE MAP

■■■■■■■■■■■■■■■■■■■

"The greatest of all epic poems . . . one of the most luminous of man's creations." That's how maps have been described.

"The most aesthetically pleasing of the sciences." That's mapmaking.

But after all the millennia of compasses and astrolabes and sextants and longitudes and latitudes and now satellites and sensors and computer graphics, there is still only one projection of the world that is absolutely accurate, and that is the globe. The flat maps we all know and use do not, and cannot, show without distortion the actual shapes and the relative sizes of continents and waters. It is also true that if you wish to study an area on a globe at a scale of one to one, you would need a sphere the exact size of Earth itself. You would need a second Earth.*

Still, as the distinguished travel writer and novelist Paul Theroux has rhapsodized, mapmaking "draws its power from the greatest of man's gifts—courage, the spirit of inquiry, artistic skill, sense of order and design, his understanding of natural laws, and capacity for singular journeys to the most distant places." Through maps, we

*An obsessive Bureau of Cartography in a Jorge Luis Borges fiction builds a map the size of the country it serves.

understand where we are in relation to other people and other places. Maps organize wonder. Confronting a well-made map has all the fascination of opening an exciting novel.

Cartographers have grappled with the challenge of rendering roughness on a flat surface since Greek antiquity and Ptolemy's declaration that the planet was an oblate spheroid. Features and phenomena cannot be flattened and placed accurately. Distortion cannot be avoided. If the shape of the continent is correct, the size of it is wrong. If the equator and the lower latitudes are projected correctly, both Poles are woefully misshaped. There seems to be no perfect solution. Accuracy is limited by technology.

Even though they're not perfectly truthful, maps convey astonishing types and amounts of information. Even a glance at a map of the United States readjusts perspectives. Atlanta, Georgia, is actually closer to Detroit, Michigan, Chicago, Illinois, and even Keokuk, Iowa, than it is to Miami, Florida, which many people believe is just "down the road a piece." The western tip of Virginia is 25 miles west of Detroit. Atlantic City, New Jersey, is south of the northern tip of Virginia. The entire state of Connecticut is south of northernmost Pennsylvania. Southernmost Canada is within 138 miles of being as far south as the Mason-Dixon Line, which became famous at the time of the Missouri Compromise (1820) as the symbolic border between North and South, politically and socially, dividing the free states from the slave states. South America is almost totally east of North America: Peru, on the west coast of South America, is farther east than Savannah, Georgia, which is on the east coast of North America.

We make do with the maps and globes we've got—they tell us where places are (approximately) and how to get there (usually). The good news is that cartographers are persistent. They continue to seek perfection.

Before there was humankind, there was geography. And before there were books, there were maps. Earliest peoples used driftwood and seashells, rock carvings and clay tablets on which to scratch maps of their vicinity. The earliest-known map, which dates from around the year 5000 B.C., was carved into a rock overlooking the former site of a neolithic village in a valley in central Italy. Regional maps

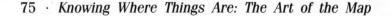

Approximate flight route

Great Circle route
(shortest distance)

IRELAND UK

LONDON

PARIS

FRANCE

A T L A N T I C
O C E A N

SPAIN

ROME

Strait of
Gibraltar

TRIPOLI

LIBYA

International boundaries are three-dimensional: from the surface of the earth to the center of the earth and from the surface of the earth to the top of the atmosphere, which is the air that surrounds the planet. An airplane can fly over a country only with that country's permission. When France refused to let American bombers use its airspace so they could fly due southeast from their bases in England to terrorist targets in Colonel Muammar al-Qaddafi's Libya in April 1986, the raiders had to head south over the international waters of the Atlantic Ocean, then east over the Strait of Gibraltar (international space by law), doubling their in-air time before reaching the Mediterranean and commencing the attack run. The maze of aerial boundaries complicates the movement of aircraft around the globe. Satellites do not need permission to pass over a sovereign state because they are so far above the atmosphere, which merges at about 310 miles with interplanetary space.

were more like an architect's rendering of a building or a community. The first maps of the "known world" displayed the Mediterranean Sea in the center or were rectangular to honor the biblical reference to the "four corners" of the Earth. Later, in the medieval era, the pious citizenry of Christendom plotted Jerusalem in the center of the world; concern was more with perfection of God's work than with physical reality. Irregularity was irreconcilable with harmony.

Early maps, like those of today, were not confined to parchment.

Stonehenge, the prehistoric circle of huge megalithic stones on the plain north of Salisbury, England, seems to be a map of celestial movements.

It was Greek reason and science that deduced that the world was round, and then devised the theory for locating places by a grid of imaginary lines based on the Poles (sites ratiocinated but still not seen) and the equator (probably still not crossed). One theory that had to be rebutted, constantly, was that if the world was indeed round, half of all the people stood on their heads.

The demand by enterprising mariners for reliable maps spurred cartography. Among the early masters of the sea were the Chinese, who were still enjoying the world's highest standard of living. By 1400, their knowledge of getting around was far ahead of Europeans', and they had invented the compass. The Khan's merchantmen sailed across the Indian Ocean to East Africa and mined jade there.

At about the same time, "Jewish cartography," which was centered on the island of Majorca, a seat of nautical sciences, was in great demand by Spanish and Portuguese navigators. "Map Jews" and "compass Jews" enlarged the boundaries of the known world and provided geographical concepts.

Portugal was driven by the lure of the sea. Prince Henry the Navigator (1394–1460), the first superstar of world discovery, went to sea only twice, but in his lifelong quest for gold he commissioned other adventurers. His fleet of caravels explored the west coast of Africa and returned to Lisbon with yellow gold and with black gold—slaves.

Development of Portugal's colonial empire, which gave rise to global prominence, was encouraged by Henry's patronage. Maps reflected geographical knowledge, and political abstracts promoted imperialism. Surveying was an instrument of authority. Maps represented the geography of power. Under pain of death, Portugal forbade selling to any foreigner maps showing its sailing routes and conquests.

Spices, especially cloves for seasoning, beckoned Arab entrepreneurs, who would ship them overland to the West. The location of the Spice Islands and their profitable aromatic-dried flower buds and tropical evergreen trees was a closely held secret. Venice, "the queen of the seas," became trade-rich, the main European go-between with Asia. Until the 16th century was on the doorstep, no one had successfully sailed from one continent to the other and made a credible record of it, and there was no Suez Canal. The oceanic link was forged by Portugal's Vasco da Gama (c. 1469–1524). With a four-vessel fleet, he sailed around South Africa's Cape of Good Hope and, with an Arab pilot taken aboard in East Africa, hove into the Indian port of Calicut on May 22, 1498. Spices now really became the spice of life.

Often I have wondered if Columbus would have sailed even once, much less four times, west to the east across the Ocean Sea for the riches of the Indies if Asia had already been reached by ship. The driven, ambitious admiral discovered a hemisphere and continents new to Europeans, reshaping their world and straining the cartographers. "The Columbian Exchange" introduced into the "new world" diseases that killed millions, and the horse, which was used to kill millions of buffalo.

Around 1402, Koreans, finishing a new map based on earlier ones, showed the Gulf of Arabia, the Mediterranean, and southern Europe accurately, and a misshapen but identifiable Africa. In 1507, the German cosmographer Martin Waldeseemuller made a map book of 12 sheets. He separated North and South America from Asia and put the name "America" for Amerigo Vespucci (1454–1512), the Italian navigator who had explored 6000 miles of South American coastline, over a patch of land he drew to the west of Africa. Ferdinand Magellan's Portuguese fleet sailed around South America's Cape Horn and went on to prove, in 1522, it was possible to circumnavigate the globe, and in only 3 years (as predicted by Spain's Council of Talavera nearly 40 years earlier).

Interaction on a global scale boomed. Spain's rapacious and cruel conquistadors plundered Central and South America. Hernando Cortez (1485–1547) defeated the Aztec empire. Francisco Pizarro (c. 1476–1541) destroyed the Inca empire and founded Lima as the center of Spanish rule in South America. Vasco Núñez de Balboa (c. 1475–1519) discovered the Pacific Ocean and claimed it and all shores washed by it for the Spanish crown. Francisco Vásquez de Coronado (c. 1510–1554) and his men were walking around New Mexico, Texas, Oklahoma, and Kansas more than a half century before the Pilgrims dropped anchor off Cape Cod. The sword and the cross established Spanish power. The inevitable skirmishes broke out over the oceans, and piracy was a constant threat.

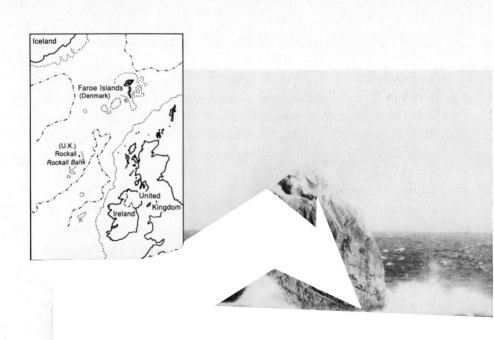

Rockall is "nothing" but a 70-foot-high, 83-foot-wide, granitelike pimple rising out of the North Atlantic about 300 miles west of Ireland. But Ireland, Denmark, Iceland, and the United Kingdom are aggressive claimants because Rockall's continental shelf, surrounding waters, and seabed for 200 miles in all directions may be mineral-, oil-, and fish-rich. A continuing puzzlement for geographers: Is Rockall by international definition an island or merely a rock in the middle of rough political seas? CENTRAL OFFICE OF INFORMATION

Holland joined the land grab. The French navigator Jacques Cartier (1491–1557), searching for a northwest passage across North America to Asia, discovered the St. Lawrence River. The English set sail bent on empire building. In 1497, their Italian-born explorer John Cabot (1461–1498) had already accomplished what Columbus hadn't, touching the mainland (Newfoundland) of North America. In the 1570s, the navigator and admiral Sir Francis Drake (1540–1596) sailed around South America and north along the Pacific coast to the present state of Washington, completing circumnavigation of the planet in 1580. James Cook's voyages around Antarctica—he never actually saw the continent, he only felt it—and throughout the Pacific in the 1770s and his charts of Australia and New Zealand provided still more clues to the shape of the planet.

Filling one of the last blank spaces on the map of the world, the English explorer Sir John Franklin (1786–1847) pushed forward knowledge of Arctic regions as he sought the Northwest Passage. It was a tragic quest. The expedition of 129 men perished. Still another English explorer, John Hanning Speke (1827–1864), unraveled the mystery of the sources of the Nile.

The American arctic explorer Robert E. Peary (1856–1920) reached, or almost reached, the North Pole in 1909. The Norwegian polar explorer Roald Amundsen (1872–1928) and the British naval officer and explorer Robert Falcon Scott (1868–1912) made it to the South Pole. The New Zealand mountain climber and explorer Sir Edmund Percival Hillary and the Nepalese Tenzing Norkay set foot on the summit of Mount Everest in 1952. In 1969, Neil Armstrong and Edwin "Buzz" Aldrin, Jr., became the first of 12 American astronauts to walk on the Moon, which is better surveyed today than Earth was a century ago.

Throughout the centuries of exploration, expunging and pushing back frontiers, maps were a paramount tool. The desire to know what was over the next mountain and beyond the next wave recalls the words of one exuberant Briton: "The force that drives men towards the summits of the highest hills is the same force that has raised him above the beasts. He is not put into this world merely to exist but to find love and happiness. Some achieve happiness best by seeking out the wildest and more inaccessible corners of the earth, and there subjugating their bodies to discomfort and even peril, in search of an ideal which goes by the simple name 'discovery,' discovery not only of physical objects but of themselves."

Maps are simplifications or models of reality. They are tools for discovering where things are and—in modern geography—how they move over space and through time. They extract specific details for inspection at a scale that is convenient. Air controllers in the United States use radar maps of the sky to keep track of the 3,000 or so airplanes that are flying over the country at any given time. Weather forecasters and fast-food marketeers couldn't exist without maps.

Even at a glance, the best new maps reveal why geography is a dynamic, process-oriented, locational art and science. Where things are is of course important, but the geographer of today looks at relative locations and changing places in his concern with the flows of capital, cold fronts, migrants, "killer bees," acid rain, and nuclear submarines. What's here today may be gone tomorrow. What's not here today may be here tomorrow. Volcanic islands pop up out of the sea: the United States' fiftieth state was once on the floor of the Pacific

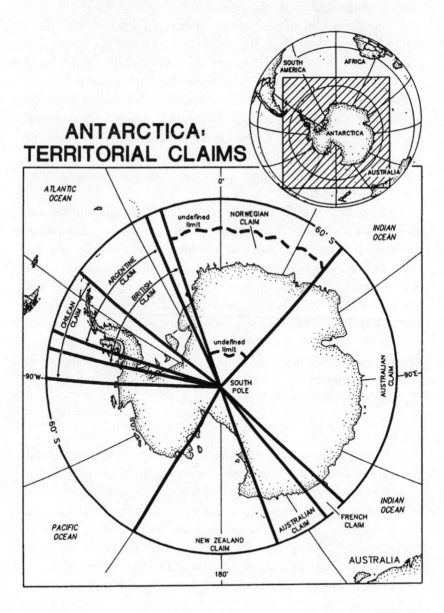

ANTARCTICA:
TERRITORIAL CLAIMS

Ocean. Towns are literally wiped off the face of the earth by natural disasters. A city is all but totaled by one bomb. Countries change their name. Iran was once Persia. In 1990, Burma became Myanmar. The king in the popular Broadway musical *The King and I* would today be the monarch of Thailand, not Siam. Cities change their name. St. Petersburg, Peter the Great's window on the West, became Petrograd for a country at war with the Germans in the First World War, then Leningrad for the heirs of the Bolshevik revolution, and St. Petersburg again, in 1991, by official decree of the Russian federated republic's parliament after the aborted coup.

In the still-familiar map projection created in 1569 by the Flemish cartographer Gerardus Mercator, and used by mariners, the world's second-largest island appears larger than South America, though Greenland's actual size (839,999 square miles) is less than one twelfth the area of South America, the fourth-largest continent. Another continent, Antarctica, wasn't even known during Mercator's day, and the North Pole isn't shown. So distorted are the distances

---

"Great God! This is an awful place," British explorer Robert Falcon Scott exclaimed at the South Pole on January 17, 1912. The island-continent, which is nearly one tenth of the world's land surface, is surrounded by a moat of treacherous, rampaging seas, and its climate is colder than are some parts of Mars. It is the coldest, windiest, highest, and driest continent in the world. There is no sunlight for six months of the year. The environment held off human presence until 1895, four centuries after Columbus's historic first voyage to the Bahamas. To this day, more people attend a Big Ten college football game than have ever been to "the Ice." But many countries, prodded by at least the prospect of lucrative commercial exploitations beneath the vast ice desert, have claimed pie-shaped sections. Acknowledging the importance globally of maintaining Antarctica's virginity, the 26 voting members of the 30-year-old Antarctic Treaty, including the United States (reluctantly), agreed in 1991 to a compromise that bans mineral and oil exploration for a half century. The United States has never staked out a claim to territory but has had bases there for decades. Antarctica, a key player in the interlocking ecology of the whole planet, is an unremitting and powerful weather-maker for the southern hemisphere. The circumpolar currents spawn enough food to feed the entire world for two centuries. Depleting ozone holes, or "donuts," above the South Pole are letting in increasing amounts of ultraviolet radiation, a deadly threat to all living things. Ultraviolet rays are penetrating deep into the ocean, causing sizeable reductions in the productivity of the single-cell organisms that form the base of the oceanic food chain, causing unpredictable effects. "I am hopeful," the American explorer Robert E. Byrd said a half century ago, "that Antarctica in its symbolic robe of white will shine forth as a continent of peace, as nations working together there in the cause of science set an example of international cooperation."

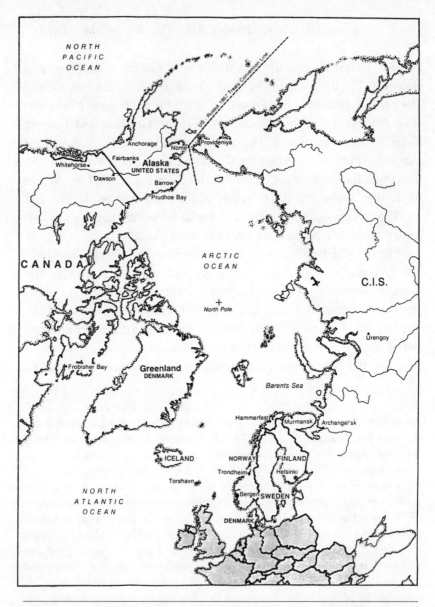

The relatively mild Arctic region—principally 5,427,000 square miles of ice water and large groups of moving ice islands—is extraordinarily important: it is strategically located and has many valuable resources, especially underwater oil deposits. Estimates of potentially recoverable reserves range up to 200 billion barrels of crude oil and up to 3,000 trillion cubic feet of natural gas. American and Russian nuclear submarines nose about surreptitiously. The United States and Canada joust over where boundaries should be located, jurisdiction over military airspace, and navigation through the Northwest Passage. Scientists are discovering all over the vast, barren area surprising concentrations of pesticides, herbicides, and industrial compounds, migrants carried north by air and water currents, some originating as far off as Southeast Asia. War may not be as much a possibility as is friction generated by native peoples protesting exploitation of resources and environmental degradation, as well as loss of their culture.

and the sizes and the shapes of the landmasses and oceans that the Mercator is unsuitable for serious use. Yet the age-old flat wall map continues to be unfurled across blackboards in American schoolrooms, its regions of vegetation colored green, its deserts red and yellow, its waters blue. I do not understand why the schools haven't pitched this archaic, false impression of the world into the nearest dumpster.

There is a flat-projection system that I regard as close to perfection as we're apt to get. This portrait of Earth was devised by Arthur Robinson, professor emeritus of geography at the University of Wisconsin and one of this country's most respected cartographers. Yes, he sees mapmaking as much of an art form as a science. There are distortions in the Robinson map. His compromises exchanged exaggeration of size for improved landmass shape. Two of the largest countries, the Soviet Union and Canada, were their truer relative sizes. In a well-known projection issued in 1922, they were more than twice their relative size.

North has not always been at the top of maps. The earliest Arab

Computerized three-dimensional mapping of the Pacific Ocean southwest of Monterey, California, reveals the previously unseen and unknown: a canyon slices through a chunk of the United States' West Coast continental shelf, and a seamount rises 6,900 feet (but not enough to break the surface). Few ocean depths have been explored. COURTESY OF THE NATIONAL GEOGRAPHIC SOCIETY maps, which were the first really usable maps, had north at the bottom. The Venetian traveler Marco Polo (1254–1324) used such a map during his 22-year sojourn through Asia. There is no north or south, no west or east, no down or up when Earth is seen from the sky and outer space. There's just in or out for our blue-green oasis floating like a delicate Christmas-tree ornament in the heavens. Given the existence of the Poles and the fact of magnetic North, which keeps shifting its position, having north at the top of maps and south at the bottom simply became a convention. It doesn't matter if Hobart, Tasmania, is near the top of the map and Hammerfest, Norway, is near the bottom. Or does it? Many of us tend to associate northernness with topness, with superiority. North has been given a psychological edge. North looks "down" on south. Temperate northern climes are associated with high achievement, tropical regions are characterized by indolence and the lack of both ambition and accomplishment. The northern hemisphere produces men and women of distinction, the southern hemisphere promotes easy living but little achievement. In the United States, northerners have patronized the south as charming but not a place celebrated for quickness and efficiency. There are distortions that should be eliminated from everyone's mental map.

We should not associate "up" only with north and "down" only with south. Up and down refer to elevation. Three of the principal rivers of Russia, for instance, flow "down north" from "up south." The elevation of the sources is higher than the mouth, water can't overcome gravity, the rivers run north and empty into the Arctic Ocean. The world's longest river, Africa's 4,187-mile-long Nile, flows downhill to the north from up south at its source, Victoria Falls. The St. Lawrence River flows down northeast from Lake Ontario and into the Gulf of St. Lawrence. Many of the tributaries of the Amazon, the world's second-longest river (only 187 miles shorter than the Nile), flow down north.

It is impossible to avoid distortions in size, shape, and direction on a flat map of the world. Professor Arthur Robinson's compromise projection has the least misrepresentations. Still, compromises were made in exchange for improved shapes of land masses. The contiguous 48 states of the United States are only 3 percent smaller than in actuality. COURTESY OF THE NATIONAL GEOGRAPHIC SOCIETY

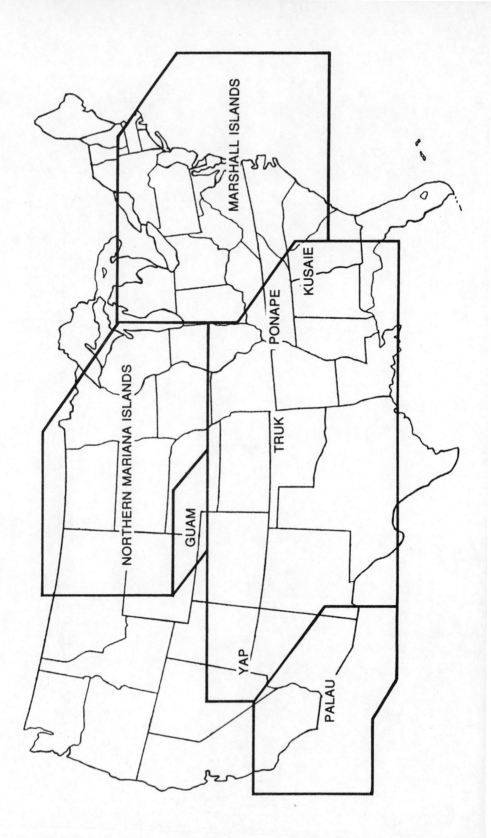

All of us are egocentric. Personal mental maps reflecting the ways we perceive our community, city, country, and the world are spatial data imprinted on the brain, many times incorrectly.

Until the 18th century, mariners had only lines of latitude for guidance, but those angular distances north or south of the equator were arbitrarily drawn. But without lines of longitude, the angular distance east or west always measured from the same point, sailors had little idea where they were. Vessels kept crashing into coastlines and islands and sometimes into one another.

Longitude was created by John Harrison (1693–1776), a self-trained Yorkshire instrument maker with an almost supernatural mechanical sense. Time is the key to measuring longitude. If the navigator knows the exact time at a particular place on the globe, he/she can calculate longitude from the time difference between that place and where he/she is. The common pendulum clock could not be used in Harrison's day because the ship's sway upset the periodic motion of the pendulum.

In 1707, a British fleet misfigured its position and tore into the rocks of Scilly Islands, off Lands End, southwest England. Four ships and 2,000 tars went down. Six years later, Parliament offered a prize for a method of determining longitude to within a half degree.

Harrison built a series of five clocks with pendulums of different metals that could handle oscillations without being adversely affected. Temperature changes expanded the metals, but the overall length of the pendulum remained the same, and the period of beat remained unaltered. The clocks kept accurate time even while being wound. They held sway more accurately than any other clock of the time. After months at sea, one of them was off by less than a minute. The modern era of naval navigation had dawned.

Mapmakers employ mathematics to determine where the equator

Examining the relative size of places sparks geographic perspective. Micronesia, widely scattered islands and coral atolls in the western Pacific Ocean east of the Philippines, is wider than the 48 contiguous states of the United States, but its land area is only 1/300th as large. Allied armies, principally the United States', shot and flame-threw their way across Japanese-held Micronesia during the Second World War.

and the North and South Poles are. The British were chauvinistic in determining where longitude zero would be. They divided the globe longitudinally into 24 equal parts. The prime meridian, or 0° longitude, would be a global north-south line through the Royal Astronomical Observatory at Greenwich, near London, in southeast England. The 180-degree meridian halfway around the world from Greenwich would therefore run through the Pacific—the International Date Line. In 1884, all the countries of the world converted to Greenwich Mean Time (GMT), or Universal Time (UT), the prime basis for standard time. Head-on railroad collisions were greatly reduced by standardization of time—everyone could be on the right track. Pizza lovers may not realize they are points of longitude and latitude for a fast-food chain using digital maps to insure quicker home delivery.

Maps are popular with just about everyone. After the outbreak of the Gulf War in 1990, one American company alone sold 4 million maps of the Middle East. Mapmaking is a $200-million-a-year industry in the United States alone. The product can be tailored to fit any interest. A book of maps of New York City includes a map of the seating arrangement in the chic Russian Tea Room and a map of public toilet facilities.

Some places are changing so swiftly that even a moderately up-to-date map is soon rendered useless. A person can go badly astray by consulting a map made in 1990 for fast-growing areas like Orlando, Florida, or Las Vegas, Nevada. Road maps are probably the least interesting, the least analytical of all maps, mainly because they display information as being static in time. Such maps are only instant photos, Polaroids; "detour" is the most familiar sign on the American road these days. The value of road maps will return when they are old enough to be an antique. There are times when I believe that the map I am looking at should carry the disclaimer "This map may lead you astray."

The planet's 200 narrow waterways are chokepoints, potential bottlenecks to the free flow of ocean traffic. Their significance is determined by their strategic importance vis à vis such factors as the superpowers, international "hot spots," trade routes, and traffic patterns.

Strategic Waterways—Stress Spots of the Globe

Narrow waterways, whether straits or canals, are potential bottlenecks to the free flow of sea traffic. A waterway's strategic significance is determined by its importance to the superpowers, international "hot spots," trade routes,

and ocean traffic patterns. Among the more than 200 international straits around the globe, about a dozen are rated as politically, economically, or militarily critical, with the Strait of Hormuz topping the list.

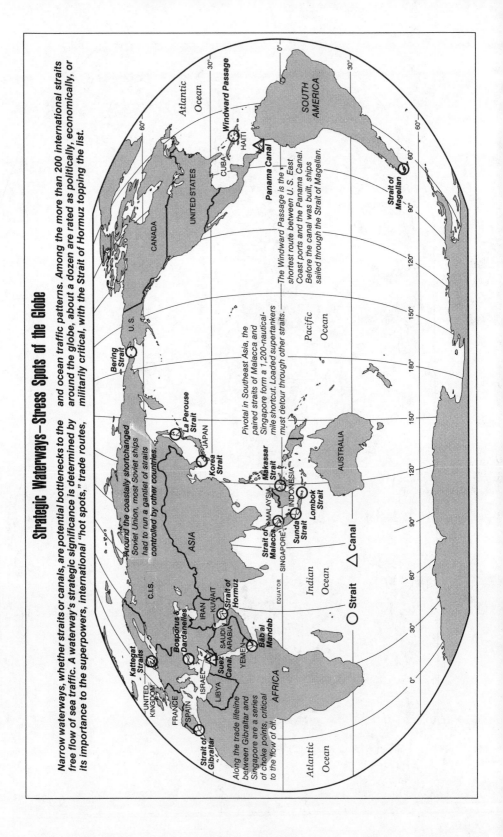

The Windward Passage is the shortest route between U. S. East Coast ports and the Panama Canal. Before the canal was built, ships sailed through the Strait of Magellan.

Pivotal in Southeast Asia, the paired straits of Malacca and Singapore form a 1,200-nautical-mile shortcut. Loaded supertankers must detour through other straits.

Around the coastally shortchanged Soviet Union, most Soviet ships had to run a gauntlet of straits controlled by other countries.

Along the trade lifeline between Gibraltar and Singapore are a series of choke points, critical to the flow of oil.

○ Strait   △ Canal

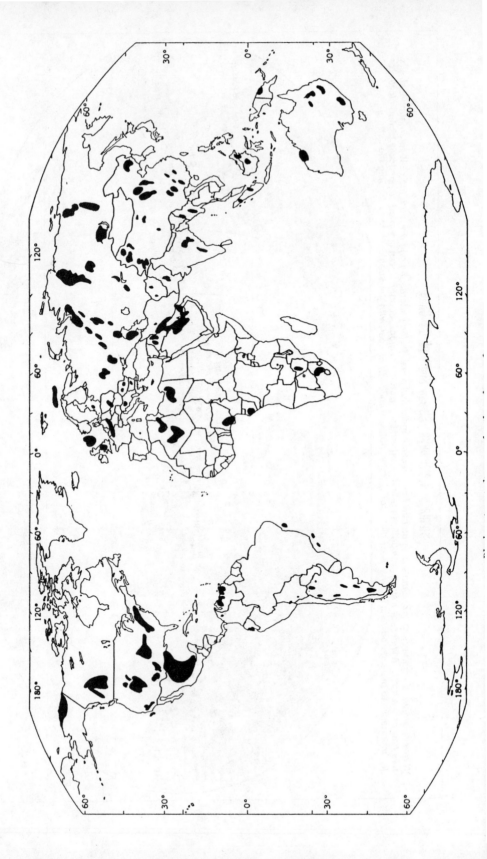

Maps are used for propaganda, for selling everything from ideologies to retirement homes. A couple planning to retire may be persuaded to move to Seattle, Washington, by the warm tones of the color maps they have been sent by the city's department of commerce and by catchy names like Carnation and Queen Anne's Estates.

I hardly need belabor the absolute indispensability and versatility of maps. But I am very impressed that the expense of moving people and goods from point to point—a half trillion dollars a year in the United States alone—is being shaved by logistical operators employing digital map systems that improve efficiency for both the trucker and the customer, proving once again that if there's movement, it can be mapped. The usual focuses of interest, of course, are land, water, and air, but geographers are also being called in to map areas as diverse and as specialized as the whereabouts of pirates—yes, there are still pirates on the high seas, as well as elsewhere—drug traffickers, terrorists, disease carriers, and refugees. It's tough to hide these days.

"A map by definition is a lie," a professor of geography once declared, explaining, "A single map is but one of an infinitely large number of maps that might be produced for the same situation or from the same data. The cartographic license is enormously broad."

And how! A few years ago, Moscow confessed that it had deliberately falsified all public maps of the Soviet Union for a half century. Correct maps were classified, practically without exception, apparently out of fear of foreign-intelligence operations and aerial bombing. Lines of longitude and latitude on public maps were positioned falsely. Highways, rivers, and railroad lines were misplaced. The size of an area or a region was added to significantly, sometimes by as much as 18,000 square miles. Towns were placed many miles distant from where they are, and on the wrong side of a river. Sometimes, towns vanished from one map and reappeared in the next printing.

I have seen ludicrous maps of Moscow. An elliptical highway was shown as either a perfect circle or a square with rounded corners.

Geologists use maps to demonstrate that fossil fuels, the world's main source of energy, are not everywhere. By the turn of the century, the United States alone will be unearthing about half of the world's coal output. The replacement of coal, oil, and natural gas with "cleaner," "safer" fuels will be a momentous issue in the next hundred years.

Maps of many cities showed only a few of the main arteries, making it difficult for visitors, tourists, and even residents alike to find their way around permitted areas. The most reliable street maps of Moscow were produced in the United States, by the Central Intelligence Agency. They were bought by Soviet authorities stationed in Washington, D.C., to check where places actually were in their own capital.

Inaccurate maps of the U.S.S.R. allowed farm managers to declare only half their useful acreage so they would have lower production targets and the opportunity to sell extra produce illegally. Hydroelectric-project builders underestimated flood plains because maps were too crude to verify information. The country's principal environmental-protection agency was accused of covering up disasters by prohibiting their mapping and concealing pollution data.

Misrepresentations were symptomatic of the paranoia I personally found throughout the Soviet Union before glasnost. President Gorbachev realized that misdrawing maps became senseless with the advent of high-resolution space photography. Other countries, working with satellite data, were making accurate maps of his country. It may be, too, that the enlightened policy on cartography was motivated by a desire to gain access to the West's advanced mapmaking technology. Redrawing the map of the disintegrating Soviet Union was the biggest revision in the map of the world since African colonies achieved independence in the early 1960s.

Maps are indispensable in coping with natural disasters. Immediately after the devastating earthquake in Armenia in 1988 (25,000 dead), the intervening agencies sought out maps of the Soviet Union in order to direct the first planeloads of blankets, medicine, and food to stricken sites. Maps were helpful in the cleanup after Hurricane Hugo had ripped through South Carolina, in 1989, in another headline-making "act of God."

My students are stunned during mapping exercises to discover the number and the variety of natural catastrophes that occur around the world in a typical year. They hardly believe the total of volcanic eruptions, hurricanes, droughts, tornadoes, typhoons, and earthquakes. They are even surprised that tornadoes are not confined to the midsection of the United States—the "Wizard of Oz" syndrome—but that the destructive, terrifying funnel-shaped whirlwinds frequently sweep through the south and southeast and north along the

Atlantic seaboard to New Jersey and even into New England. The students become alert to phenomena whose causes and origins—and spatial patterns—even experts still can only guess at. They readily see that the biosphere is under severe strain.

Talking about *The Wonderful Wizard of Oz:* maps have always served as guides and stimuli to the literary imagination. Before he composed *Treasure Island,* Robert Louis Stevenson sat down and drew a map of the imaginary site. He recalled that "it was elaborately and (I thought) beautifully coloured: the shape of it took my fancy beyond expression; it contained harbours that pleased me like sonnets; and with the unconsciousness of the pre-destined, I ticketed my performance *Treasure Island.* . . . as I pored over my map of Treasure Island, the future characters in the book began to appear there visibly among imaginary . . . quarters, or they passed to and fro, fighting and hunting treasure. The next thing I knew I had some paper before me and was writing. . . ." This was before Stevenson had ever left home in Edinburgh.

The Barsetshire novels of Anthony Trollope recreated a panorama of rural England. He used a map drawn from memory to chart the figures in his literary landscape. Thomas Hardy drew a map to help readers picture the locales in his classic Wessex (Dorset) novels. The American Nobel Prize winner Sinclair Lewis created detailed maps for the settings of his novels, locating homes and streets and buildings involving his characters before he began writing. The Polish-born English novelist Joseph Conrad had a genuine enthusiasm for "the geography of open spaces and wide horizons built up on men's devoted work in the open air," while deploring his teachers whose geography was "very much like themselves, a bloodless thing with a dry skin covering a repulsive armature of uninteresting bones."

The most remarkable map and unusual geography lesson in English literature may occur in W. H. Hudson's mystical novel, *Green Mansions,* which is set in the forests of Guyana. To illustrate his explanation of the world to the forest girl Rima, Abel constructs a "map" on the ground, using stones to mark places he mentions. Full of awe, Rima tells the spirit of her dead mother, "On the mountaintop he marked out and named all the countries of the world, the great mountains, the rivers, the plains, the forests, the cities."

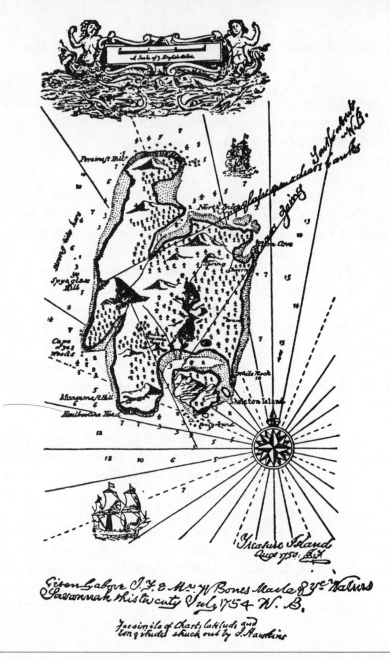

Authors make maps as guides into and through their work. The Scottish novelist Robert Louis Stevenson sketched a map of Treasure Island before launching the adventure novel that brought him renown and his own loot. The creators of the television series "Twin Peaks" evolved a blueprint of the fictional town of Twin Peaks in the tall-tree country of the United States Northwest before they could summon up their characters: "We drew a map. We knew it had a lumber mill. We had to know the place before we could make up a list of suspects." Places formed by imagination, such as Shangri-la, El Dorado, Lilliput, and Utopia (Greek for "nowhere"), tend to be magnetic. The novelist Eudora Welty is persuaded that there is no fiction without place. Space is essential in the real world, as well as in the imagination.

Samuel Taylor Coleridge has been described as an ardent student of geography, a tireless and insatiable reader of travel books and voyages of discovery. "Through wood and dale the sacred river ran,/ Then reach'd the caverns measureless to man,/And sank in tumult to a lifeless ocean," the opium-addicted English poet wrote in "Kubla Khan."

Many ascribe the sleuthing genius of Sherlock Holmes to his profound knowledge of geography. After his supposed death-plunge in the arms of Professor Moriarty over Reichenbach Falls, he traveled to some of the most exotic places on earth and learned the customs. India's deadliest snake (the swamp adder), discussions with the head Lama in Tibet (which barred other foreigners until 1920), medicinal cures from Sumatra, the aborigines of Adaman Island, a visit to the Khalifa at Khartoum—all supply clues and knowledge to solving a mystery that would have eluded any other detective. And how casually, on return to London, Holmes described his travels to Dr. Watson: "I then passed through Persia, looked in at Mecca. . . . Returning to France, I spent some months in a research into the coal-tar derivatives, which I conducted in a laboratory at Montpellier. . . ."

The American novelist Pat Conroy, in the popular *The Prince of Tides,* page 1, lines 1 and 2, writes, "My wound is geography. It is also my anchorage, my port of call."

Scale is an important issue in cartography. In an atlas, the scale on a single 8½-by-11-inch page is 1:140,000,000; one inch on the map represents 140 million inches on the actual surface of Earth. A map of the United States would be 1:20,000,000. The usual street map of a city is at a scale of 1:50,000—about one inch to three quarters of a mile. The amount of detail that can be included depends on the scale: the larger the scale, of course, the more detail.

Changing the scale changes the concept of distribution. A room looks different to an ant, a dog, a child, an adult, a giant. I have seen a slide show that demonstrates differentials brilliantly. First we see a picture of a man close up. Click. We see the man on a beach as the camera pulls away from the site. Click. The man is on a beach in a region. Click. The man is on a beach in a region in a country. Click. The man is on a beach in a region in a country in a hemisphere.

Click. The man is on a beach in a region in a country in a hemisphere on a planet. Click.

The art of mapmaking has been revolutionized by computers, which can create an exhaustive world-data base in accuracy and detail. Distortions are reduced, new calculations pour out at the speed of light. Computers keep up with spatial processes that are changing a place's contents and connections. They are gathering, processing, and storing information on relief contours, ethnicities, economics, waterways—everything geographical, no matter how large, no matter how minuscule. Layers upon layers of maps can be called up simultaneously to give the geographer real-time data. The work is complicated, difficult, expensive, and time-consuming, but it is also accurate, rewarding, and current.

For the 1990 census, the United States Census Bureau created a computerized map of the whole country. Woven into the tapestry was every street, every residence, and every political boundary. But when the checkers went into the field, they found that streets and intersections were no longer there, or they were directed to lakes or uninhabited lots. Such errors mounted up to 20 percent of the geographical data, raising fears, particularly in the major cities, that the census could not be counted on for accuracy because the computer had worked with information 10 years old. The time lag had not taken into account the new geography—the developments of growth and recession, the changing landscape. A topologically integrated geographic encoding and referencing file can locate every political subdivision, every street, and every road—even what kind of street, what kind of road—every bridge, every tunnel, then store the data in digital form for updating day by day. The file *is* versatile. Its commercial uses are incredible. This is why corporations have begun to appreciate the practicality of mapping and the dynamism of place.

Computer techniques produce real-time animated maps. Generating "live," swirling color displays, the maps project the positions and the movements of major storms you see breaking out and gyrating on television news programs. They help to focus on relevant processes. Analytical geography requires flow maps. Photographing the planet from above provides an excellent perspective on deforestation, desertification, mineralization, and the greenhouse effect—serious problems that must be addressed by professionals from geologists to generals determining military strategy. (But I see that General

H. Norman Schwarzkopf, commander of coalition forces in the Gulf War, complained that his headquarters was unable to obtain reconnaissance information—updated maps, presumably—that was less than a day old.)

But to most of us, average map users, a map is still a guide to the directions and the distances between places. Inuits before "civilization" looked at those labeled distances on their maps entirely by the amount of time it took to travel from one place to another. Alaska's Point Barrow and Nome were not 525 miles apart, they were 10 days apart. It is indeed easier to give directions for getting somewhere if a sketch, a map, can be supplied. Most people have more faith in such information when it is presented cartographically rather than orally. Rodney Dangerfield tells of being directed along a road in a town new to him: "You turn left at the church that isn't there anymore," he is told, "and go past the billboard that was taken away by the local do-gooders . . ."

IN SEARCH OF THE ONCE AND FORMER SOVIET UNION

▪▫▪▫▪▫▪▫▪▫▪▫▪▫▪▫▪▫▪▫

Comrade, believe.
She will rise, the star of captivating happiness;
Russia will arise from slumber,
And they will write our names on the ruins of
    autocracy. —PUSHKIN

Who of us can say anything today about the disintegrated Soviet Union that will be certifiably true as late as tomorrow? Can we even say that there will be anything we can lastingly call a union or a confederation or a commonwealth? One thing is for sure, there will always be a Russia.

But whatever the reforms and changes, I shall always remember Russia and the various parts of the Soviet empire as a geographer's paradise. My more-than-30 visits to the world's largest country gave zest to my own life as a geographer. It was my privilege to explore more and more of its seemingly endless expanses and its contrasting cultures, climates, and terrains, and to make the acquaintance of many of its fascinating heterogeneous peoples. And, of course, to contemplate the why of that extraordinary where and the spatial processes spanning 11 time zones and remarkably varied places—cities, republics, steppes, inland seas, and regions.

Not even the keenest of us old "Soviet hands" foresaw the complete collapse of Communism there. It had been "74 years to nowhere." In hindsight, we now wonder that it didn't happen sooner. The Gorbachev reforms, including glasnost and perestroika, clearly sounded the death knell for central planning, with its five-year plans, and the bureaucratization that had failed to give the people a tolerable standard of living, let alone the most basic freedoms. We observed the inefficiencies of a rigid system and accepted them as standard and enduring. We witnessed nationalism and ethnic identification in some of the republics and the yearning for religious and intellectual freedom of expression—and understood its suppression by a coercive system. How shocked the world was when repression was removed and the country became a map of conflicts: places and nations set against one another.

On my first visit to the Soviet Union, more than three decades ago, I became aware of the almost fanatical importance of place and space in the Russian psyche. This primacy of space reflects attitudes and behaviors on the part of the people. At the outset, I was deeply, and negatively, impressed by Soviet borders and boundaries. They were everywhere. Miles of barbed wire, growling dogs, mine fields, armed and angry guards on the boundaries and even guards at the entrance to the hotels and the university, all contributing to an instilling of irrational feelings of guilt. The Soviet Union was fanatical about protecting its space and, within the country, its spaces and places. When one intended to leave the country from any port, one understandably became fearful that permission would be denied. The bark was always the same: "WHAT ARE YOU TAKING OUT?"

Early into conversation, new acquaintances would inquire, *"Otkuda vee?"*—"From where are you?" (One word you will rarely hear again in greeting is *"Tovarishch"*—"Comrade.") This question can lead to warm and emotional exchanges about the special characteristics of one's home or region. Citizens of St. Petersburg wax poetic about "white nights," the canals, the Hermitage, Stravinsky, Pushkin, Dostoevsky, their window on the West, and the 900-day Nazi siege. Irkutsk residents are reduced to tears contemplating the beauty of Lake Baykal or the deliciousness of Siberian pelmeni (a meat dumpling dish). Literature, art, music, and folklore reflect regions and places—the steppes, the Volga, the taiga. The word for homeland is *rodina,* a feminine noun that rolls off the tongue lovingly and can be loosely translated as "place of birth."

There are many positive aspects of this Russian/Soviet attachment to place and space. Wherever one goes, there seem to be local and regional museums that reflect history, natural wonders, and artistic endeavors of the area. There is an enormous richness in regional variations of accent, food, folklore, customs, music, languages, and manners.

To the first-time visitor, Moscow appears somber and forbidding. But if one arrives by railway, one might still find the station a colorful

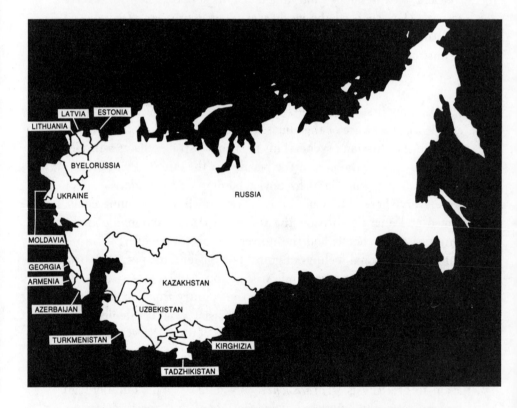

The Union of Soviet Socialist Republics, when it consisted of 15 republics following the forceful incorporation in 1940 of the three Baltic states, Estonia, Latvia, and Lithuania, was for a half century the world's largest single united political unit by far. Several former republics, including the three Baltics, successfully tested Mikhail Gorbachev's glasnost in the wake of the attempted coup in August 1991, agitating for independence from the far-flung empire at the very time that Western Europe's many countries were coming together as an economic powerhouse, the EC. Save this map.

thoroughfare of Russian peasants, Caucasian peddlers, Central Asian cotton workers, Siberian miners, and Arctic reindeer herders. The cavernous waiting room may be filled with idlers and children and crowds of transients. The Babel of languages creates a cacophonous symphony. Sad and shabby prostitutes (who didn't exist officially in the 1960s) ply the crowds with rates painted on their shoes.

That was the Russia I encountered, "Mother Russia" in all her diversity. Such a kaleidoscope of impressions only whetted my appetite to take my explorations to the farthest corners of the empire.

When I first arrived there, Khrushchev had long since castigated the murderous regime of Stalin. But the Communist system had changed little. There still existed the slow, smothering bureaucracy, suspicion, and xenophobia, and the impervious and ominous supremacy of the Party. It was a society with few consumer goods, with an immense housing shortage, long lines for everything, and an ardor for socialism.

I traveled and I marveled at the people and their environments and the spectrum of geographic variations. The three Baltic republics were Western in culture and held an intense hatred of Moscow and everything Russian. By contrast, Samarkand, Tashkent, and Bukhara in Central Asia were rural and superficially Islamic and Turkic. When I went to Tbilisi University in the then-republic of Georgia, where Stalin was born, I was met at the railway station and welcomed as if I were arriving in a friendly and especially blessed country. My hosts greeted me with flowers in February and the words "Welcome to Georgia, you have left the Soviet Union." Almost treasonous words at that time! The Siberians and their environment were tough and indominitable. The "old-timer" residents of Siberia had an openness and guilelessness that clearly distinguished them from recent immigrants, usually officials who had brought Moscovite and European attitudes with them. They proudly referred to themselves as Sibiriyaks.

Such astonishing differences within the U.S.S.R. reflected the geographic mosaic that made up the empire (once Tsarist, then Soviet, eventually who knows?). Most of this empire was made up of pieces of territory belonging to various nations with their own cultures, languages, and ethnicities. The empire juxtaposed more than a hundred nationalities speaking in tongues incomprehensible to their neighbors. Before the First World War and the birth of the Soviet

The end of the road. A bust of the founding father of Soviet Communism and first premier of the Soviet Union (1918–1924), Vladimir Lenin, is carried out on a stretcher from a military school in Moscow in late summer 1991 as the death knell for the Party reverberated throughout the expiring Soviet Union. THE BETTMANN ARCHIVE/REUTER

Union, Russia had been annexing the territories of its neighbors at the rate of 18,250 square miles a year for a period of four centuries. (Three geographic factors led to Moscow as capital: its location in the mixed forest region protected the people against mounted enemy attacks; its location at the convergence of three rivers made trade and communication with others easier; its location in woodland provided building materials.)

What went so disastrously wrong? Generalizing, roughly, observers point out that a rigid ideology was superimposed with vicious coercion over geography, over people, over ideas, and over the economy. The resultant ideological and political systems were such that they made the country extremely inefficient and seriously flawed.

The metamorphosis of the Soviet Union from a rural society into a highly urbanized one occurred with extraordinary speed, a speed possible only in a forced economic system. A series of five-year plans initiated by Stalin emphasized heavy industry and technology over the needs of the citizenry: consumer goods. They propelled people by the hundreds of thousands out of villages and into cities.

Economic development created wholly new places. More than 1,200 were built from scratch in the years between the Revolution and the advent of the Gorbachev era in 1985. In 1926, two cities in the U.S.S.R. had populations of over a million people. Today, more than 20 cities have populations of that size, and some 300 cities have over 100,000 citizens. The urban population exceeds 70 percent. However, the country is so large that many regions are devoid of cities and are out of the mainstream of development; for example, northeast Siberia.

The redistribution of population and the growth of cities were not part of a master plan. The serious problems aside, urban growth was hailed as good spatial process and a towering achievement. There are those who claim that the achievement reflected the views of Marx and Engels in *The Communist Manifesto:* rural life is "idiocy." However, many of the poorly housed and deprived Soviet urban population may have preferred rural idiocy to their urban barrenness.

Although Engels deplored the misery of life in capitalist cities, he noted that "only as uniform a distribution as possible of the population over the whole country, only an intimate connection between industrial and agricultural production, together with the extension of the means of communication made necessary thereby—granted the abolition of the capitalist mode of production—will be able to deliver the rural population from the isolation and stupor in which it has vegetated almost unchanged for thousands of years." Seventy-four years of Marxism produced few of Engels's goals, but they did produce an antithetical solution—a search for a free marketplace.

Americans read about a bountiful harvest in the Soviet empire but kept seeing television news programs showing Soviet residents lining up and waiting hours for a few potatoes and carrots and a slice of sausage. "How come?" I have been asked. Easy: there was a breakdown of the food growing and distribution systems. Farm regions, even whole republics, were hoarding produce for local consumption or losing it to waste and corruption. Most of the abundant harvest of 1990, for example, never made it to market. Built-in weaknesses became exacerbated in a period of social breakdown: ethnic unrest, the collapse of discipline, the erosion of all economic and distribution arrangements. The food industry learned that a system based on a command system works only as long as people obey.

Losses of 15 percent of the grain and 30 percent of the vegetables along the way from soil to store was considered acceptable "spillage." More than half the losses occurred at the farm itself because of inadequate machinery and tools (including fuel and batteries) and manpower. Few farms have enough trucks to take care of a big harvest. There is a shortage of freight cars.

On the other hand, in Moscow and in most other large cities excellent produce and meat and fish are available at the numerous "free," "capitalist" food markets, which have sprung up in profusion. But the new entrepreneurs, heady with the first whiff of private enterprise, know how to fix markups to insure a profit; such outlets have become too pricey for many.

One of the many indignities heaped on the citizens and the republics was the arbitrary or politically motivated redrawing of regional boundaries or erasure of boundaries surrounding the space of distinct groups. Such callous decisions rendered by Stalin and his henchmen haunt the new leadership.

Uzbekistan, a Central Asian republic populated by Islamic Turkic groups, was never satisfied with Russian rule. It issued a declaration of economic independence, demanding control over the republic's resources. The Crimean Tatars and the Volga Germans lost their space and their special political administrative units after Stalin accused them of collaboration with the Nazis during the Great Patriotic War.

Muslim Central Asia and the Islamic part of the Caucasus were conquered during the 19th century, after throwing up fierce resistance. These regions were carved up to form six republics. This "sovietization" was decreed by the Bolshevik government during Stalin's reign of terror. Kazakhstan, a republic with nearly as many Russians as Kazakhs, harbors many complaints as the result of Russian, and later Soviet, colonialism.

More than 60 spatial/place injustices caused by old boundary claims, forced dispersals, and environmental disputes have flared into open conflict. To understand one of the main political/economic problems of post-Brezhnev U.S.S.R., one must know its ethnic geography and the changes imposed on the original spatial arrangements. In short, place and space are intimately entwined with ethnicity and history.

In the secessionist fever that began gripping the republics of the Soviet Union, Georgia was among the most clamorous. Georgians proudly remember their heritage as a kingdom that sought protection in 1801 from Tsarist Russia. In an effort to squelch dissidents, Soviet troops in 1989 armed with shovels and tear gas killed people attending peaceful demonstrations in the capital of Tbilisi. The tactic only made the activists more radical and led to a Georgian declaration of independence in 1991.

Ukraine, which has had religious disputes between the once-oppressed Ukrainian Greek Catholics and the Ukrainian Orthodox, is no less fiercely bent upon a drive to reestablish its identity. Ukraine has a population of 52 million and boasts a fifth of Soviet industry.

Lenin once confided to his fellow Bolsheviks: "If we lose the Ukraine, we will lose our heads." Stalin tried to remove the threat of Ukrainian secession once and for all by breaking the spirit of the Ukrainian nation. In the 1930s, he created an artificial famine to crush the resistance of Ukrainian farmers to collectivized agriculture.

The history of the empire is also one of migrations. All major cities in the largest republics have large and often majority Russian populations. In all, 1.3 million people left what used to be the Soviet bloc in 1990, the year after political upheaval swept away Communist governments across the region.

In the ethnic bazaar that was the Soviet Union, the growing nationalism of some of the republics stimulated similar feelings among some of the Russians themselves. Many conservative Russians would like to see a reincarnation of the age of great Russian literature and architecture and art and music. Many feel that all that made Russia distinguished and unique was destroyed by the Revolution and the atrocities committed by a Communism promising a better life for all citizens. From 20 million to 50 million Soviet people had their lives snuffed out by government-sponsored terrorism, and millions more as an indirect result of callous Soviet policies and practices.

Severe environmental problems, attributable largely to a Moscow-directed economic system, are everywhere. They have led to vociferous protests. Regions and ethnic areas have risen in anger to block nuclear-plant construction and other major projects. They have forced "unsafe" plants to close and demanded local control. It is estimated that 15,000 "green" groups have sprung up across the country.

One of the current environmental sagas is a tale of two lakes.

Lake Baykal, in the southern Siberian mountains north of Mongolia, is the longest (395 miles) freshwater lake in Eurasia and the world's deepest lake (5,715 feet). Its staggering volume of 5,500 cubic miles contains a fifth of all the world's fresh water. At stake was the very conservation of the old lake itself. Baykal harbors a unique ecosystem, including 1,500 kinds of plants and animals found nowhere else. Baykal, declared a Siberian activist, contains "such pulchritude as to be unimaginable this side of paradise." Despite the designation of a protected area around the lake, there were efforts to create polluting enterprises in the basin. Geographers effectively warned that paper mills and mining and processing of nearby phosphorous-containing ore, including the storage of its waste products, would lead to irreparable damage of the environment. The Baykal victory united the input of geographers, biologists, ecologists, economists, and geologists.

There is not a happy ending to the story of the inland, salty Aral Sea. It has been devastated in a horrendous man-made disaster. Its main feeder rivers, the Amu Dar'ya and Syr Dar'ya, were largely diverted to irrigate Central Asian agriculture, particularly rice and cotton fields. The Aral receded 50 miles, then 60 miles, then 70 miles, until the sea became a virtual desert facing extinction. It is a grotesque sight, a graveyard of ships, rotting, capsized fishing vessels in a field of cotton, rusting docks tens of miles from the water's edge.

The incredible shrinking Aral once supplied 6 percent of the fish for the entire country. The fish are now essentially gone. As the sea contracted, the salinity increased. Meteorological events came into play and exposed the sea bottom; salt deposits devastated the area. Peasants are losing their crops and being forced to move away. Chemicals and pesticides add to the damage. In its dwindled state, the Aral, which once supported a fishing industry employing 60,000 people, no longer exerts a moderating influence on temperature. Growing seasons have been reduced. Soon, it will be a lifeless brine lake. The dry salt is already whipping around in the atmosphere. General health is deteriorating, typhoid is rampant, tuberculosis cases are doubling. Infant mortality has soared. Livelihoods have been lost. The sea is dead for all practical purposes. The Chernobyl catastrophe took place in a minute. The Aral disaster, a sustaining disaster, is described as "creeping Chernobyl."

No official heed was taken of the alarms. The Stalin mentality of man conquering nature via a state plan imposed by Moscow has had tragic consequences. No one thought about ecology and the consequences of actions. Ecological situations in the former U.S.S.R., as they are elsewhere, result from human disrespect for nature and from interference in processes over space.

Finally recognizing that concern about the environment is not a matter of ideology and empty claims, steps are being taken to stop the catastrophic damage. New governmental committees, new legislation, and the formation of thousands of ecological groups are signs of a changed mentality. Correcting mammoth environmental problems will be enormously difficult because of the huge costs involved in reversing destructive processes and the breakup of the Union. As for the Aral Sea, its liquidation is a symbol of mismanagement, a memorial to years of ideological excess.

Even before the 1991 collapse, the all-encompassing Communist Party had begun to lose its special status and credibility as the "leading" institution in society. There were at least 25 new political parties in the Soviet Union, ranging from Christian Democrats to Anarcho-Syndicalists and Monarchists. There are over 60,000 informal organizations as diverse as National Fronts and reactionary clubs, such as the anti-Semitic Pamyat, and Russian monarchists who would like to restore the Tsar, who would resurrect the *modus vivendi* that was terminated so abruptly in 1917. The once monolithic national trade union, to which all workers belonged, fell into complete disarray.

During the early '90s, cities and political units throughout the Soviet Union were demanding that shoppers in local stores prove residency by showing identification. Many areas, regions, or republics were refusing to ship products, especially foodstuffs, across their boundaries. Others, acting cooperatively, were signing agreements with one another to assure supplies of all kinds of goods.

New labor unions emerged. Most groups threw out their party affiliation and launched strikes. Strikes, once illegal, became characterized in *Izvestia* as "the most common form of communication with the Authorities." In a few months in 1989 there was an average of 50,000 workers on strike every day.

In the spring of 1991, the Soviet government transferred control of the Siberian coal industry to the Russian republic. It was a milestone, representing as it did the first transfer of a major industry from the national government to that of a republic. (Russian president Boris N. Yeltsin had successfully complained to the Kremlin that the republic of Russia—which contained two thirds of the Soviet Union's land area and half its population—effectively controlled only 16 percent of its wealth.) The transfer of control had been one of the principal demands of strikers who had sharply reduced output at vital mines in Siberia.

Marx said that Communism wouldn't catch on in Russia, but it was a Russia that was agrarian. The irony is that Marxism was forced on a nonproletarian Russia in 1917. In 1991, an industrial proletariat was rejecting Marxism. The conclusion, apparently, is that new ideas can be imposed on a weak, destitute, rural population more readily than on an educated, urban-industrial proletariat.

I did not hear of, nor see evidence of, starvation in the Soviet Union on my visit there in the summer of 1991. There was some homelessness, some unemployment. For democracy to work, and to pay the rewards of a free-market economy, time and a commitment are needed. The transition will be painful, long, and could go in any direction.

In the United States, people blend, but they haven't been assigned specific pieces of space. Russia and the Soviet Union expanded over peoples who were already in places they had bonded to and considered their sacred hearth. They conquered those peoples or spread their sovereignty over them. The empire with all those ethnic and nationalistic units was a hodgepodge, manipulated and treated with little attention to ethnic sensitivity. Countless nightmares must be resolved as a consequence of such a spatial, political policy. One could not understand the complex problems of a dissolving U.S.S.R. and the creating of a set of new states without understanding the

The policy of a state lies in its geography.—NAPOLEON

"The Colossus of the Nineteenth Century," as the brilliant Corsican-born French emperor with multiple sobriquets was hailed, was routed on the Russian front by geography: a huge spatial expanse with disrupting processes that cross it—arctic fronts and blistering winds. The Grand Armée's lines of supply and communication were much too long, and an unforgiving winter decimated the troops (as they exacerbated Hitler's assault more than a century later). Only 10,000 of Napoleon's force of 422,000 soldiers escaped death or capture. On the Italian front, "The Little Corporal" had manipulated time and space with an ingenious semaphore telegraph that carried messages from mountaintop to mountaintop Paris-Rome, Rome-Paris in four hours each way, giving the "Gallic Caesar" an enormous advantage over his foes. The spaces of Russia were less amenable to manipulation with the technology of the time. THE BETTMANN ARCHIVE geography of ethnic groups, the geography of ethnic political units, the geography of ethnic administrative politics, and the historical geography of the regions. The Soviet empire, in whatever form it eventually takes, is a fascinating set of geographic puzzles.

For mainly economic reasons, women have long been members of the work force, and have felt the pressure to keep families small. There is the problem of space deficiencies also to be reckoned with. Abortion is the principal method of birth control. Most women there have multiple abortions, from six to eight. More fetuses are aborted than are born. The country, in fact, has the highest abortion rate in the world—more than 7 million abortions a year: that's 131 for every 100 births. Among the republics, the highest rates were in Russia. (They also were high for the former Baltic republics of Estonia and Latvia; they were not high for Lithuania, which has essentially a Catholic population.) The lowest rates have been among the traditional non-Western cultural populations in the Central Asia region and the Caucasus.

On the general subject of health among the Soviets, it could be said that the combination of rapid industrialization, inadequate medical services, and environmental degradation had a devastating effect on life and death. Infant mortality and adult male death rates soared. A male infant born in, let's say, Moscow had a life expectancy of only 64 years, which is 8 years less than a white male born in the United States. Urban, industrial, Western life-styles seemed to be worse for Soviet males than rural, agricultural life-styles; the earlier death rate for males was highest in the European section of the empire than in the southcentral section, further evidence that some places are better for promoting longer lives.

From a geographical point of view, one of the empire's historical difficulties was its vastness and the difficulty of connecting the many places with a communication-transport system that would truly unite the state. The steppes, or great plains, are surrounded by ice on the North and rugged mountains in the South. (The steppes on both flanks of the country allowed incursions from the Swedes, Germans, and French from the West and Mongol hordes from the Southeast.)

In the North, the land is protected from invasion; no enemy is about to cross the Arctic ice and waters except with great effort and high technology. In the South, high mountains; in the East, the Pacific. Surrounded by these barriers of nature, the people were isolated from an important series of cultural events in Western Europe; among them, the Renaissance and the Reformation. I often think of Russia-U.S.S.R. as having missed much of the "civilizing" forces from the West, though 70 percent of the population lived in the European part of the empire. There is an exotic aura about the whole country, as reflected in an old saying, "Scratch a Russian, find a Tatar."

There is, as well, the remoteness of mentality, a remoteness self-imposed out of paranoia, suspicion, and a profound fear of the rest of the world. The Russian bureaucracy earlier and the Soviet more recently distrusted its own people, keeping them separate from others both at home and abroad, and controlling information flow. I recall being in Moscow during the Cuban missile crisis of October 1962. While my countrymen back home were having a "nervous breakdown," thinking they were on the verge of atomic doomsday, we in Moscow were blissfully unaware of the situation and didn't learn about it until much later.

As I write several months after the failed coup ostensibly brought the curtain crashing down on Soviet Communism, the future of the individual republics was problematical. The greatest threat to civil liberties is no longer Communism but violent nationalistic passions unleashed by the overthrown system. Many republics, now new states, are already quarreling over the spoils of the old empire.

The three Baltic countries declared and won their independence, but with all the interdependencies between them and the former Soviet Union, especially Russia, over decades, it may be years before they are truly free.

There were few concrete steps being taken toward power sharing among the former republics to prevent a chaotic disassembling of the Soviet empire. Privatization of industry and housing and land seemed to be in the works, though no one had come forward to take charge of implementing economic planning.

Outwardly, the coup and its aftermath seemed not to have

changed much. There still were long lines, and shortages remained acute. Most predictions were that things would only get worse. Most leaders were generally assumed not to have the expertise or the support to come to grips with the economic and political crises they faced. Many reformers, intellectuals, nationalists, and religious leaders had fallen out, after they had lived to witness the death of Communism. I doubt that any of us professional Soviet-watchers has a quick fix for the complexity of woes burdening that puzzling and conflicted land.

Tottering on the brink of economic collapse, the entity once known as the Soviet Union is a colossus that has never had a significant entrepreneurial or capitalist culture. The old system became bankrupt, completely bankrupt. The transition to an open-market economy will be a long, painful process.

I believe that in the long run aid from the West, Japan, and the Asian Tigers should take the form of advice, shared expertise, and resources available through numerous contacts, joint commercial ventures, and exchanges at regional and local levels. Some of these contacts already exist in the form of sister-city programs and university exchanges. Regions, cities, and even local groups are seeking help directly. It is in everyone's best interest to respond at levels where impacts will be more direct, and certainly the United States will want to do all it can to spare people the ravages of famine.

It may be that the republics of the old empire will evolve into something resembling the contemporary European community—that is, a confederation or commonwealth based on egalitarian principles and voluntary membership. Whatever the outcome of the intricate political and economic processes during the period of great transition, be assured that Russia and a large part of the old union will survive and play their rightful role, cultural, political, and economic, in what we can all hope will be a "warm peace."

CULTURAL INTERCOURSE

AND

LANDSCAPES

There was a time when the very word "culture" would bring a glaze to people's eyes. Soviet authorities spoke of Soviet culture, but it was a "culture" dictated by the rigid boundaries of party lines. In the United States a few years ago, a book was published that proposed to be a guide to "cultural literacy." Its listing of 500 things that "literate Americans know"—abominable snowman, Muhammad Ali, fireside chat, lyric, Dolley Madison, Tin Pan Alley, "You Are My Sunshine"—propelled the book to the top of bestseller lists. It is hardly the last word on the issue, or even a good word. And it is truly arrogant to think one can limit such a list to a finite number— and why only Americans?

Culture in its broad application is of course too complex a process to be containable in a few hundred facts easily committed to memory and dispensed in tables like latitude and longitude. Humankind throughout history has left its impression on the planet in ways that vary over time and space because people are agents of change. Our structures and artifacts progressively transform natural landscapes into cultural landscapes.

Cultural geographers are interested in the way humankind uses its natural resources—the way it moves from place to place, produces food, builds houses and roads, and puts products into circulation. The process is called land use. Geographers examine land use of a particular region, which is usually defined as a physical area occupied by a specific social or cultural unit.

There are some cultural geographers, however, who regard it their duty to pinpoint the state of every Doric column or lighthouse that has graced—or disgraced—the surface of the planet. To me, this is much ado about trivia and best left to experts in other fields. Cultural geographers have a role to play in solving problems that matter. We should focus on such fundamentals as language, religion, and customs, so that we can bring an informed understanding to the perplexities, conflicts, and travails of peoples adrift in a contentious world.

Cultural landscape has been defined as the composite imprints superimposed on the entire physical landscape over the globe by human activities. Earth is dynamic because it is in constant change, albeit in a framework of time much greater than the human life span.

It is almost universally accepted that humans and the environment are co-equal partners in an interacting unity, with humankind the agency of the systematic modification of the physical landscape. Nature, however, periodically reacts angrily when mistreated and abused.

Cultures are patterns of learned behavior that leave permanent traces and are transferred from generation to generation, from group to group. Studying such activity is the role of the cultural geographer. Human cultures are many, diverse, complicated, and ever changing. Every generation hands along to the next the sum of everything it knows and has lived by. There are cultural differences within countries and spatial variations within cultures. Processes of migration change cultural levels and cultural types. The Latin appellation for the sacredness of a place is *koltus*—culture.

The cultural geographer examines all the elements that have gone into the making of a culture. These include religion, language, folklore, art, family structure, sexual mores, education, political systems, and economic structures. They can be reflected in our tastes and fashions, our housing and gadgets, our school curricula, our customs, our artistic expressions and entertainments, and our attitudes toward work and leisure.

Language is one of the great linking devices in human culture. As we all know, it is also a great divider. Language can be viewed at many scales and has a varied spatial scale and distribution. Excluding dialects, there are more than six thousand languages in the world. Indo-European languages, some global, some local, are spoken by half the world's population. In terms of spatial extent, they have great variations. English and French are global languages. Mandarin happens to be the language spoken by most people in the world, but it is spatially concentrated, limited mostly to mainland China. In the Caucasus of the former Soviet Union, hundreds of languages are spoken in a small space, an example of tremendous complexity and variation in a circumscribed area.

Man-made landscapes of fear are everywhere, from drug-ridden neighborhoods of murder in inner cities in the United States to famine-plagued Sahel areas of Africa. In Cambodia, Pol Pot's Khmer Rouge regime eliminated perceived opposition and tortured and butchered upwards of 2 million people. Skulls were cleaned, numbered, and stacked on the killing fields. WIDE WORLD PHOTOS, INC.

In the fastest and widest expansion of prehistoric times, the Austronesian languages of the Pacific spread across 6,000 miles of coastline and sea within 1,500 years. They were carried by expanding agricultural populations into regions that were either empty or sparsely inhabited by groups of foragers.

Some languages are threatened with extinction. The Gaelic language is kept alive almost solely by the passionate efforts of Irish patriots and Scottish Highlanders. The Welsh language has its devoted adherents keen on protecting it from the almost universal acceptance of English. Cornish, once the language of Cornwall, in the far southwestern corner of England, is no longer spoken by many. In Africa, nearly a thousand languages are spoken, but they do not travel from one tribe to the next. Many languages have expired altogether.

Language, like other cultural factors, is usually in flux. That is because language moves over space and interacts with other cultures. Only isolation "freezes" this evolutionary process. Many South American Indian tribes are a case in point; they have had little or no contact with outsiders since the invasion of the Spanish conquistadors nearly five centuries ago. The speech of people in some remote settlements in the Appalachias reflects their place of origin in the British Isles centuries ago.

It is hard to exaggerate the role of religion in the geography of cultures. Religions were among the first major sources of human identity. In time, nationalism was often superimposed on religion, but it never replaced or buried it. Some religions today are more passionately and widely embraced than they ever were.

Differences in religion have been the harbinger of more armed conflicts and bloodshed, misunderstanding, and eternal enmity than any of us care to think about. How sad and ironic that a spirited force that should have elevated us and brought us closer together has so often torn us asunder. Religious wars through the centuries have taken as heavy a toll as all the other forces that drive men to kill one another.

Religions take root, spread, and breed schisms. Christianity, born in Palestine two millennia ago, has found a home in every continent. In Northern Ireland, two Christian religions have been fighting to death over a small territory; it seems to be a hopeless conflict, causing a continuous flow of blood among peoples who should be living side by side in a harmonious community. (I know, there are those who argue that this is essentially a political struggle, not a religious one, with the minority Catholics wanting complete freedom from British-Protestant rule, but I believe it is more complex.)

Islam began seven centuries or so after Christianity. It was born in the Middle East, then cut a wide set of arcs eastward, northward, and westward, eventually crossing the ocean to North America. As it moves across space and time, Islam wears many faces, often a warring one.

The miserable, historical, violent, and blood-soaked split between India and Pakistan is based on nothing more than the conflict between the Hindu and Islam religions. Pakistan used to be one country divided east and west. East Pakistan was primarily Hindu, and West Pakistan Islamic. Today, East Pakistan is known as Bangladesh and West Pakistan is simply Pakistan, and their relations are hardly brotherly.

In 1948, the world peace-seeker Mahatma ("Great Soul") I. K. Gandhi was assassinated by a fellow Hindu for his efforts to resolve the strife between Hindus and Muslims. Mrs. Indira Gandhi, India's first woman prime minister, agreed to considerable autonomy for western Punjab's 7 million Sikhs, but her concessions fell far short of demands for a Sikh Khalistan, and she was shot dead— riddled, as noted earlier—by Sikh members of her personal bodyguard in 1984. The assassination in 1991 of Mrs. Gandhi's son, apparently by Tamils, who have been seeking their space in Sri Lanka, seemed to be more factional than religious. Hinduism in India cannot subdue a multitude of different peoples and groupings in contention with one another. The assassination of Razhiv Gandhi, who was seeking re-election as prime minister, may well set back India's drive to a genuine democracy.

Judaism is the oldest religion that worships a single godhead (with the possible exception of Zoroastrianism). Jerusalem, the capital of Israel, and the "umbilical cord of the world," may be the most religious city in the world. Its uniqueness is that it offers a spiritual home for Jewish, Muslim, and Christian faiths alike. But peace and tolerance do not always prevail there.

Cultural geographers deal with such oddities of religion as why northern Sudan is Islamic and rural and southern Sudan favors

Cultural geography! The premier of the People's Republic of China, Deng Hsiao-ping, in a 10-gallon hat waves from a stagecoach in a Texas rodeo during a tour of the United States. The Kipling refrain of "East is East" notwithstanding, the East-West twain has had marked meetings for ages. East and West are relative and arbitrary descriptions, coined by Europeans creating longitude with the zero-degree line through Greenwich, England. There'd be different terms if the global grid system had been initiated in Delhi or Toronto.
WIDE WORLD PHOTOS, INC.

Christianity and animism; the schism compounds overriding problems of drought and starvation.

Cultural regions of the United States have been examined. The conclusion, sometimes, is that political boundaries between the states don't make a lot of sense. Citizens of neighboring states often share, because of cultural factors, a closer relationship than they feel with other residents of their own state. Idaho is a good example. Northern Idaho feels a closer kinship to the state of Washington than it does to southern, Mormon Idaho. The coal miners of western Pennsylvania and eastern Ohio are more allied than people living in other parts of each state.

It has been argued by some political scientists and cultural geographers that if dividing the United States into "appropriate" units were the goal, cultural regions would be the touchstone instead of the historical and near-meaningless political lines that divide the states. How much sense did it make to divide the former Dakota Territory into North Dakota and South Dakota, both of which share a remarkably similar terrain and agricultural output? Kentucky and Tennessee, midsouthern neighbors, are sufficiently similar to be one political unit (state).

Our culture is reflected in the sports we play and who plays them. Skyscraping basketball players tend to be black, many from the inner sectors of New York, Philadelphia, Chicago, Detroit, and Los Angeles. Baseball favors warmer climes; there have been twice as many professional players from the southern United States as from the north. Hockey, on the other hand, was exclusively an outdoors diversion in the cold winter blasts of Canada until ice-making machines were developed for use in indoor arenas. (The medium was once again the message.)

Western Pennsylvania was home to an extraordinary number of young men who went on to become outstanding quarterbacks in the National Football League. Joe Montana grew up near Monongahela, a coal-mining town 25 miles south of Pittsburgh. Before him were Joe Namath of Beaver Falls, Johnny Unitas of Pittsburgh, and George Blanda of Youngwood. Dan Marino and Jim Kelly, current NFL luminaries, also grew up in western Pennsylvania.

Another aspect of culture is music. It is often said that jazz is the one unique art form America has given the world. Jazz evolved from the blues created by blacks giving expression to their hardships and griefs. Country music was born in the isolation of Appalachia, with its origin in Anglo-Celtic balladry. It became localized to express, simply and directly, human themes that touched a common response: love and hate, happiness, heartache, jealousy, infidelity, desertion, death. Country and folk became infused with western to mesh into a kind of Everyman's language communicating universal messages. Most of the world's extraordinarily great violinists have come from eastern Europe.

The ravages wreaked upon the African continent by European businessmen severely disrupted indigenous cultures. Europeans carved the continent into bits and pieces and threw up illogical boundaries. These were nothing more than land grabs based on resources and the claims of explorers. Boundaries cut willy-nilly across tribal units, which had always been the social core of most of Africa. With

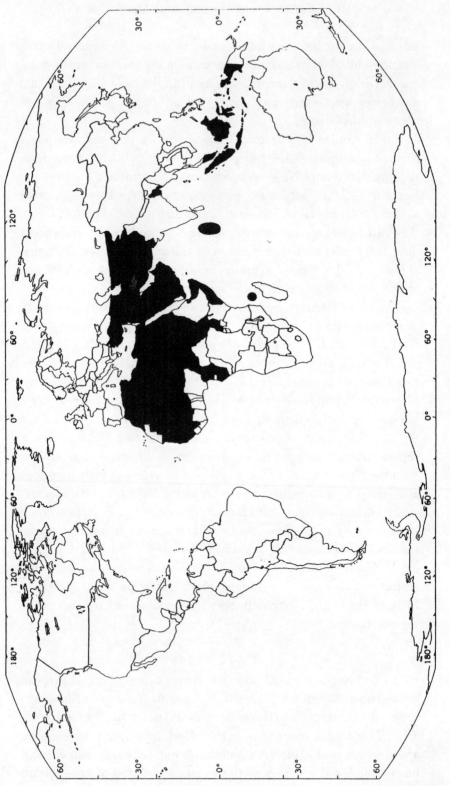

Black regions indicate the Muslim world.

improbable boundaries in place, cultural ties were violated and inefficiency reigned. Railroads were built from the resource to the coast. The tracks were like so many spokes converging upon a port, where the raw material, including the manpower, was loaded onto ships and sent to ports abroad.

Communications in Africa worked pretty much the same way. At the beginning of the decolonization process, in the 1950s and '60s, many newly independent countries could not make a direct call to a neighbor 15 miles away; the call had to be routed through Paris or another European capital. Someone could drive a car to the nearby coast but not into the next country. There were few systems of contact even within the country. These exploited peoples inherited their own country but were untrained for government. Little wonder they have had trouble developing, and assuming the burden of independence. Many of the countries are actually too small to sustain themselves. Animosities have erupted across contentious boundaries. Civil wars, rooted in tribal rivalries and cultural differences, do not point to an early resolution of the sub-Sahara's latent conflicts and open disunity.

All changes, as I have noted, can be traced through space and time. The very food we eat can have the most remote origins. A grain of rice grown in a Far East delta scatters its seed around the tropical-temperate world to become a diet staple of hundreds of millions of people. Corn (maize), the staff of life for Native Americans for centuries, is today grown on several continents, though there are some people, notably the French, who disparage corn as hog food and not people food. The increased movement of people throughout the world

---

A remarkable cultural diffusion process originating in Medina and Mecca, the two most important holy cities of Islam, on the Arabian Peninsula, led to the "Muslim region" of the globe. The spread was sometimes aggressive, sometimes repulsed (e.g., by the monarchs of Spain in 1492). Muslims, who profess belief in Allah as the sole deity and in Muhammad as the prophet of Allah, have been extraordinarily effective in imposing a lasting culture over millions of square miles and many generations of humans. Muslims accounted for nearly 20 percent of the population of the former Soviet Union. The overwhelming victory of the Islamic fundamentalists in Algeria's first free parliamentary elections could presage Islamic-world democratic elections yielding Iran-style dictatorships: can God be voted out of office? The geography of religion permeates the globe. American preachers have been indoctrinating Catholic Latin America with Protestant Fundamentalism.

has broadened gustatory horizons by diffusing cuisine over space and incorporating once-exotic foodstuffs into our own culture. A dinner served in a Boston restaurant can be every bit as Thai-authentic as one served in Bangkok.

"The Columbian Exchange" is a sample of such interaction. The Admiral of the Ocean Sea's four historic voyages west across the Atlantic, in 1492, 1493, 1498, and 1502, initiated an exchange of microbes, plants, and animals. The Europeans took to Native Americans death-dealing invisible microorganisms (pestilences), a new sense of time, the gun, the wheel, the plow, swine, cattle, sheep, the domesticated dog, reindeer, wheat, barley, ironworking, the honey bee, glass, and stringed instruments, and reintroduced the horse, which had become extinct in North America about 9000 B.C. The Native Americans in turn introduced the Europeans to corn, potatoes, tomatoes, oranges, tobacco, and other agricultural produce.

"Made in America" may have lost some of its luster, but it is impossible to calculate the prodigious impact that artifacts of our culture have had on other cultures around the world. Our movies, our pop music, our television shows are devoured from Warsaw to Bogota and often slavishly imitated. Is there any doubt that it was television o'erleaping the Iron Curtain that ignited Eastern Europe's revolution and threw off the Communist yoke: those millions wanted what folks in the West were consuming. Hollywood has often been a fomenter of anticolonialist revolutions. *Variety,* the show-business newspaper, headlined, "Ice Boxes Sabotage Colonialism." President Sukarno of Indonesia confessed, "The motion-picture industry has provided a window on the world, and the colonized nations have looked through that window and have seen the things of which they have been deprived. It is perhaps not generally realized that a refrigerator can be a revolutionary symbol to a people who have no refrigerators. A motor car owned by a worker in one country can be a symbol of revolt to a people deprived of even the necessities of life. . . . [Hollywood] helped to build up the sense of deprivation of man's birthright, and that sense of deprivation has played a large part in the national revolutions of postwar Asia." Our ingenuity has given the world such wonders as blue jeans, Big Macs, Kentucky Fried Chicken, transistor radios, Coca-Cola, Hula Hoops, and Monopoly. Trivial as some of these contributions may be, they imprint part of American culture on other peoples. They look in the mirror and see us.

The tunnel under the English Channel connecting Calais and Folkstone will open in 1993 over the passionate protests of many Englishmen. The English may love the foods and the wines and the scenery of France, but many of them just don't like the French people. "The channel may be only 30 miles long," one young British woman told me, "but culturally no tunnel is going to bring us any closer together. We'll always be oceans apart." It is a mutual antagonism (the French tend to sneer at the British as being an effete, repressed people) which goes back to the Norman conquest. This joining of the British Isles with continental Europe will indeed bring about profound cultural changes along with uniting processes.

Cultural observers of Australians describe the Australian "genius" as being "one of the mob," of accepting the environment and getting on with the job (but not too strenuously), enjoying friendly relations with their fellow humans and prizing the value of leisure time. The geography of magnificent beaches and a benign sun seems to promote a joie de vivre "down under."

The people least accessible to the world at large are the Japanese. But it is the industry and efficiency of the Japanese that has been the most dazzling since the end of the Second World War—phoenix rising from the ashes. How did they do it? Culturally, the Japanese are more group-oriented than individualistic. Employees are likely to mingle only with colleagues from their own company, both at work and at play. This social conformity and single-mindedness and almost fanatical dedication to work did much to help Japan achieve economic supremacy. The Japanese are far and away the most homogeneous of all First World countries; intermarriage with "outsiders" is rare.

It is interesting to contrast how the two defeated powers of the Second World War have dealt with their humiliating defeat. The Germans are still agonizing over what went wrong, while most Japanese (it is said) never give the experience a second thought; tomorrow is another day. The German character is to brood over history, the Japanese character is to look to the future. Put another way, the Germans are trying to learn from history and the Japanese are striving to bury it.

And so we are in flux, all of us, people interacting with other people, ideas and images lifting their wings to be borne afar by the winds that make the world go round.

5.3 BILLION

—AND

MULTIPLYING

■ ■ ■ ■ ■ ■ ■ ■ ■ ■ ■ ■ ■ ■ ■ ■ ■ ■ ■

Every time the second hand on your watch ticks—and it ticks 60 times a minute, 3,600 times an hour, 86,400 times a day—3 babies are born into the world. That's more than a quarter-million babies a *day*. That's more than 100 million babies a year. That's tantamount to adding another China to the family of mankind in this decade alone—and China already has as many people as Europe, North America, and South America combined.

By the year 2030, it is projected that the world will have a population of 10 billion people—nearly double today's population. But, like everything on Earth, there are wide differentials among places around the world. (About 40,000 infants die every day, the vast majority of them in Third World countries.)

Africa's population is exploding. Kenya is one of the fastest-growing countries in the world, with a 4 percent annual increase in population; the average Kenyan woman bears seven or eight children in her lifetime. By the year 2025, it is projected that Africa's population will be greater than the combined populations of Europe, North America, and South America. Forty-five percent of the present population of Africa is less than 15 years of age. One can only imagine the burden this huge, young population will place on struggling governments.

Most growth in population has been recent. When Jesus was born, the world was a cozy little planet. Global population is thought not to have reached 1 billion until the 1820s. It took only another century for the population to zoom to 2 billion. Between 1930 and 1975, the total doubled again. We are now at 5.3 billion, more or less—and multiplying. Ninety percent of all people who have ever lived are alive today.

One of the challenges for geographers in this supercharged world, of which a booming population is one of the foremost, is to determine how many people are too many. "Too many people" may affect the environment by consuming nonrenewable resources or they may be negligent or ignorant about renewing tapped resources. But "few people" can also despoil the environment by high consumption and by high demand for energy and industrial processes. Per capita, environmental damages are higher in the developed and presumably enlightened world than in the developing poor countries.

The relationship between the rising demands of people and threats to the environment is best shown by the example of the automobile. Since 1950, the human population has doubled but the car population increased sevenfold. Today, there are 400 million cars; the projection is that there will be 700 million in the year 2010.

More babies are born every year in India (which has a population of more than 850 million) than there are people in the entire island-continent of Australia (17 million). India is on its way to becoming the most populous country in the world. It also has the oldest government-sponsored family-planning program in the world. The United States, with severe problems of homelessness, joblessness, and hunger, may double its population to a half billion in the next century.

During the 1970s, Third World cities had to house 30 million new people each year. This figure will double during the 1990s. The United Nations predicts that there will be a couple of hundred cities each with more than 2 million citizens by the year 2025, 80 of them in Third World countries. Some, like Mexico City, will be over 30 million!

These are daunting statistics, and very geographic. Urbanization

THE WORLD'S MOST POPULATED CITIES

Rank in 1985	Agglomeration	Country	1950	1955	1960	1965	1970	1975	1980	1985	1990	1995	2000
													(millions)
1	TOKYO/YOKOHAMA	JAPAN	6.74	8.59	10.69	12.60	14.87	16.38	17.67	19.04	20.52	20.93	21.32
2	MEXICO CITY	MEXICO	2.88	3.77	4.93	6.58	8.74	11.05	13.97	16.65	19.37	22.02	24.44
3	NEW YORK	USA	12.34	13.22	14.16	15.18	16.19	15.88	15.58	15.62	15.65	15.67	16.10
4	SÃO PAULO	BRAZIL	2.75	3.60	4.71	6.16	8.06	10.05	12.50	15.54	18.42	21.16	23.60
5	SHANGHAI	CHINA	10.26	10.60	10.67	10.81	11.41	11.59	11.80	12.06	12.55	13.39	14.69
6	BUENOS AIRES	ARGENTINA	5.13	5.86	6.69	7.46	8.31	9.06	9.88	10.76	11.58	12.35	13.05
7	LONDON	UNITED KINGDOM	10.25	10.48	10.72	10.66	10.55	10.44	10.44	10.49	10.57	10.68	10.79
8	CALCUTTA	INDIA	4.45	4.95	5.50	6.16	6.91	7.88	9.00	10.29	11.83	13.71	15.94
9	RIO DE JANEIRO	BRAZIL	3.45	4.13	4.93	5.89	7.04	7.96	8.98	10.14	11.12	12.07	13.00
10	SEOUL	KOREA, REPUBLIC OF	1.02	1.55	2.36	3.45	5.31	6.80	8.28	10.07	11.33	12.25	12.37
11	LOS ANGELES	USA	4.05	5.15	6.53	7.41	8.38	8.93	9.51	10.04	10.47	10.77	10.91
12	OSAKA/KOBE	JAPAN	3.83	4.73	5.75	6.73	7.60	7.96	8.71	9.56	10.49	10.86	11.18
13	GREATER BOMBAY	INDIA	2.90	3.43	4.06	4.85	5.81	6.85	8.05	9.47	11.13	13.12	15.43
14	BEIJING	CHINA	6.64	7.06	7.30	7.63	8.29	8.91	9.10	9.33	9.74	10.43	11.47
15	MOSCOW	CIS	4.84	5.54	6.29	6.81	7.11	7.62	8.21	8.91	9.39	9.81	10.11
16	PARIS	FRANCE	5.44	6.27	7.23	7.92	8.33	8.62	8.68	8.75	8.75	8.76	8.76
17	TIANJIN	CHINA	5.36	5.74	5.97	6.29	6.87	7.43	7.68	7.96	8.38	9.02	9.96
18	CAIRO/GIZA	EGYPT	2.41	2.99	3.71	4.61	5.33	6.08	6.93	7.92	9.08	10.38	11.77
19	JAKARTA	INDONESIA	1.73	2.17	2.73	3.44	4.32	5.29	6.42	7.79	9.42	11.26	13.23
20	MILAN	ITALY	3.63	4.05	4.50	4.99	5.53	6.12	6.77	7.50	7.90	8.32	8.74
21	TEHRAN	ISLAMIC REP. OF IRAN	1.04	1.40	1.87	2.51	3.29	4.27	5.55	7.21	9.21	11.27	13.73
22	MANILA/QUEZON	PHILIPPINES	1.54	1.87	2.27	2.83	3.53	5.00	5.96	7.09	8.40	9.88	11.48
23	DELHI	INDIA	1.39	1.78	2.28	2.85	3.53	4.42	5.54	6.95	8.62	10.57	12.77
24	CHICAGO	USA	4.94	5.44	5.98	6.34	6.72	6.75	6.78	6.84	6.89	6.95	6.98
25	KARACHI	PAKISTAN	1.03	1.38	1.85	2.41	3.13	3.97	4.95	6.16	7.67	9.45	11.57

#	City	Country											
26	BANGKOK	THAILAND	1.36	1.71	2.15	2.58	3.11	3.84	4.75	5.86	7.16	8.63	10.26
27	LAGOS	NIGERIA	0.29	0.47	0.76	1.24	2.02	3.29	4.45	5.84	7.60	9.86	12.45
28	LIMA-CALLO	PERU	1.01	1.31	1.69	2.19	2.84	3.57	4.41	5.44	6.50	7.63	8.78
29	HONG KONG	HONG KONG	1.75	2.13	2.59	2.98	3.40	3.90	4.48	5.16	5.44	5.78	6.09
30	ST. PETERSBURG	CIS	2.62	3.02	3.46	3.78	3.98	4.32	4.70	5.11	5.39	5.65	6.84
31	MADRAS	INDIA	1.40	1.54	1.71	2.24	3.03	3.60	4.19	4.87	5.69	6.68	7.85
32	MADRID	SPAIN	1.55	1.86	2.23	2.74	3.37	3.82	4.30	4.83	5.06	5.26	5.42
33	DACCA	BANGLADESH	0.42	0.52	0.65	0.97	1.50	2.28	3.29	4.76	6.40	8.54	11.20
34	BOGOTA	COLOMBIA	0.68	0.94	1.30	1.79	2.37	3.14	3.91	4.74	5.59	6.38	6.94
35	BAGHDAD	IRAQ	0.58	0.72	1.02	1.61	2.11	2.74	3.54	4.39	5.35	6.44	7.66
36	NAPLES	ITALY	2.75	2.96	3.19	3.39	3.59	3.80	4.02	4.26	4.33	4.39	4.46
37	SANTIAGO	CHILE	1.33	1.65	2.03	2.41	2.84	3.24	3.70	4.23	4.70	5.16	5.58
38	PHILADELPHIA	USA	2.94	3.28	3.64	3.83	4.02	4.07	4.12	4.18	4.23	4.29	4.33
39	SHENYANG	CHINA	2.22	2.34	2.47	2.75	3.14	3.53	3.82	4.11	4.46	4.91	5.50
40	PUSAN	KOREA, REPUBLIC OF	0.95	1.05	1.15	1.37	1.81	2.42	3.12	4.02	4.75	5.34	5.82
41	DETROIT	USA	2.77	3.14	3.55	3.76	3.97	3.89	3.81	3.83	3.86	3.89	3.92
42	BANGALORE	INDIA	0.76	0.95	1.17	1.38	1.62	2.11	2.81	3.73	4.86	6.19	7.67
43	ROME	ITALY	1.57	1.91	2.33	2.70	3.07	3.30	3.48	3.67	3.72	3.77	3.82
44	SYDNEY	AUSTRALIA	1.70	1.91	2.13	2.39	2.67	2.96	3.28	3.64	3.79	3.92	4.06
45	CARACAS	VENEZUELA	0.68	0.93	1.28	1.63	2.05	2.47	2.94	3.51	3.96	4.38	4.79
46	WUHAN	CHINA	1.25	1.86	2.17	2.42	2.73	3.01	3.21	3.40	3.66	4.00	4.47
47	LAHORE	PAKISTAN	0.83	1.02	1.26	1.58	1.97	2.39	2.85	3.40	4.08	4.91	5.93
48	KATOWICE	POLAND	1.69	2.02	2.41	2.58	2.76	2.94	3.15	3.36	3.58	3.73	3.88
49	GUANGZHOU	CHINA	1.43	1.68	1.93	2.17	2.50	2.82	3.07	3.33	3.62	4.00	4.49
50	SAN FRANCISCO	USA	2.03	2.23	2.44	2.71	2.99	3.09	3.20	3.30	3.39	3.47	3.53

THE WORLD'S MOST POPULATED CITIES

Rank in 1985	Agglomeration	Country	(millions)										
			1950	1955	1960	1965	1970	1975	1980	1985	1990	1995	2000
51	BARCELONA	SPAIN	1.56	1.74	1.95	2.27	2.66	2.87	3.07	3.28	3.31	3.34	3.38
52	BELO HORIZONTE	BRAZIL	0.47	0.64	0.87	1.17	1.59	2.00	2.52	3.17	3.81	4.43	5.01
53	TORONTO	CANADA	1.07	1.37	1.74	2.09	2.53	2.77	2.96	3.17	3.32	3.47	3.61
54	MELBOURNE	AUSTRALIA	1.54	1.70	1.87	2.07	2.33	2.56	2.79	3.03	3.10	3.18	3.27
55	BIRMINGHAM	UNITED KINGDOM	2.50	2.58	2.67	2.73	2.80	2.86	2.92	2.97	3.02	3.07	3.10
56	AHMEDABAD	INDIA	0.86	1.01	1.18	1.41	1.69	2.04	2.45	2.95	3.55	4.26	5.09
57	HYDERABAD	INDIA	1.12	1.18	1.24	1.46	1.75	2.08	2.47	2.94	3.49	4.16	4.94
58	ISTANBUL	TURKEY	0.97	1.25	1.45	2.00	2.78	2.87	2.91	2.94	2.97	3.08	3.27
59	ANKARA	TURKEY	0.28	0.44	0.64	0.95	1.27	1.67	2.20	2.91	3.62	4.40	5.19
60	WASHINGTON, D.C.	USA	1.30	1.54	1.82	2.14	2.49	2.63	2.77	2.91	3.03	3.13	3.19
61	ALEXANDRIA	EGYPT	1.04	1.25	1.50	1.75	1.99	2.24	2.53	2.87	3.28	3.76	4.29
62	MONTREAL	CANADA	1.34	1.65	2.03	2.37	2.68	2.79	2.82	2.85	2.85	2.88	2.93
63	HOUSTON	USA	0.71	0.90	1.15	1.40	1.69	2.03	2.43	2.83	3.19	3.46	3.62
64	HO CHI MINH VILL	VIETNAM	0.87	1.07	1.32	1.63	2.00	2.23	2.48	2.78	3.17	3.70	4.42
65	ATHINAI	GREECE	1.35	1.56	1.81	1.96	2.08	2.28	2.51	2.76	2.90	3.03	3.15
66	CHONGQOING	CHINA	1.54	1.88	2.15	2.28	2.46	2.59	2.65	2.72	2.86	3.08	3.42
67	YANGON	MYANMAR	0.67	0.81	0.98	1.18	1.43	1.75	2.18	2.71	3.18	3.75	4.45
68	BOSTON	USA	2.24	2.33	2.42	2.54	2.65	2.67	2.68	2.71	2.74	2.77	2.81
69	ALGER	ALGERIA	0.44	0.62	0.87	0.93	1.20	1.59	2.11	2.70	3.43	4.27	5.16
70	CHENGOU	CHINA	0.70	0.93	1.12	1.29	1.58	2.00	2.35	2.69	3.06	3.49	3.98
71	CASABLANCA	MOROCCO	0.71	0.89	1.10	1.29	1.51	1.82	2.21	2.69	3.26	3.92	4.63
72	PORTO ALEGRE	BRAZIL	0.67	0.82	1.01	1.24	1.52	1.84	2.22	2.68	3.11	3.53	3.94
73	RECIFE	BRAZIL	0.82	1.00	1.21	1.47	1.78	2.04	2.34	2.68	2.98	3.28	3.57
74	DALLAS	USA	0.86	1.00	1.16	1.54	2.03	2.23	2.46	2.68	2.87	3.02	3.11
75	GUADALAJARA	MEXICO	0.40	0.60	0.88	1.16	1.51	1.86	2.28	2.66	3.06	3.48	3.89

	City	Country											
76	HARBIN	CHINA	0.93	1.28	1.55	1.74	2.00	2.24	2.44	2.63	2.86	3.16	3.56
77	KIEV	CIS	0.74	0.94	1.19	1.43	1.66	1.93	2.24	2.60	2.90	3.17	3.39
78	KINSHASA	ZAIRE	0.17	0.28	0.45	0.79	1.37	1.75	2.24	2.57	2.99	3.57	4.35
79	SINGAPORE	SINGAPORE	0.95	1.08	1.22	1.38	1.56	1.94	2.42	2.56	2.70	2.84	2.95
80	MANCHESTER	UNITED KINGDOM	2.51	2.52	2.52	2.52	2.52	2.52	2.52	2.53	2.55	2.57	2.60
81	TAIBEI	CHINA	0.59	0.76	0.97	1.19	1.50	1.84	2.17	2.52	2.89	3.30	3.78
82	MONTERREY	MEXICO	0.36	0.56	0.88	1.04	1.23	1.57	2.02	2.43	2.88	3.32	3.75
83	ZIBO	CHINA	0.48	0.64	0.80	1.01	1.30	1.65	2.03	2.41	2.82	3.27	3.76
84	TORINO	ITALY	0.88	1.05	1.25	1.43	1.62	1.83	2.07	2.34	2.50	2.65	2.77
85	SURABAJA	INDONESIA	0.61	0.76	0.95	1.18	1.47	1.74	2.01	2.32	2.71	3.16	3.67
86	XIAN	CHINA	0.94	1.16	1.34	1.51	1.73	1.95	2.12	2.28	2.48	2.74	3.08
87	HAMBURG	GERMANY	1.79	1.94	2.10	2.16	2.20	2.24	2.24	2.24	2.24	2.24	2.24
88	BUCHAREST	ROMANIA	1.18	1.28	1.38	1.50	1.67	1.87	2.05	2.23	2.34	2.45	2.55
89	LIUPANSHUI	CHINA	0.59	0.78	1.19	1.44	1.66	1.87	2.04	2.20	2.40	2.66	3.00
90	MUNICH	GERMANY	0.96	1.13	1.34	1.52	1.71	1.92	2.08	2.20	2.28	2.33	2.33
91	SALVADOR	BRAZIL	0.45	0.57	0.71	0.90	1.14	1.42	1.76	2.18	2.60	3.00	3.39
92	NANJING	CHINA	0.89	1.19	1.45	1.60	1.78	1.94	2.05	2.16	2.31	2.53	2.83
93	TASHKENT	CIS	0.61	0.78	1.00	1.21	1.41	1.61	1.85	2.12	2.34	2.54	2.70
94	KITAKYUSHU	JAPAN	0.94	1.12	1.31	1.46	1.59	1.74	1.91	2.09	2.28	2.34	2.39
95	WEST BERLIN	GERMANY	2.15	2.17	2.19	2.16	2.12	2.08	2.08	2.08	2.08	2.08	2.08
96	BUDAPEST	HUNGARY	1.62	1.71	1.81	1.88	1.95	2.06	2.06	2.07	2.08	2.10	2.15
97	LEEDS-BRADFORD	UNITED KINGDOM	1.91	1.92	1.93	1.95	1.98	2.01	2.04	2.06	2.09	2.12	2.14
98	MEDAN	INDONESIA	0.35	0.40	0.46	0.53	0.61	0.88	1.34	2.05	3.00	4.14	5.36
99	NAGOYA	JAPAN	0.96	1.18	1.50	1.72	1.85	1.94	2.00	2.05	2.11	2.11	2.11
100	POONA	INDIA	0.59	0.66	0.73	0.89	1.10	1.34	1.64	2.00	2.44	2.95	3.56

reflects the distribution of population, economy, and culture—the geographic location of its people. The statistics are particularly awesome for the population geographer whose job it is to figure out why these people are where they are and how and where they will change. Population distribution—the process of redistribution over space—is conceivably the single most important and complex of all geographic phenomena. The way people are grouped in space depicts all the accommodations they have made to their total environment. At this juncture, humankind has clustered in four principal concentrations: East Asia, South Asia, Western Europe, and Eastern North America.

How can the destitute, who make up the majority of humankind, especially in Third World cities, be fed, employed, and spared life-threatening illnesses? How can skilled manpower be attracted to places where it can be employed in development? How can population flows be "controlled" to bring a population into balance with a country's needs and facilities?

Can the factors that encourage low fertility (technology, medicine, education, urban living, modernization) be exported? Not very easily, because these changes are slow in coming. In lieu of quick modernization, does family planning offer the solution? It could, but people must want smaller families to relax their resistance to family planning, and they must alter their attitudes regarding family size.

Only half, probably less, of the world's population is fed adequately, though all of us *could* be fed adequately if the world's supply of food were equitably distributed irrespective of cost, politics, hatreds, and corruption. Straightening out the system and ending political infighting, disparities in income, and other problems will not occur in our lifetime.

The rapid growth of population in some areas of the world relates to a number of problems, be they economic inequality or environmental deterioration or the rise of political tensions or aggression between countries. It is economic ineptitude, as well as a soaring population and population maldistribution—not the 1.7 million Palestinians—that is the root of Middle East instability. In several of the 20 Arab countries, many of them fabulously wealthy, there have been bread riots. Arab potentates chronically fail to provide their subjects with enough food, health care, education, and jobs. In many Arab countries, the population increases at rates ranging from 2.5 to

Cities with Populations of 2 Million or More in:

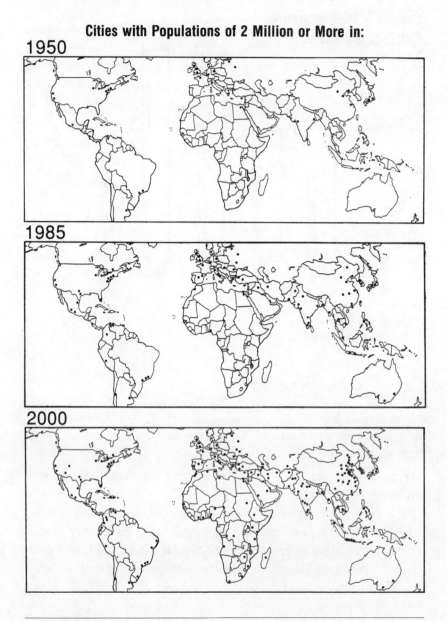

1950

1985

2000

Major spatial flows—capital, technology, sustenance, to name three—can create very high-density places (cities) through migration. Changes in the flows of population distribution can be predicted, enabling policymakers to minimize strains. Worldwide at midcentury, each of around 25 cities had a population of 2 million. By the year 2000, the cities with populations exceeding 8 million will include Seoul, Dhaka, Bangkok, Manila, and Madras.

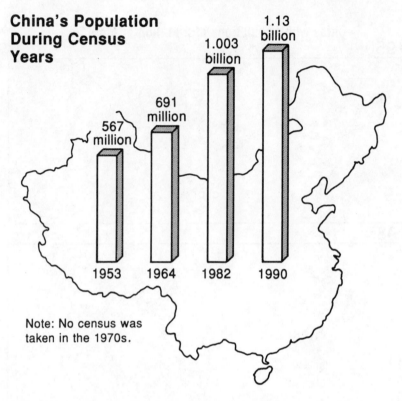

China's Population During Census Years

- 567 million — 1953
- 691 million — 1964
- 1.003 billion — 1982
- 1.13 billion — 1990

Note: No census was taken in the 1970s.

The Wall Street Journal *reported that by mixing IOUs, IUDs, and the profit motive, the outback of the world's last major Communist country is being penetrated with an innovative family-planning program—contraceptive capitalism.* U.S. CENSUS BUREAU

4.5 percent a year. At least 40 percent of the Arab population is under 14 years of age.

Population is the concern of economists, sociologists, and behavioral scientists, as well as geographers. The geographer studies the subject as a set of problems in particular places and in processes connecting places. He or she examines natural events such as birth, death, and disaster, and the ceaseless movement of peoples in various regions and states.

The fundamental bases of the study of population are fertility and mortality (the number born and dying, and where) and mobility (the human flows between places between birth and death).

The world is not growing at the same rate everywhere. Rates and numbers vary even within the same country. Urban populations in developed countries tend to have fewer children than do rural families. When a population is not growing fast enough to fill the needs of a society, efforts are made to increase the rate of fertility.

A case in point is the former Soviet Union. For years, Soviet women were given incentives to produce more children; there was so much work to be done nationwide, and many women worked rather than bear children. The Great Patriotic War claimed tens of millions of Soviet lives, especially young males'. But inadequate housing and other conditions discourage large families—the average woman has six to eight abortions, as noted earlier. The Soviet population grew very slowly except in the rural, non-Western republics, where fertility rivals that in the Third World.

We usually measure a country's mortality rate by the number of people per thousand who die in the population. The very youngest and the old, we know, have the highest death rates; in between, the causes of death can relate to negative factors in the culture or in the environment or in the quality of the genes. No one would argue that extension of the life span is a constant and common goal of humankind. Some places are considered preferable to others simply because they offer the hope of living longer there. Even in the United States, people in some places live longer than in others. There is a geography of life. Whatever crosses we may bear in this vale of tears, we would rather be alive than dead.

Mobility in geographic terms refers to the impact people who move have on space or places. When people leave one place for another, it always makes a difference. Taxes, social services, schools, and local economies are affected in the places they left and the places they newly occupy. The migration of seasonal workers leaves its imprint by boosting sales and the demand for services in a community.

Adding the three factors together—births, deaths, and migrations—provides a profile of the rate of population growth and the change in population distribution—the geography of population.

I wish I could convince more people that the population problem is a complexity of issues that cannot be stated in simple terms. There are those visionaries, I know, who say the planet can accommodate 10 billion or more, it is just a matter of more efficient management and distribution. More important than such aggregate perspectives is the *real* issue: some places are growing too fast and have too many people, and others are growing very slowly and face declining populations. The population problem *cannot be understood* or solved without a geographic perspective. Even more important problems include maldistribution of people.

The People's Republic of China has given us the most evidence that the population can be brought under control in a developing country. I was there in 1988 and observed firsthand what China was doing to keep its population from exceeding 1 billion. It was an eye-opener, and chilling. Each family in most areas was restricted to one child. Rural families were allowed two children if the first born was a daughter. It was like observing the inner workings of Big Brother, Chinese-style. Somehow Beijing was managing to keep tabs on most families. Officials visited houses and factories, checking on pregnancies, asking intimate questions. A woman who conceived a second child and failed to abort could bring down on her family economic reprisals, such as the loss of her job and subsidized housing and medical care. It is suspected that the barbarous custom of killing first-born daughters is still operative in China. Elsewhere in Asia, especially in the Koreas and India, the aborting of female fetuses is on the increase.

Back in the 1950s, Indian men were offered transistors if they would submit to a vasectomy. In a single year, 2.5 million vasectomies were performed. But interest abruptly declined. In the early 1970s, India's population increased by 50 percent. War and drought did not interrupt the boom.

Indian children of poor parents are not looked upon as future consumers, people who will grow up to live independent lives and acquire possessions. They are seen as co-workers or co-seekers in the basics of survival. They are, more importantly, viewed as social security for parents who have many children—just in case. If the cycle of poverty can ever be broken, the poorest peoples of many Third World countries will by choice have fewer children, as has happened in the industrialized societies. Economic development and modernization will change attitudes toward smaller families.

The main concern of the peasant in the world's poorest lands is getting enough to eat. Children may be more mouths to feed, but they can also help in the quest for food, thus becoming a kind of security net.

Notable starts toward family planning have been made in several Third World countries. (Family planning is a useful way of reducing fertility, but it must be preceded by attitudinal changes by which fewer children are perceived as better than more.) Cuba's birthrate has dropped from 52 to 15 per thousand. Colombia and Mexico have made some progress, possibly because both countries have strong, government-encouraged family-planning programs. Of all the Latin American countries, Brazil seems to be progressing the most methodically. In one generation, Brazil's birthrate has been cut almost in half. As one expert has put it, "The country's population bomb has been deactivated." The sharp decline is attributed to the rapid spread of contraceptives, economic stagnation, and universal access to television—but the truth is much more complex.

Japan has the lowest birthrate of the advanced countries. In 1989, the number of births fell to 1.24 million, or a rate of 10.1 per thousand people. The total fertility rate—the average number of children a woman bears in a lifetime—has fallen well below the 2.1 needed to maintain a stable population level. There are more deaths than births. Young men are putting off marriage because Japan's corporate culture demands long working hours and there is little leisure time for social life. The average age at which a Japanese man marries has risen to 28.2 years. Many young women are hesitant about marriage, because they fear it would deprive them of some of the new-found freedoms they are merrily savoring; married women are expected to stay home and start raising children. There is still the sense that an unmarried woman in her late twenties is an "old maid."

Japanese authorities are fearful that after the turn of the century the declining population will cause a labor shortage, a stagnant economy, and higher tax burdens so that social services will have to be provided. So great is the concern that the Japanese government has proposed baby bonuses. But Japanese women, enjoying their unprecedented independence, have served notice that they are not interested in becoming "breeding machines," no matter how many yen are involved.

We social scientists contend that even the most primitive societies evolve toward modern, developing societies, however slowly. Slowly, I would say, is the operative word. Most important factors in the transition that most affect attitudes toward smaller families and fewer children are increased education (especially of females), increased incomes, and the freedom of women to make decisions. Typically, a rural backward agrarian society gives way gradually to an industrialized society of high technology focused in cities. In an urban society, attitudes about children change: they are no longer seen as economic assets, the better to help with menial chores and ultimately to be the chief support of aging, infirm parents; rather, they are seen as dependents whose number must be controlled if the family is to prosper and enjoy a high standard of living. Some countries that have modernized recently and rapidly—Singapore, South Korea, Malaysia, Thailand—have at the same time had dramatic drops in birthrates.

One of the greatest threats to the environment and to human-kind's well-being is the rapid population growth in developing countries. The only sane choice for the 1990s is to slow population growth, decrease poverty, and guard the environment. Failure to do so will mean that we hand to our children a world that will be dangerous, if not largely uninhabitable.

The booming population in developing countries will reach a crisis stage in the early 21st century. To avoid disaster, methods will have to be found to manage urban growth, the pollution of land and water resources, the economically driven destruction of forests, and the buildup of greenhouse gases that produce global warming.

At any level of development, we can say the more people, the more pollution. But equally significant is the high rate of consumption in the rich world. Both worlds must change habits.

No less than 95 percent of the growth in population throughout the world in the next 35 years will be in the developing countries in Africa, Asia, and Latin America. Africa is expected to more than double its present population (650 million) and reach 1.6 billion. The United Nations chart on pages 126–129 reveals how many of the Third World cities have already outstripped the largest cities elsewhere in population growth.

Unless the rate of urbanization slows, more than half the population of the world will be living in the largest cities. Cities serve as a link in the economic interaction of different areas of a country. Their development in the modern era is distinctly linked with specialization in production.

Short of mobilizing for war, the taking of the census decennially has been called the United States government's single most ambitious undertaking. It is called for in the Constitution: "Representatives and direct taxes shall be apportioned among the several States which may be included within this Union, according to their respective numbers, which shall be determined by adding to the whole number of free persons, including those bound to service [a synonym for indentured servants] for a term of years, and excluding Indians not taxed, three-fifths of all other persons [a euphemism for slaves]. The actual enumeration shall be made within three years after the first meeting of the Congress of the United States, and within every subsequent term of ten years, in such manner as they shall by law direct." The United States population at the time of the Federal Convention (1787) was about 3 million whites and 50,000 free blacks—and 600,000 slaves.

The first census was taken in 1790. Only male heads of households were asked to supply data. The men were asked as to the number of women and children in their household. Slave owners disclosed their inventories of "human property," with each slave counted as three fifths of a person. No effort was made to count Native Americans, except for those few who had joined the white society and were paying taxes. (Today, those who pay no taxes and are wards of the government are still not counted in the apportionment of representatives.)

The accuracy of the census is always in contention. The current policy is to count illegal aliens. In 1980, the Census Bureau counted 2.1 million of them, though the actual number residing in the United States at the time was thought to be much higher. There are those who believe that the bureau should count only people who live there legally.

For the first time in the history of the United States, more than half the population lives in cities of more than a million. Forty-four percent of the population in 1950 lived on farms or in small towns. That segment has dwindled to 23 percent. (In ten years, incidentally, the median age of the United States' population will be 36, three years older than it is now.) In 1990, one of every four Americans belonged to a minority group—Hispanic, Asian, African, or Native American.

Booming immigrant populations, largely Hispanic and Asian, render the future of California politics uncertain. The Texas electorate is already one-third Hispanic and black. Florida, with eight of the nation's fastest-growing metropolitan areas, owes its rapid growth in the last decade to older Americans seeking more benign climes and to the burgeoning Hispanic communities, particularly Cuban in

Miami. There has been a general hollowing out of the interior of the country.

The 1990 census showed that in the years 1950 to 1990 the United States population grew about 65 percent. Nevada showed the greatest percentage growth of any state, 650 percent. In numerical growth, California led the country with the addition of more than 19 million people.

The Census Bureau estimates that 9.7 million to 15.5 million people were not recorded in the 1990s' canvas. The official United States population, excluding military and federal employees stationed overseas, was 248,709,973 rather than the projected (and apparently more accurate) 254,000,000.

The Secretary of Commerce, Robert A. Mossbacher, who oversaw the census, decided not to use statistical methods to adjust the undercount, because, he said, it would be difficult to determine where in the country people-missed should be allocated. Large cities and states contend they are being deprived of billions in aid in a politically-biased call.

In 1990, New York City again found itself undercounted, by several hundred thousand this time. Whole apartment houses were ignored. Political skulduggery could be at play. While blacks and whites were fleeing to the hinterlands, the Hispanic population grew substantially and the Asian population more than doubled.

So many whites have moved out of New York City that for the first time in history whites account for less than half the population. However, so many immigrants—mostly "people of color"—moved into the city in the 1980s that they not only replaced the numbers of departed Caucasians but increased the population by 300,000 over what it was in 1980.

In 1980, the census failed to count 6 percent of New York City's population; in other words, it failed to count 450,000 people, a good-sized city in itself. That meant a loss of $1,500 in federal aid for each person not counted, a total of hundreds of millions of dollars. If the undercount had not occurred, the city would also have gained an additional Congressional seat, one state senate seat, and 3½ assembly seats in the state capital.

Considerable fanfare preceded the 1990 census. The counting would supposedly be error-free because computers would be extensively used. But census officials declare that the 1990 census was the most difficult census in the agency's 200-year history. Only 63 percent of households returned their questionnaires; neither the rich nor the poor seemed to have any interest in answering questions about their families.

Since 1980 the Hispanic population of the United States has grown at a rate five times as fast as the non-Hispanic population. It totaled around 20.1 million in 1990. By the end of the decade, 46 percent of the population of California will be Hispanic, Asian, and black, and 92 percent of all Californians will live in counties where the "minority" population is more than 30 percent. In San Francisco County, the figure will be 65 percent—mostly Asian—and in Los Angeles (sometimes called Central America North or Asia East) the count will be 60 percent. From much experience in the field, let me say that the threat to the country, and to California in particular, would be the failure to educate its rising population groups and to produce a growth economy.

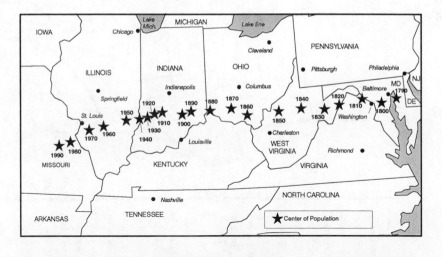

The population center of the continental United States has moved ever-westward 818.6 miles from the first designated center, 23 miles slightly northeast of Baltimore, calculated after the country's first census, in 1790. The center has also moved in a southerly direction recently, reflecting the push toward environmental amenities, benign climates, and increasing economic opportunities. The center is calculated according to where the country would balance perfectly if it were a flat surface and all 248,709,873 Americans had identical weight. The present center is at latitude 37°52'20"N and longitude 91°12'55"W, a spot of heavily wooded land in Crawford, Missouri, which is about 65 miles southwest of St. Louis. The center, which moved 39.5 miles in the last decade, first moved west of the Mississippi River in 1980. Nine decades ago, the center of population was six miles southeast of Columbus, Indiana. Continual flows and changes in linked processes (people, jobs, capital) equal continual changes in geography. U.S. CENSUS BUREAU

Apportionment of the U.S. House of Representatives
1990 Census of Population

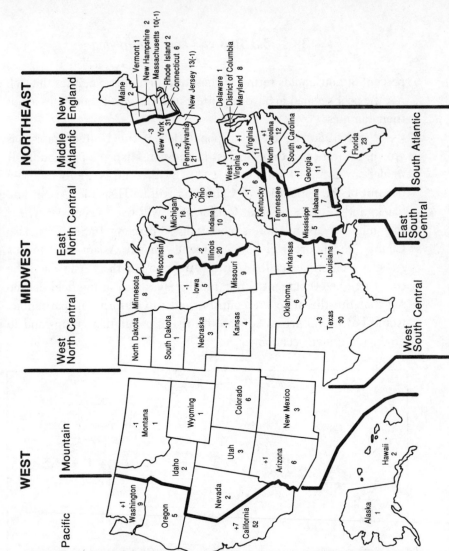

WEST

Pacific | Mountain

Washington +1 9
Oregon 5
California +7 52
Nevada 2
Idaho 2
Montana -1 1
Wyoming 1
Utah 3
Colorado 6
Arizona +1 6
New Mexico 3
Alaska 1
Hawaii 2

MIDWEST

West North Central | East North Central

North Dakota 1
South Dakota 1
Nebraska 3
Kansas -1 4
Minnesota 8
Iowa -1 5
Missouri 9
Wisconsin 9
Illinois -2 20
Michigan -2 16
Indiana 10
Ohio -2 19

NORTHEAST

Middle Atlantic | New England

New York -3 31
Pennsylvania -2 21
New Jersey 13(-1)
Maine 2
Vermont 1
New Hampshire 2
Massachusetts 10(-1)
Rhode Island 2
Connecticut 6

Delaware 1
District of Columbia
Maryland 8
West Virginia -1 3
Virginia +1 11
North Carolina +1 12
South Carolina 6
Georgia +1 11
Florida +4 23
Kentucky 6
Tennessee 9
Mississippi 5
Alabama 7

South Atlantic

East South Central

West South Central

Oklahoma 6
Arkansas 4
Louisiana -1 7
Texas +3 30

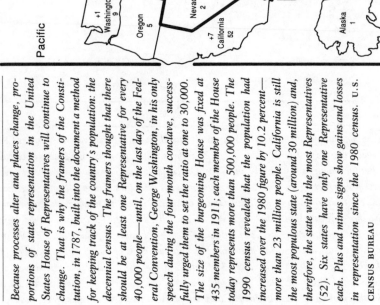

Because processes alter and places change, proportions of state representation in the United States House of Representatives will continue to change. That is why the framers of the Constitution, in 1787, built into the document a method for keeping track of the country's population: the decennial census. The framers thought that there should be at least one Representative for every 40,000 people—until, on the last day of the Federal Convention, George Washington, in his only speech during the four-month conclave, successfully urged them to set the ratio at one to 30,000. The size of the burgeoning House was fixed at 435 members in 1911; each member of the House today represents more than 500,000 people. The 1990 census revealed that the population had increased over the 1980 figure by 10.2 percent—more than 23 million people. California is still the most populous state (around 30 million) and, therefore, the state with the most Representatives (52). Six states have only one Representative each. Plus and minus signs show gains and losses in representation since the 1980 census. U.S.

CENSUS BUREAU

Any census yields interesting findings about populations and the changes they bring:

- The exploding population of Florida cities and environs is changing that state from largely agrarian to vast suburban sprawl, threatening waterways and endangering species.
- Finland and Japan have the fewest infant deaths per 1,000 births in the first year of life of any country in the world. Among the higher number of deaths in industrialized societies occur in the United States (18th in the world); poverty deprives many expectant mothers of proper prenatal care, and vast rural parts have few or no facilities for helping these women.
- Japanese women have a life expectancy of over 82 years, the highest of either gender of any country in the world.

Population growth and maldistribution figure in all the ills of the world. In this last decade of the millennium, I would say the problem confronting the human race is not so much natural resources per se but rather their misuse and overuse and rate of consumption. Any way you examine it—be it by the criteria of technology or consumption or waste or varying degrees of plenty or poverty—the size of the population and its appetite determine the toll on the environment.

There are those who make the point, harsh as it may be, that when prosperous countries send food to a starving population that has already grown beyond the environment's ability to sustain itself, they become a partner in the devastation of the land. Gifts of food to an overpopulated country encourage the recipients to demand more food; in the long run, this will only lead to increased starvation. Foreign aid may not really be an aid but a disfavor if certain African countries in particular become permanently dependent on handouts. The goal should be increased land productivity with agrarian policies and practices, as well as reasonable population control.

"It is well to remember," the philosopher-economist Kenneth Boulding observed, "that the symbol of Christian love is a cross and not a Teddy bear." Help for disasters aside, the affluent should stop subsidizing poor peoples whose multiplying populations are rapidly increasing, and the affluent must curb their appetites and share their wealth. Otherwise, Earth's prospects are slim.

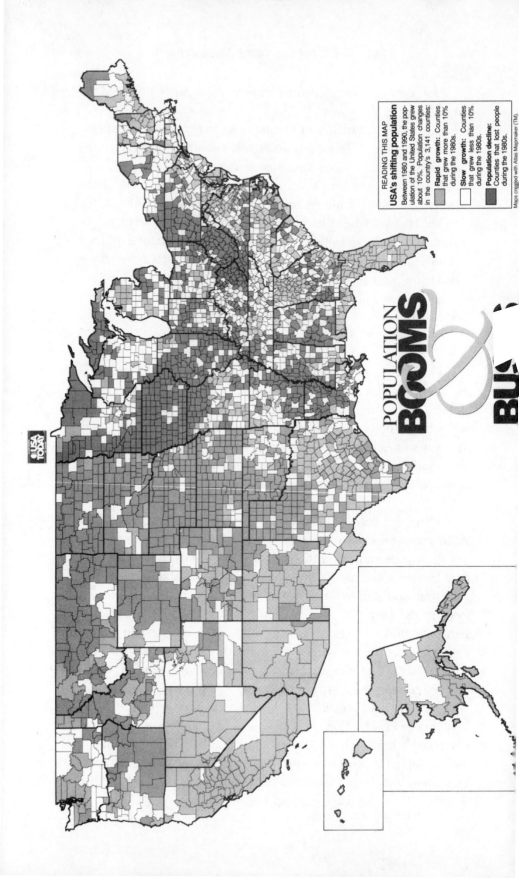

READING THIS MAP
USA's shifting population
Between 1980 and 1990, the population of the United States grew about 10%. Population changes in the country's 3,141 counties:

Rapid growth: Counties that grew more than 10% during the 1980s.

Slow growth: Counties that grew less than 10% during the 1980s.

Population decline: Counties that lost people during the 1980s.

Maps created with Atlas Mapmaker (TM).

POPULATION
BOOMS
BU

USA TODAY

The role of the population geographer is to monitor the distributions of population flows and related impacts and use all the measures within his or her competence to discourage negative place impacts, to improve health, and to identify and predict the dangerous trends over space. Marketeers are watching demographic shifts in population in order to adapt to changing population patterns. Such trends as a population growing older, the growth of minorities, and evolving family structures are studied carefully. Products and services are being adapted to accommodate these changes.

In summary: Population from a geographer's perspective is concomitant with space. Some places on the planet have populations growing at unacceptable rates and others at rates that presage decline. Birth rates have been plummeting in eastern Germany, Romania, Ukraine, and parts of Russia and continuing their steady declines in Hungary, Poland, and Czechoslovakia. Other places are cursed by maldistribution of population. Third World cities are growing so large so rapidly as to become unmanageable, and soon. Some places are emptying—the upper, rural Midwest of the United States, for one. Some places are scenes of horror or political coercion, driving thousands of people across boundaries as refugees seeking asylum in places in which they are not welcome. Some places give rise to unemployed people crossing boundaries to other places looking for succor—some places accept them, others reject them. Some places are connected by flows of tourists, job seekers, and salespersons. Demography, population, and the fundamental processes associated with them—begetting, dying, and moving—are inextricably tied to place and cannot be understood apart from place.

USA's Shifting Population
Between 1980 and 1990, the population of the United States grew about 10 percent. Population changes in the country's 3,141 counties:
█ **Rapid growth** *(like Florida): counties that grew more than 10 percent during the 1980s*
☐ **Slow growth** *(like most of New York State): counties that grew less than 10 percent during the 1980s*
█ **Population decline** *(like most of North Dakota): counties that lost people during the 1980s*

THE

CENTURY

OF MIGRATION

W̲e are smack in the throes of what may be the greatest migratory thrust in the history of mankind. Most of the aspirants want to go in one of two directions, either northward or westward. Latin Americans want to flee conditions of economic hopelessness or the threat of persecution for political beliefs. The collapse of the Soviet empire and the unleashing of its satellite countries have lifted the lid on nationalistic and ethnic squabbling and displayed the poverty and squalor of Eastern Europe—igniting a mass movement toward Western Europe, which is not able to absorb all the peoples pressing to move in.

Since time immemorial, or at least since the Garden of Eden and the first food gatherers and hunters and the historical flight of Moses and the enslaved Israelites out of Egypt and into the land of "milk and honey," people have moved from place to place. Locally, regionally, nationally, globally. Temporarily. Permanently. By choice. At the point of a gun. Alone. With families. Tribes. Whole cultures.

Population distribution is one of the most revealing manifestations of geography. Inevitably, humankind has become more mobile with modernization and development. It may be that modern societies move so much that many of us have lost our sensitivity to space and place.

Everywhere emigrants go, it is usually in search of new lives. And everywhere they go, they influence geographic processes: population distribution, urban size, wealth, politics, culture, education, ethnicity, social services, even language. Population distribution is one of the most fascinating revelations of modern geography.

Every non–Native American has an immigrant somewhere on his family tree. Presumably, every Native American does, too, from Asia via the land bridge across the Bering Sea. Many nonaboriginal Australians probably have a convict on their genealogical chart; that continent was "settled" by English prisoners, who were shipped "down under" beginning in the late 18th century.

When the Spanish invaded the American Southwest, they took along with them horses, which had an enormous impact on Native American culture. When hundreds of thousands of Cubans sailed into Florida, they encouraged sections of the state to adopt bilingual modes, and in Miami they established Little Havana. When well-off residents of Hong Kong, fearing the consequences of the takeover of the British crown colony by the People's Republic of China, emigrated to Vancouver, British Columbia, they bought up property and boosted rents (earning them the sobriquet "yacht people," as opposed to the impoverished "boat people"). When Vietnamese moved into Louisiana, they were perceived as a threat to the livelihood of local fishermen who had been working the Gulf of Mexico for generations. When oil was discovered in the Middle East in the 1930s, engineers and toilers moved there *en masse*, turning the forbidding Arabian desert into towns and cities boasting extraordinary wealth and splendor. When 10 million or more Africans were forcibly moved to the New World, their descendants, still enslaved, were an inciting cause of the four-year United States Civil War; in the largest migration in United States history, 6 million poor, uneducated blacks moved from the South to the North over five decades beginning in 1940, helping to establish the black middle-class, working-class, and lower working-class presence in northern cities. When the Berlin Wall fell to sledgehammers in 1989, tens of thousands of East Germans became immigrants, so to speak, in their own country; before the 28-mile-long barrier was constructed in 1961, an estimated 2 million East Germans had already fled to the West. When 110,000 Bangladeshis lost their jobs and fled Kuwait and Iraq during the Gulf War, a million dependents in Bangladesh lost their sole source of income, further taxing the poor country's social structure and economic

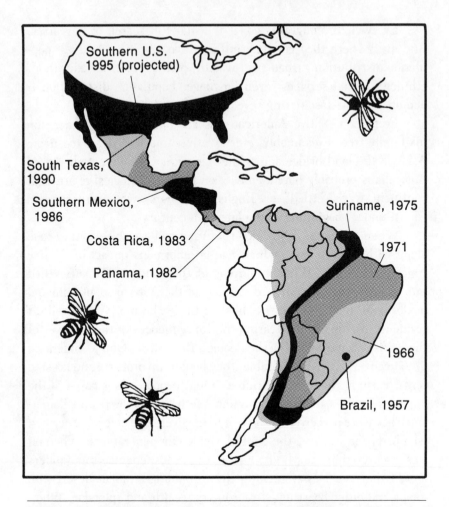

Southern U.S.
1995 (projected)

South Texas,
1990

Southern Mexico,
1986

Costa Rica, 1983

Panama, 1982

Suriname, 1975

1971

1966

Brazil, 1957

Tracing movements over and through space and time—ideas, fashions, so-called "killer" bees, for example—is graphic geography, a dynamic process. The northward migration through the western hemisphere of the genus of almost-pure honeybees, driven by an unbroken line of African queens, is an episode in population biology. The bees are notorious, because they aggressively protect their nests and pursue intruders for up to a half mile. Their sting is said to be no worse than that of a more tranquil European bee, but it is best to run as fast as possible downwind to prevent the bees from following any scents. The bees were imported from Africa in 1957 for a breeding experiment in Brazil. Queen bees escaped and bred, and have been blamed for about 600 deaths in South America. The wild population entered the United States near Hidalgo, Texas, in October 1990, and scored their first major attack 11 months later, laying 300 stings on a 65-year-old man in McAllen, Texas. Long winters could inhibit their circulation. The economic attraction of these bees is their superior productivity, with a yield several times greater than European bees'. These are pests whose location is worth knowing.

growth. When Russian Jews emigrated to the United States at the turn of the century, they retuned American entertainment from Tin Pan Alley to Hollywood.

The immigrant Asian population in the United States expanded 70 percent in the 1980s, making it the country's fastest-growing minority. More than a third of the new immigrants, who are mostly from South Asia and Southeast Asia, settled in California. The Vietnamese population in Orange County, California, could be as high as 150,000, and around 700,000 for the country as a whole. New York City is now home for at least a half million Asians. The gritty Massachusetts city of Lowell is the "Cambodian capital" of the United States; most of the 20,000 Cambodians living in that once-prominent textile center are refugees from civil wars back home.

Asians were the most undercounted minority in the 1990 census. As much as 20 percent of New York's Chinatown population was uncounted. Myriad Asians, Chinese particularly, reside in the United States illegally, and also have a distrust of American ways. Many fear eviction or stiff rent increases by landlords who discover that more people are living in their properties than they had been told. In the last 30 years only 500 Tibetan refugees have entered the United States, largely because Washington was reluctant to displease the People's Republic of China by giving Tibetans either refugee or immigrant status. (China claims historical sovereignty over Tibet, a stance reinforced by military occupation since 1950.)

Counted or not counted, the Asian influx is having a profound effect on the American scene. Some of the manifestations are obvious. Vietnamese business signs glitter in chrome and neon in California: Van Hoi Xuan Chinese Herbs, Kim Son Fabrics. Another manifestation is the enthusiastic welcome for a cuisine that is extraordinarily varied, imaginative, and healthful, influencing dietary habits (less fat, less sugar, increased fiber). The successes of Asian enterprises is a reminder of the rewards of hard, unremitting work. Young Asians are winning scholastic awards in disproportionate numbers. They have a hunger to learn, to be all they can be—a motivation that must be encouraged at home by parents who care about the value of education. But they are not yet a political force.

When I lecture to a new audience, I pass around a flat map of the world and ask everyone to mark on it where their four grandparents were born. At a teachers' conference in Manhattan, I wasn't surprised to see that most of the marks covered Africa or the "Jewish pale" in Russia and eastern Poland. The genealogical exercise leads to exciting dialogues on the conditions that prompted emigration. It is a personal, painless way to learn about spatial patterns, such as transportation, and about political and economic conditions, and much more. More reasonable geographic knowledge about places and their content and conditions and connectivity can be learned this way than by hours of old-fashioned memorization.

Refugees are spatial flows from places of danger, poverty, and drought to places of security and economic opportunity: massive geographic displacements, or flows, from bad places to good places or places perceived as good. It is pure human, political, economic geography.

In the former Soviet Union, large masses of people were moved from one place to another. The deportation of whole groups became just one more harassment of "hostile elements." About 400,000 Poles were moved to "special settlements." In 1944, about 650,000 people were deported from the Crimea and the northern Caucasus.

Migrant Muslims now constitute about two thirds of Europe's immigrants. Their presence, along with their different religious, political, and cultural traditions, is raising issues of national identity that Europe had skirted in the Cold War era. Are European societies destined to follow the United States and become multiracial and multiethnic?

Emigration, forced or chosen, across national frontiers or from village to metropolis is an all-too-frequent phenomenon of our time. A map of refugee flows in southern Africa is crisscrossed with traffic patterns: 70,000 Namibians have moved to Angola; 295,000 Angolans to Zaire; 65,000 Mozambiques to Zimbabwe; another 225,000 Mozambiques to South Africa; 4,000 Zimbabweians to Botswana; 1,000 South Africans to Lesotho; another 7,000 South Africans to Swaziland. When the oil boom went bust in Nigeria in 1983, about 2 million immigrant workers were expelled, left to wander homelessly across the continent.

Many countries are erecting barriers: there are electrified fences on African and European borders, armed guards and a heavy metal barricade at the United States–Mexico line—stanching the influx of people, establishing immigration and asylum policies to check the flows.

Immigrants tend to move around by choice. Refugees, on the other hand, usually flee in peril for their lives. They flee for protection in a foreign land, especially from war or from political or religious persecution. Refugees became an international issue in the wake of the Second World War.

Refugees are found on every continent, except Antarctica, and are of every persuasion and income level. Their only common denominator is that they live in a country other than where they were born and they have a well-founded fear of returning to their homeland because of their political, religious, racial, national, or social group.

Refugees today are primarily a product of political instability in Third World countries; unfortunately, most flee to poorer countries, where resources to support them are few. New York City, between 1980 and 1989, took in 854,000 immigrants from Latin America, Asia, and the Caribbean. Until the mid-1970s, most Third World refugee flows resulted from colonial struggles; many of the people were able to return home after independence. Nowadays, an increasing number of refugees are from independent countries that are less likely to undergo the radical political changes necessary to ensure their safe repatriation.

Beginning with Iraq's conquest of Kuwait, which took only four hours to accomplish, the Gulf War in 1990–1991 ignited what was regarded as one of the largest, fastest, and most widespread flights of population. It was a world on the move, a tide of misery. The United Nations Disaster Relief Organization and the United Nations High Commission for Refugees estimates that at least 5 million people from more than 30 countries were temporarily or permanently displaced or resettled: 450,000 Iraqis (mostly Kurds) to Turkey; another 400,000 Iraqis (mostly Kurds) internally displaced in northern Iraq; 7,000 Egyptians (from Iran and Kuwait, many via Jordan); 250,000 Jordanians and Palestinians (220,000 to Jordan, 30,000 to Israeli-occupied territories); 1.2 million Iraqis (mostly Kurds) and 100,000 Shiites to Iran; 70,000 Iranians (mostly Shiites from Karbala and

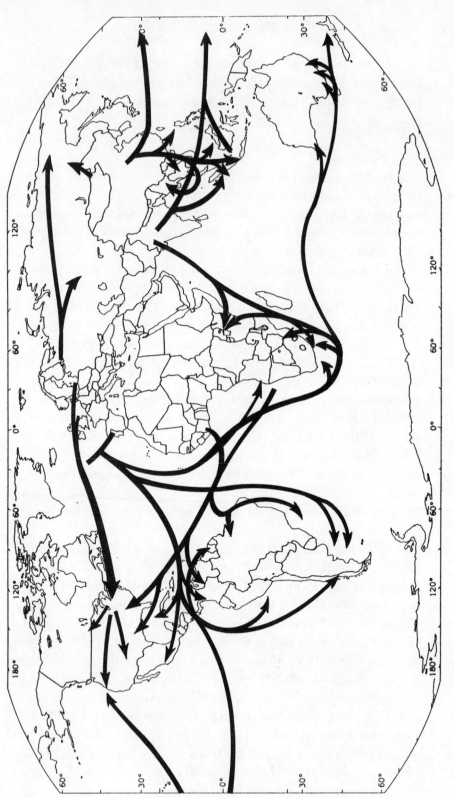

People on the move—by the millions.

Najaf); 380,000 Kuwaitis to Saudi Arabia; 45,000 Yemenis (from Iraq and Kuwait); another 1 million Yemenis (from Saudi Arabia); 21,800 Sudanese (from Iraq and Kuwait); 50,000 Syrians; 8,780 Europeans; 2,200 Americans (living in Iran and Kuwait); 67,600 Pakistanis; 150,000 Indians (most through Jordan); 85,000 Bangladeshis (most through Jordan); 85,000 Sri Lankans (most through Jordan); 9,900 Thais; 16,000 Vietnamese; 30,000 Filipinos.

It is probable that never before had so many people moved in so short a time. Consequences included loss of back wages, loss of jobs not replaceable back home, abandoned savings, and altered populations in terms of ethnicity, number, and skills. Relief organizations had to establish tent cities and transportation facilities in a desert environment to assist the millions coping with inadequate food, housing, and medical care.

More than 80,000 impatient, frustrated Albanians, desperate for political freedom, stole into the night, fleeing across the Adriatic Sea to Italy in the first half of 1991; children as young as 10 fled with classmates without telling their parents. Italy set up tent cities as clearing centers for those who would not be granted political asylum. It was a social earthquake. "If we don't send them all back," warned a prominent Italian, "all of Albania will be in Italy. I have been there, it is a horrible country. Many people would walk on water to escape it." Eventually, most of the Albanians were sent home. Italy was already supporting 400,000 illegal workers, second on the continent only to Germany's 600,000. If one tries to solve population-distribution problems, one does it by attacking the geographic process that causes it: migration and its roots.

A recent map of Haitians on the move reveals that 600,000 have migrated to the United States, 40,000 to Canada, 25,000 to the Bahamas, 15,000 to Cuba, 150,000 to Haiti's neighbor the Dominican Republic, 20,000 to Martinique and Guadeloupe, 1,000 to Venezuela, 5,000 to Suriname, 20,000 to Guyana, 3,000 to Africa, and 7,000 to France. In December 1991, tens of thousands of Haitians tried to

---

Human migrations around the world in modern times—spatial flows of people altering population distribution and spreading second-order consequences, such as enlarged pools of cheap labor for the host countries. New immigration policies have slowed permanent moves.

flee the latest of oppressive regimes; many were consigned to a kind of limbo in the United States naval station at Guantánamo Bay, Cuba.

As of May 1991, the Palestinian diaspora of 5.2 million people was roughly like this: 650,000 were living in Israel; 1.1 million in the West Bank; 700,000 in Gaza; 1.5 million in Jordan; 35,000 in Lebanon; 300,000 in Syria; 150,000 in Kuwait; 250,000 in Saudi Arabia, the United Arab Emirates, Qatar, and Oman; 150,000 in other Arab countries; 150,000 in the United States; and 175,000 in other non-Arab countries.

The military government of Myanmar (formerly Burma) forcibly removed hundreds of thousands of citizens from their homes in the cities and relocated them in new "satellite towns" in the outskirts.

Pakistan in 1990 became home to the most refugees of any country in the world—about 3 million Afghans who had escaped the decade-long Soviet invasion.

In the century between 1835 and 1935, about 75 million Europeans emigrated, fleeing famine, oppression, war, horrendous living conditions—and a hopeless future. Millions and millions swept into the United States, most around the turn of the century. Eleven million Europeans have found new homes on continents other than North America since 1945. In the 1960s, Latin America replaced Europe as the leading source of immigration into the United States.

The United States is clearly influenced by the spatial interaction of migration. Today, the United States is attractive even to its neighbors to the north: Canadians who move freely across the 4,000-mile-long unguarded border to escape high taxes and the high price of goods. From the United States perspective the flood of weekend shoppers from Canada is an economic advantage; but to Canada, which experiences a permanent exodus of factories, jobs, and even corporate headquarters, the consequences can be bleak. On the other hand, some Americans have fled the politics of the United States to Canada. The Vietnam War and the United States military draft engendered a steady emigration of United States young men to Canada, to the point that the Canadian government protested the untoward job competition and the influence of United States' youth culture on its own.

Geographers track emigrations and immigrations. They analyze the impact that immigrants have on the environment of places left and arrived at. As we all know, a heavy tide of migration funneling into one place changes the character of that place. Growth in population and its attendant needs—housing, schools, jobs, welfare benefits, health-care services—are the most obvious. In the eastern United States, Korean newcomers often gravitate to open-air greengrocer markets, Indians and Pakistanis move into the photocopy and newsstand businesses, Chinese open restaurants. Whole new communities arise as if transplanted, with mores intact, from the "old country." Once the newcomers feel comfortable in their new environment, they begin to introduce and champion attitudes they knew in their native land, further altering the geographic character of the region.

Europe employs more legally-resident foreigners than the United States or even the Arab oil-producing states. Among Western Europeans, only the French believe in assimilation. Europeans reluctantly acknowledge a discomforting paradox: immigrants may not be wanted, but they are needed. Many perform jobs Europeans no longer are willing to accept. Western Europe's population is dwindling and rapidly aging. France will need 315,000 new immigrants a year to maintain its current work force. It is said that neighborhoods, suburban areas, and even cities like Marseilles could pass for non-French.

Migration, as opposed to refugee influx, is the norm. Movements are episodic and finite, eternal and worldwide, goaded on by hunger or impoverishment, hope, and play. The Mexicans who cross the Rio Grande to harvest citrus crops in Texas or sugar beets in Minnesota are migrant workers. The thousands of southeastern Europeans, mostly Turks who were imported by West Germany to fill menial jobs after the Second World War, were, or became, migrant workers. The Okies who abandoned their Dust Bowl farms in the 1930s found a day's work in the lush fields of California. They were joined on the road to the West Coast by fellow migrants from Texas, Arkansas, Missouri, and Kansas who could no longer eke out a living on their own drought-ravaged farms. California presented them with the difficulties of an unfamiliar way of life—seasonal hiring for minimal wages and a nomadic existence. Nevertheless, the hearty endured, even prospered.

In 1989, about a million immigrants were caught trying to enter the United States. Experts guess that the number of illegal immigrants who successfully move to the United States annually is between

200,000 and 250,000. Immigrants from Mexico account for the largest number—about 40 percent. The solution to illegal Mexican immigration is not to be found in better fences, wider moats, or guard dirigibles. The solution in the long run will be for the two governments, working cooperatively, to develop employment opportunities in Mexico that will discourage illegal migrations. That process seems to be under way—but not to the satisfaction of American workers, who cannot compete with the lower-paid Mexicans and may have to accommodate themselves to lower wage demands.

There is a new type of migrant: the environmental refugee. More and more people are fleeing from threatening environments, from man-made disasters and growing pressures on ecosystems or predicted natural disasters, such as earthquake zones, volcanic eruptions, and typhoon routes. The exodus from Chernobyl, which was obligatory, left behind the first modern peopleless city. In the sub-Sahara, millions of Africans flee from famines exacerbated by political decisions and warring factions.

Israel, through moral obligation, takes in and shelters every Jew from anywhere, without regard for the consequences. In the last few years, more than 1 million Soviet Jews emigrated to the "promised land." Was life in the Soviet Union so intolerable that even professional peoples would fly off to a tiny land where they find themselves performing menial jobs? Were many seduced by promises of challenging employment and an improved life-style that are not available? Jerusalem, for its part, is confident that the extraordinary influx will infuse Israel with strength, wealth, and increased importance on the world stage. The ramifications of this saga continue to unfold.

Nomads include people who know how to lead their flocks to good grazing land. They extend hospitalities to others they meet along the way. They carry with them everything they need. Typically, their movement is a seasonal transhumance. It might be well for us if we recovered some of our atavistic nomadic spirit.

"I was six years old when Lindbergh flew to France in 1927," the columnist James J. Kilpatrick has recalled. "It seemed an unbelievable adventure. Now the unbelievable becomes routine. Satellites and supersonic planes have turned strangers into neighbors. Our children and grandchildren ought to know them better. The globe dwindles. The planet shrinks."

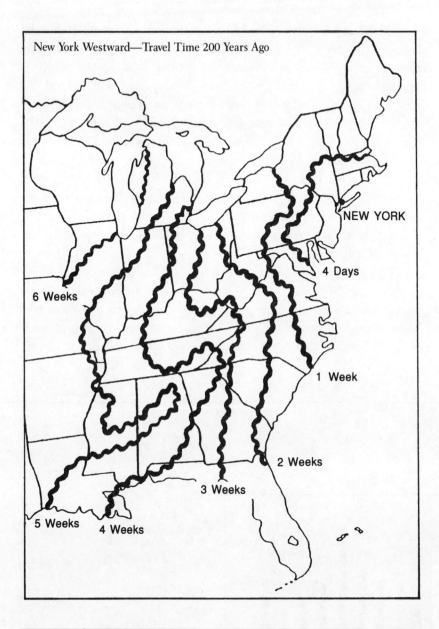

New York Westward—Travel Time 200 Years Ago

NEW YORK

4 Days

6 Weeks

1 Week

2 Weeks

3 Weeks

5 Weeks   4 Weeks

Technology has transformed "distance"—space, time, cost, difficulty—as measured between places. In covered-wagon days, migrants needed a half year to cross the North American continent; the railroad reduced the time to about three days. Electronic technology, the medium or process of our era, continues to reshape and restructure patterns of travel— space and location—all spatial interactions. Jet time between New York and Paris or London, via the Concorde, is three hours. In 1986, President Ronald Reagan proposed development of a hypersonic Orient Express space plane that could make the Washington–Tokyo flight, in space orbit during midcourse, in 120 minutes.

The United States continues to be the first choice for emigrants from all corners of the globe: "Give me your tired, your poor, your huddled . . ." Congress's immigration bill of 1990 permits 750,000 immigrants to enter the United States every year. Well over a half million of these are admitted because they have family members here already. Especially welcome are immigrants with specific skills needed in the States. This is a newly stated preference. Yes, it invites charges of elitism. "More people do not necessarily make for a better society," Richard D. Lamm, the controversial former governor of Colorado, asserts bluntly. "Better people do." The number of immigrants the United States accepts is about twice that of the rest of the free world but still a far cry from the words expressed in the Statue of Liberty's welcome.

The United States is quite specific in its definition of a refugee. A person is not a refugee if he returns to his native country or is granted citizenship in a country of first asylum or in a third country. Because of this, people who leave their country as a result of such events as famine or natural disaster are normally excluded from ref-

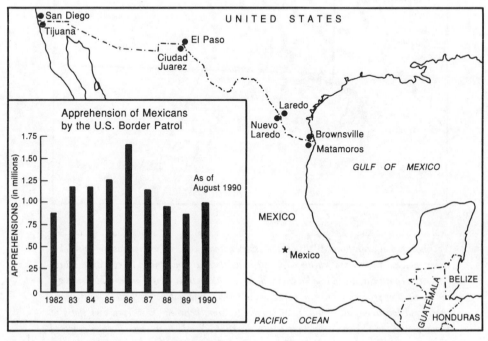

SOURCE: IMMIGRATION NATURALIZATION SERVICE STATISTICS BRANCH

ugee counts and may not be considered refugees for purposes of gaining legal admission to another country.

After the United States barred immigration of Orientals in the 1920s, Brazil emerged for a time as the largest destination for Japanese emigrants. Brazil today has 1.2 million people of Japanese descent,

There are around 17 million refugees worldwide. Unlike migrants, who voluntarily shift from place to place, usually in search of sustenance, refugees leave their homes involuntarily, usually out of fear of persecution, to seek the protection of elsewhere; many settle permanently in their place of asylum, further changing societal geography. About 110,000 "boat people" and other refugees fled the Communist conquest of South Vietnam, then fought forced repatriation, fearing political harassment. "Yacht people," on the other hand, are selective migrants, flying from the Orient, especially Hong Kong, to a prosperous life in Canada, far from the feared consequences of the negotiated takeover of the British crown colony by the People's Republic of China in 1997. There are 3 million Afghans (a quarter of Afghanistan's population) in sanctuaries in neighboring Pakistan and Iran, having escaped the Soviet Union's failed invasion of their country in the 1980s. WIDE WORLD PHOTOS, INC.

slightly more than does the United States. Brazil, in fact, is home to more Japanese than any other area outside of Japan. The city of São Paulo has been a particular magnet.

The European Community, which denounced the Iron Curtain for four decades, is anxious that millions from the former Soviet bloc don't decide to migrate west with their new freedom of movement. In 1990 alone, the hordes nurtured the speculation that upwards of 20 million East Europeans and former Soviets might flee hardship and environmental pollution and seek refuge in sovereign states in Western Europe. "Immigration will be the big issue of the 1990s," a senior official of the European Community's headquarters in Brussels has said. "There are, in fact, signs of a new xenophobia in Western Europe, but it is not a question of border controls. We can't build a Fortress Europe or put up our own 'Berlin Wall.' But I can't tell you how many times I've awakened in a cold sweat in the middle of the night, having had the nightmare of 100,000 Eastern Europeans and Russians walking west—all at the same time."

The pervasive fear that host societies will be profoundly changed by massive immigration is not entirely unfounded. "Even if we adopt and enforce quotas, take only those we need, these people, too, will probably change our societies," a European friend told me. "There is a fear that a white, wealthy, and Christian 'fortress Europe' will be pitted against a largely poor, Islamic world."

Throughout Western Europe, there is a perception that immigrants—Arabs in France, Turks in Germany, Pakistanis in Britain, and so on—are "taking over" schools, housing complexes, and neighborhoods. This in turn has led to the movement of European families. The entire process has become known as the ghetto phenomenon.

The challenge to Western democracies will be to help long-subjugated countries get to their feet, clean themselves up, and establish sturdy economies that will keep most people employed and living prosperously and in good health in their native environs. Solutions lie in helping the former socialist East build healthier places, places they like and know but must bond with in a natural way—without piles of toxic waste, without polluted water and the repressive regimes that forced their populations to remain in place.

Mobility—migration, commuting, tourism—is a mighty set of

spatial processes that connect places, change places of origin, alter places of destination, affect flows of everything—ideas, disease, ideology, money—and change our world and the geography of every place.

Every time a person changes his living arrangement—and the average American moves his residence 18 times in his or her life—the places and spaces he or she leaves and moves into are altered, sometimes dynamically. Spatial interaction, be it money, an air mass, a disease, always interfaces with the environment. That's the nature of geography. Some people can't bear to leave a familiar surrounding, so they take it along, even if it's a nine-room house. WIDE WORLD PHOTOS, INC.

WATER, WATER . . .
WELL,
NOT EVERYWHERE:
CONFLICTS IN
THE MAKING?

The geography of water is of enormous importance. The planet's flowing waters and oceans are networks connecting many of Earth's places and promoting trade and tourism. The maldistribution of water over Earth and the variations over space of consumption levels are leading to shortages, tensions, economic stress. Understanding the spatial dimensions of water will be a key to peace in the next century.

Colorless, tasteless, odorless, calorie-free . . . and absolutely essential to all life.

It's water—two atoms of hydrogen, one atom of oxygen—the most common, the most abundant substance on the face of the planet. Yet it is water, or at least the maldistribution and shortage of water, that may breed violent disputes. Water is life, wealth, and power. It is far more rare, vulnerable, and precious than oil.

The Euphrates River flows from Turkey to Syria to Iraq. Ankara has built dams along its course, virtually allowing Turkey to control the flow. Syria, distressed by cuts in water flow, has lodged diplomatic protests to Turkey. But Turkey, like most upstream countries, insists it is under no more obligation to give away water than, say, Saudi

Arabia is to give away oil. This is but one of many, *many* international water disputes looming.

Seven tenths of Earth's surface is covered with water, but 97 percent of the water is saline and unusable, leaving a meager 3 percent to nourish and sustain all terrestrial life. The availability of usable water is often a life-and-death issue. On an international level, rivers flowing across more than one country can initiate difficult political problems; one country can control the flow into another. Within countries, disputes over water are arising among states, regions, and communities. Water, in a very real sense, is a geographic issue at many scales and in many ways: pollution, irrigation, and so on.

It is by no means an exaggeration to say that water occupies a central place among resource issues. Population growth and economic and cultural development are simultaneous global phenomena, interdependent in a plethora of regional contexts. Ignorance, poverty, and poor agricultural practices have gravely endangered water resources and resulted in severe shortages.

Every 20 years, the volume of water-use worldwide about doubles, and increasingly rivers and reservoirs become more polluted. At least 80 countries, supporting 40 percent of the world's population, suffer from serious water shortages. The most seriously affected areas are in the arid climate zones, the Middle East, Saharan Africa, the American West, and Central Asia. In many areas, particularly monsoon Asia, the problem is precipitation, unreliability, and timing. By the end of the decade, 29 developing countries are expected to be without adequate renewable water resources for domestic and irrigation use. In developing countries, 1.2 billion people are without safe water. In some parts of Africa, women must walk 10 miles to reach an acceptable water source and carry it home in jars on their head.

In the Middle East, demand and consumption have grown remarkably, and competition for water grows daily. The region relies on irrigation to grow food, coaxing bigger harvests out of the land. Only two countries, Egypt and Turkey, have reasonably adequate supplies of water within their borders. Much of the water for Jordan and Israel passes through, and is diverted farther upstream by, Turkey, Syria, and Lebanon. The major water sources for Iraq are the historic Euphrates and Tigris rivers. Desalting plants, processing water financed by oil wealth, produce most of the fresh water in the Gulf States, especially in Kuwait and Saudi Arabia.

My colleague Brad Thomas, a water specialist, believes that self-sufficiency in water seems an elusive goal in Israel, where water is very scarce. Not one drop of Yarmuk River water ever reaches the sea. Two consecutive years of abnormally low rainfall prompted Je-

Rivers, liquid highways to the sea, and places between, are vital for the planet's water cycle involving land, ocean, and atmosphere. They can inspire music and cast an aura of romance, but they also can be vicious killers. Repeatedly destructive floods have earned the People's Republic of China's nearly 3,000-mile-long Yellow River, or Huang Ho, the lamentable nickname "China's Sorrow." Rivers can be adjuncts to the waging of war. To stem the tide of the Japanese military invasion in 1938, General Chiang Kai-shek's retreating Nationalist Army dynamited dikes along the Yellow River, flooding 20,000 square miles and hundreds of cities and villages, but also drowning nearly a million innocent Chinese. In Egypt, the Aswan High Dam at the First Cataract of the Nile expands both agricultural opportunity and hydroelectric power, at the same time depleting farmland of silt and the eastern Mediterranean Sea of nutrients. Aswan backs up so much of the world's longest river that if the 375-foot-high structure were breached, about 50 million Egyptians would be drowned. Such a possibility may have stimulated Cairo to sign a peace treaty with Israel. The world's largest hydroelectric project is Paraguay's 60-story-high Itaipu Dam on the Parana River, which is second only to the Amazon as the longest river in South America. (The Grand Coulee in the state of Washington had been the world's largest dam.) Eighteen turbines, shared with Brazil, which arranged the financing, produce six times as much electricty as Egypt's Aswan. WIDE WORLD PHOTOS, INC.

rusalem to implement a 60 percent cutback in water for agriculture. Even with normal rainfall, the sharp rise in municipal and industrial demand, driven by population growth, required taking water away from agriculture.

The water level of Lake Kinneret, which is the source of a third of Israel's water, has been near a record low. Pumping from the lake was stopped so the remaining liquid would not become too saline. The annual artificial winter recharge of the two principal groundwater aquifers—the source of another major share of the water supply—could not take place. With no artificial or natural recharge, continued drawdowns left the aquifers increasingly saline from seawater intrusion.

Farmers adjusted to the cuts by reducing production of water-intensive cotton and other field crops and by irrigation of the remainder with treated urban wastewater. Use of expensive desalinized seawater requires a shift to higher-priced market vegetables. Raising water prices—now heavily subsidized in Israel—would encourage further conservation.

Israel receives some of its water from outside its borders. The "Unity Dam" proposed for the Yarmuk River between Jordan and Syria might regularize flows to downstream Israel but would not add significantly to the supply. Israel manages 97 percent of the natural-water resources available to it; only erratic desert flash floods have not been harnessed. The urban population continues to make increasing demands on limited sources; reduced agricultural water allotments will no doubt become the norm.

Egypt, to ensure its long-term water security, has supported the efforts of a group of Nile riparian states to cooperate in the development of the river, on whose banks and in whose delta more than 50 million Egyptians live. Any future plans by Ethiopia to develop the Nile's headwaters could become a *casus belli* for Cairo. It is said by military historians that every war in the Nile basin since the turn of the century has been fought at least partly over water.

The plan to turn Siberian rivers around to flow south into Central Asia rather than down north into the Arctic Ocean has been halted. This was to have been the "project of the century." Russian friends have told me that ethnic disputes were important in stopping the project, as well as the lack of knowledge of the potential ecological impacts. Costs, too, played a large role in the decision. Central Asia

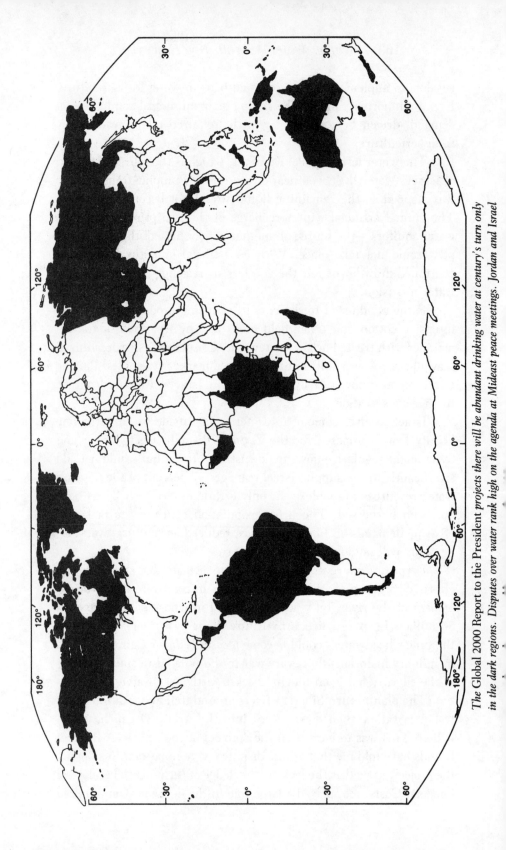

The Global 2000 Report to the President projects there will be abundant drinking water at century's turn only in the dark regions. Disputes over water rank high on the agenda at Mideast peace meetings. Jordan and Israel

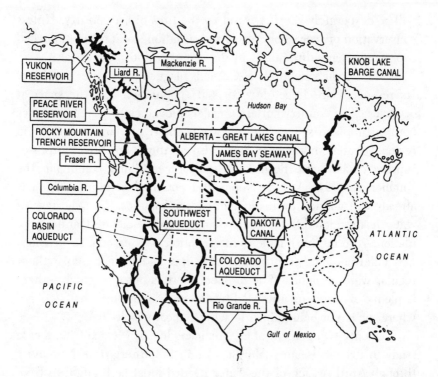

Water, water, but not everywhere—therein lies a life-and-death spatial problem. Diversion to parched lands is controversial, expensive, and ecologically risky. The North American Water and Power Alliance (NAWPA) was a monumental scheme to head off a calamitous shortage: tapped Alaskan and Canadian waters would be diverted southward through British Columbia and the prairie provinces to 33 U.S. states, including the desiccated Southwest, and Mexico. Diversion of at least 110-million acre feet of water annually would keep 337 cities the size of New York in potable water. (An acre foot is the volume of water that would cover one acre, or 1/640th of a square mile, to a depth of one foot.) Scientists studying NAWPA feared the dangers posed by tons of water flowing across the continent's unstable seismic zones. Another scheme would have 2,000-mile-long underocean pipelines carry fresh water from one or two of Alaska's great rivers to California, which has been acting like a thirsty vampire; it would be like putting a faucet in a blood bank. The Soviet Union's plan to divert some of the flow of its Arcticward rivers to arid and semiarid southern regions was shelved after 20 years of development because of the country's weak economic situation, ecological concerns, and ethnic tension. In Central Asia, the future of agriculture may well depend on more efficient water use: conservation. The next century will be a period of great tension around the world over water—its supply and quality. Over 200 rivers flow across international boundaries; they will become areas of conflict unless we learn to manage water properly.

will be desperately short of water by the turn of the century. Unless conservation or technology alters the situation, local populations will experience many difficulties in the economy and public health.

The People's Republic of China is plagued with water problems. Its resources are huge, but so is its population and the space it occupies. The country suffers a major imbalance in water resources both in time and space; in other words, there are both seasonal and regional water shortages. There also at times are huge surpluses causing catastrophic floods, the most recent killer coming in the summer of 1991. Any developmental potential of North China is already constrained by inadequate water supplies. The shortage of water has rendered industrial production in the city of Tianjin (population: 7.76 million) liable to interruption and seasonal suspension.

Palermo, Sicily's principal city, has had to make do without regular water supplies. The problem there is not how much—there is plenty of water on the Mediterranean island—but that it is not where, or is not being piped to where, most people live.

Twenty-two percent of all Mexicans live in Mexico City, a case study in urban disaster. Almost no rain falls there from November through April. Much of the water needed must be brought in from the outside.

By the year 2000, there may well be a total of three quarters of a billion people without an adequate water supply. The average American's way of life "requires" 2,000 gallons of water every day, 87 gallons of it for personal use, but only two of these gallons to meet the basic needs of drinking and cooking.

Four trillion gallons of precipitation fall on the United States every day, but there are shortages in many places. The East Coast receives an average of 40 inches a year; Iowa gets 31 inches; western Nebraska, 18 inches; Nevada, the driest state, a mere 9 inches. Sixty percent of the country's real estate is in the West, but only 25 percent of annual rain and snow falls there. Seventy percent of the rainfall in California is in the north, but 80 percent of the demand for water in the state is south of Sacramento.

Cities like Las Vegas are beginning to confront limits to growth set by the shortage of water. The fastest-growing major city in the United States, the Nevada metropolis may outstrip its water supply by 1995 unless it finds new sources and conserves what it has.

To manage water, the United States has rearranged its landscape on a colossal scale: it has built innumerable dams, irrigated 60 million acres, drilled millions upon millions of wells. Roughly half the irrigated land is in the Great Plains, most of it water from the Ogallala aquifer.

The last quarter century has seen a rapid increase in the scale and the scope of North American water development. Where projects costing a half billion dollars and involving the construction of dams over 500 feet in height were once considered gigantic, they are commonplace today, and necessary. Some areas have been drying up, extinguishing important fish supplies, turning timberlands into tinderboxes, and forcing populations to relocate to assured water resources. Gigantic water-development projects have become a must. The drier states of the southwest have relied for decades on an irrigation-ditch system manned by local farmers who buy into consortiums.

On the drawing boards was the biggest engineering project ever attempted: the diversion of water from northern North America to the south, from a Canada "drowning" in water to a parched United States and Mexico. The plan called for the annual diversion of at least 110 million acre feet of water, probably enough to keep 337 cities each the size of New York (7 million people) in potable water.

The proposed monumental, continental water development scheme was known as the North American Water and Power Alliance—NAWPA. Briefly, it called for the damming of the headwaters of the Yukon and Tanana rivers; the lifting of this water into the already-existing Peace River Reservoir; the transferring of the water in part to the proposed Alberta Great Lakes Canal to serve the Prairie Provinces and the Missouri and Mississippi river systems; its pump lifting in part into a reservoir created in the Rocky Mountain Trench by damming of the headwaters of the Columbia, the Fraser, and the Kottenay rivers—the reservoir would extend for some 500 miles into British Columbia; distribution from the reservoir would be into the southwestern United States and into Mexico. NAWPA's estimated cost was $100 billion.

Because water shortage is expected to become disastrous in the western United States in the 1990s, NAWPA was a serious proposal. The ecological problems that would develop by running tons of water across an unstable seismic zone are sizeable and not entirely predictable. The project lost its credibility with most technicians early on. Most American and Canadian scientists have debunked it. But the urgency of the water shortage indeed calls for big thinking.

Alternatives to water diversion can be divided conveniently into two categories: technical solutions involving the provision of additional sources of supply, such as desalinization, weather modification, and groundwater exploitation; and nontechnical or institutional alternatives involving the regulation of demand such as water rationing, the use of pricing mechanisms, and the reallocation of use—for example, decreasing consumption, especially waste.

Water can indeed unite or divide a region. In India, in the summer of 1991, a dispute over river-sharing between two southern states threatened the government coalition. Water can be a catalyst for cooperation and peace or for conflict. International cooperation in the integrated management of water resources can defuse incendiary situations.

DOTH
THE EARTH
ABIDE
FOREVER?

■■■■■■ ■ ■ ■ ■ ■ ■ ■ ■ ■ ■ ■ ■ ■

*Paradise is a toilet! It's as if the planet had aged four
billion years in the last two centuries. If this had
happened to a human being, you'd say, "Jesus, what's
he been up to? He looked like Apollo yesterday, now
he looks like Methuselah."*
—from *London Fields*, by MARTIN AMIS

Here on Earth, God's work must truly be our own.
—JOHN F. KENNEDY,
Presidential Inaugural Address, 1961

The environment has become a prime geographical issue because
we live in an ecologically imperiled global society in which the deg-
radational processes move across national boundaries with no regard
to laws, places, or ideologies. The shipping, trading, and movement
of toxic wastes, for example, is an international problem with enormous geographical implications. It is literally as down-to-earth as
geography can get—environmental threats and spatial processes that
must be solved by global negotiation and treaty.

Worldwide environmental problems are so severe, so critical, that a distinct branch of geography whose purview is the interfacing of humankind and the environment was created even before public awareness of the crisis existed. In a less sensitive period, it was called the man-land school of geography. Today, it is generally referred to as the society/environment approach. We must accustom ourselves to an accelerating ecological interdependence between man and the biota, the animal and plant life of a region or period. Ecology and economy, of necessity, are becoming ever more interwoven—locally, regionally, nationally, globally—a seamless net of causes and effects.

In managing the global commons and resolving environmental problems, we must take into account the following:

• The growing pressure to implement environmental policies is often muted by traditional vested political and economic interests and by conflicts over new regulatory standards and enforcement procedures. This must change!

• Most environmental nongovernmental organizations are focused at the national and local levels rather than at the global, holistic level.

• Perhaps the most influential supranational agencies on the environmental front are multilateral development banks and other leading nongovernmental institutions that view the problem from a global perspective.

• Innovative alliances sometimes build hybrid institutions that can creatively tackle international environmental programs otherwise bogged down by government red tape, lack of funding, or poor scientific support.

• Environmental politics can be viewed in a hierarchical framework with five layers that represent local, subnational, national, international, and global ecopolitical scales or stages. They must work together from local to global levels.

• Concerns about the global implications of once-local problems can cause or worsen international disputes and stimulate competition among politicans to appear environmentally sensitive. These days, being green is politically correct, but . . .

• Both the scientific community and the public are demanding that a wide range of transboundary environmental issues be raised in

"Hard to believe this was all rain forest just fifteen years ago."

international forums. Public education is a vital component in solving ecological problems.

These ideas and trends clearly demonstrate the significance of geographic awareness: the connectedness of place and the increasing awareness of the public to the environmental problems plaguing Earth. To solve such problems, we will need governance systems— treaties, agreements without regard to boundaries—and not just actions by individual governments and national policies.

All life exists in the biosphere—the intersection of the atmosphere (the air that surrounds the Earth), the hydrosphere (the water both on the surface of the planet and in the atmosphere surrounding it), and the lithosphere (the crust of the planet). We have not been very good as custodians of our living planet. We are choking the biosphere to death. We are stripping the gears of the ecological systems that support civilization. We are committing suicide. As I write, I hear on the radio that a satellite passing over Antarctica measured the lowest stratospheric ozone level on record. It is the fourth severe "donut" to develop since 1986, another ominous indication of potential global health risks. Once again human activity is the villain: the release of chlorofluorocarbons into the atmosphere, substances widely used as refrigerants, solvents, and foaming agents in insulating plastics. Destruction of the ozone shield could be a devastating event beyond imagination.

We slash and burn the trees of the world. We overgraze grasslands. We overplow croplands. We crisscross the landscape with freeways. We have taken little care about the dumping of toxic wastes. (There are Love Canals everywhere.) Increasing immolation of the Amazon rain forest—pharmaceutical laboratory, flywheel of climate, library for an ecosystem of 15 million plant and animal species—is one of the great threats to humanity.

Pollution, such as acid rain, knows no boundaries. It doesn't stop at the corner down the street or at the nearest border. Nor does it recognize hemispheric differences. It attacks all ecosystems. Pollution not only permeates the places of origin but inexorably becomes part of spatial interactions (the air and water systems) and diffuses over the planet. Clouds of arid smoke from the decimated Amazon rain forest—170,000 forest fires were burning at the same time in the western Amazon, settlers turning the woods to ash in the name of progress—floated to the southern tip of South America, 3,000 miles away. Volcanic ash from Mount Pinatubo eruptions in the Philippines that killed 700 people in 1991 collected in the southern polar region. Months later, flash floods and mudslides after a typhoon killed 6,000 Filipinos. An Exxon tanker spilled 10.5 million gallons of crude oil into Prince William Sound, Alaska, damaging 1,224 miles of shoreline. In that same year, 1989, there were 527 other oil-tanker mishaps worldwide. (Had the Alaskan oil spill occurred on the eastern seaboard of the United States, the 1,200 miles of affected shoreline would have stretched from Cape Cod to the Outer Banks of North Carolina.)

One hopeful development in coping with United States pollution is the requirement by the Environmental Protection Agency that manufacturing concerns must report on a yearly basis the amount of toxic chemicals they emit into the environment. The offenders are identified, and there is encouraging evidence these corporations are making progress in reducing emissions.

Deforestation of the Himalayas disrupted flood control in Bangladesh. The smoke from the torching of more than 700 oil wells in Kuwait by retreating Iraqi troops in 1991—a mad act of vindictiveness—desecrated the snows of Switzerland and could deflect vital monsoons across the Indian Ocean. Many lakes and rivers are a witch's brew of chemicals. Acid smog eats the marble of the Acropolis in Athens. Black fumes besmirch the marvelous vistas of Istanbul, once the brightest jewel in the eastern Mediterranean.

Polluted air flows lead to changes in the weather and sometimes even in the climate. If they lead to global warming, as it appears likely, mountain ranges of polar ice would melt and bring on life-threatening sea-level rises around the planet. Seashores and ports will be negatively, and expensively, affected.

Everyone who breathes knows that the physical geography of the atmosphere is becoming unhealthy and even dangerous. The air in Mexico City is so foul, it is often visible to the eye and dangerous to the lungs; the smogs of Los Angeles, Denver, and Seattle are frequent and asphyxiating. Atmospheric processes have specific, mappable patterns of flow.

Villagers of Broughton Island, 70 miles north of the Arctic Circle, have higher levels of PCBs in their blood than any known pop-

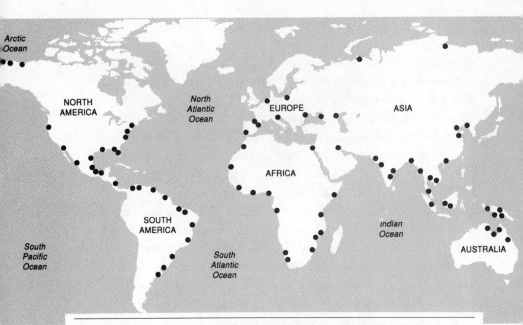

The dots represent the areas of vulnerability of sea-level rise. But be of good cheer: the waters will rise slowly; there will be time to head for the hills.

ulation. All across the Arctic, scientists are turning up surprising concentrations of pesticides, herbicides, and industrial compounds used much farther to the south, some of them banned years ago in Canada and the United States.

Economic development in all societies has been a rapacious process. Supercomputers continue to collect critical masses of environmental data, evidence of environmental batterings and abuses around the globe, which are shared in international conferences. But not much can be accomplished if the major powers do not cooperate. About 65 percent of the carbon-dioxide emissions originate in the industrialized sovereign states in the northern hemisphere. As we entered the last decade of this century, more than 60 countries in conference in the Netherlands agreed that by the year 2000 carbon-dioxide emissions—a primary source of the so-called greenhouse and global-warming effects—would be stabilized, and that by the year 2005 they would be reduced. The United States, the Soviet Union, and Japan, the three most industrialized countries in the northern hemisphere, refused to sign the declaration. "The conference was a total failure, the U.S. was saboteur," an American environmentalist cried. In Washington, George Bush, chief executive officer of the country that spews more pollutants than any other, defended his opposition to limiting the greenhouse effect: the extremist position could limit industrial development!

Third World countries are industrializing. They, too, want to escape poverty and to provide education and a reasonable life for their citizens. They are finding it expedient to ignore problems of pollution. They resent being told that pollution is bad for the environment, that they should not build a smokestack economy. They argue that the West had its turn, now it is theirs: "We have been deprived long enough." The truth be told, *we* are not interfering in their internal affairs. Pollution affects everyone.

Brazil resents being told what to do with its rain forests. International negotiations may be scheduled to offer differing views on the topic. How much can be cut? How much compensation should be paid for not cutting, and by whom? "Chopping down trees is our livelihood," Brazilian laborers hired by agribusinesses declare. "How else can we create new farmlands and grazing areas?" International law is based largely on long-standing notions of autonomy and state sovereignty. Clearly, solutions that will be complex and costly must be found to protect the global environment, allow the Third World to develop without destroying global resources, and find enormous sums of money to pay for appropriate practices and policies. It will call for imagination to resolve the problems of disparate places and the processes that connect them.

Greenhouse gases are atmospheric gases that include carbon dioxide, nitrous oxides, methane, and CFCs. They allow light to reach the surface of the planet. But they also block the return of the infrared rays to space, thereby retaining heat. These gases are analogous to the glass windows of a greenhouse. Carbon dioxide was once regarded as a harmless product of clean combustion. It is now considered highly dangerous. This greenhouse warming confronts humankind with the prospect of trying to maintain agricultural and other systems in a protracted period of climatic change historically unprecedented in speed and magnitude. Geographers and other scientists are striving to understand the regional, spatial, and climatic processes in order to be able to predict the where of these changes and, subsequently, *the why of where.* This is an instance where the global processes are better understood than the regional, that is, the regional *spatial* processes. Everyone knows about global warming, but the more important question is, Where will it get warmer and what impact will warming have in one place as compared to another? Warming in Russia, let us say, may be beneficial warming; in the United States, it may be harmful. The *where* of global warming is critical to economics and lives.

The term "acid rain" was introduced as far back as 1872, in an article discussing the increasingly sooty skies of industrial Manchester, England, and the acidity found in precipitation. Acid rain has a number of geographic dimensions: where it is created; why it is created there; what paths, or tracks, over space it follows from origin to destination; what its impact is. Emission sources, winds, rainfall patterns, and soil sensitivity combine to determine the impact of acidity on the environment. In the United States, acid rain is created in the upper Midwest, especially in Ohio, where high-sulfur coal is burned in power stations. Particles shoot into the atmosphere, mix with moisture in the clouds, and are blown northeast by westerly winds. They combine with other chemicals and form an acid mixture, which falls over the eastern U.S.

(In the United States, the highest amounts of toxic waste are in Texas, Louisiana, Tennessee, Virginia, Illinois, Indiana, Ohio, Pennsylvania, New Jersey, Massachusetts, Connecticut, and Rhode Island. The water is most tainted in Louisiana, Alabama, Tennessee, Indiana, Ohio, South Carolina, Virginia, Delaware, New Jersey, Massachusetts, Connecticut, and Rhode Island. The three leading polluters are Du Pont (petroleum and coal refining, chemicals), Monsanto (industrial, household chemicals), and American Cyanamid (chemicals). The three leading water polluters are Arcadian Corporation (liquid nitrogen, phosphates), 3M (chemicals, tapes, automotive products), and Freeport McMoran (minerals, oil and gas refining). The three leading air polluters are Renco Holdings (part of the Renco Group, which is an iron and steel processing business), 3M, and Eastman Kodak (photographic chemicals, household products).

Countries perceive that external constraints are being imposed on their national sovereignty in the name of saving the global commonweal. Sovereign states behave pretty much like individuals when their turf is invaded. They need to be assured that restoration of the planet calls for global bargains, compromises, and trade-offs. Most modern processes, industry and transportation, for example, are driven by energy (coal, natural gas, oil, wood—solar power captured in convenient packaging), hence the existence of air-carried pollutants. Should a paper mill in Maine be closed to save the environment in the northeastern United States and neighboring Canada—and effect massive unemployment? We are told that controlling industrial output could limit development. I believe, however, we should consider drastically altering our life-styles to reduce at least the rate of environmental deterioration. Changing life-styles means consuming less, sharing more, and recycling.

One can read the news stories and witness signs of ecological changes. They herald and reflect the evolution of the globe as a set of regions or landscapes of fear. They do not occur in isolation but

Agro-Ecological Zones In Sub-Saharan Africa

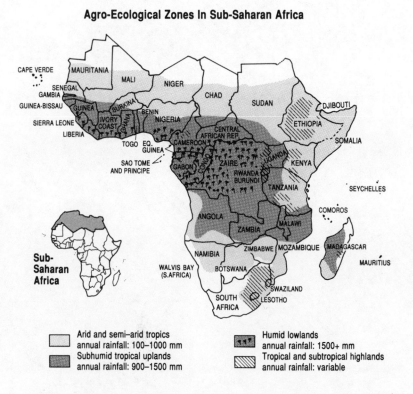

Sub-Saharan Africa is vulnerable to the complex processes of drought, famine, and human exploitation, not to mention savage, long-running civil wars. Millions in Sudan, the continent's largest country, and the Horn of Africa are threatened with starvation by internal forces all too frequently. Two drought- and heat-resistant cereals that thrive on dry, withered soils are mainstays: millet and its cousin sorghum.

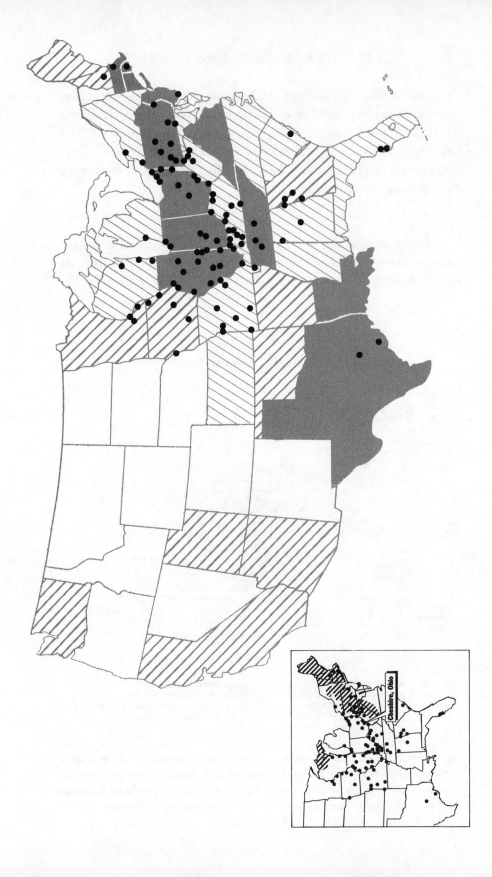

Cheshire, Ohio

rather as events that connect places and people in those places. Is the following litany our epitaph?

Indoor Pollution Leads to Sick-Building Lawsuits

Antarctica's Ozone Hole Stretches Toward Chile

Chronic Smog Blankets Eastern Europe After 40 Years of Inept Economic Management

The Everglades, "A River of Grass," Shrinks by 50 Percent

Miners Pour Two Pounds of Toxic Mercury into the Environ- ment for Each Pound of Gold Extracted from Rivers of Bra- zilian Rain Forests

It Takes 20 Trees to Keep a Baby in Disposable Diapers for Two Years

Factory's Gas Leak Kills 3,000 in Bhopal, India

Nuclear-Bomb Tests Render Sites Uninhabitable

Disappearance of Biotic Diversity Escalates

Air in Cubatao, Brazil, Is So Bad, If You Go Outside You Will Vomit

Pollution in Eastern Europe Is Destroying Cities, Rivers, Seas, and Killing People

Czechoslovakia's Air Pollution Termed "Worst in the World"

Carbon Emissions from Fossil Fuels Reach Billions of Metric Tons

Expert Says Pesticide Killed All Life in California River

Turning Rain Forests into Hamburgers Turns Out to Be a Poor Trade

New Threat to Maya Ruins: Acid Rain

Los Angeles's Air Dirtiest in U.S.; Mexico City Restricts Driving

Air Fit to Breathe a Growth Industry in Mexico City

The dots represent the sites of the power plants in the United States that produce the highest emissions of sulfur dioxide, which are inexorably blown eastward by the wind and reportedly spread acid-rain damage (the slanted lines in the box). United States–created acid rain has polluted even European woodlands. Damage tends to occur when high levels of acid rain combine in areas where soil and bodies of water are low in natural buffers that neutralize acidity. Western states have the lowest amount of toxic waste, determined by pounds released per square mile into the air and water, on land, underground, in public sewage, and off-site. Amounts are highest in the blackened states; above average in the states like Florida with right-slanted lines; average in the states like California with left- slanted lines; lowest in the all-white states.

Vermont's Sugar Maples Sicken Under Acid Rain's Pall

Destruction of Habitat Kills Off Koalas

China Resists Signing International Agreement to Save Earth's Ozone Layer

Ninety Percent of California's Wetlands Have Disappeared

Bamboo Forests Are Cut Down; China's Giant Pandas May Starve to Death

Last Great Stand of Teak Forest Falls to Finance Military in Myanmar

Arctic Haze Rivals Pollution Levels in Large Cities

Miles of Shoreline Are Vanishing

Virtually Every System of Coral Reefs Is Suffering

Japan Wantonly Destroys Tropical Rain Forests in Southeast Asia

One Hundred Species of Plants and Animals Become Extinct Every Day, Their Genetic Material Lost Forever

Endangered Species Now Include Snow Leopard, Zebra, Wallaby, Humpback Whale

The headlines are not isolated facts. They are further evidence of what's going on. They are a profile of a planet in jeopardy, because all places are connected.

We know more today than any of our predecessors ever did about the environment. How best can Earth be made safe for healthy habitation? Would it be feasible to call for a worldwide time-out for industry and agriculture? What is the best strategy for reconciling economic development with a habitable environment? Is the rapid rise in skin cancer a consequence of the forming of those ozone holes over the South Pole?

We should not forget that large parts of Earth were at one time very different from what they are now. We simply don't have the data from the past to know how much of what we are experiencing is cyclical and how much we are altering the ecological systems. Are we experiencing for the first time our interference with the atmosphere, our assaults on nature, or are we merely repeating ourselves?

We do know that environmental change has been the norm. We know that continents have shifted, shorelines have changed, seas have disappeared. We know that temperatures affect the wind, the wind affects the ocean currents, offshore temperatures affect here, not there, then there and not here, all influencing rainfall patterns, evaporation rates, vegetation patterns, drought areas, growing sites— all enormous interactions of processes over space, causing prodigious geographic variation.

Mutilation of the physical environment has been going on since the dawn of humankind. There is an area in north-central Pennsylvania that looks like a vast virgin forest. Dusty photographs of the site taken in the late 19th century reveal thousands of tree stumps. It turns out that thousands of trees were cut down there to make supports for tunnels in the nearby bituminous coal mines. Humans, alas, have always had difficulty in refraining from consuming the bounty of nature, and often have been ravenous in the process.

A few years ago, I visited a community that was being carved out of remote northeast Siberia, a storehouse, a vault of natural resources. It was a Soviet "company town" in the worst sense of the description. The region was being decimated. The earth was being gouged asunder. Forests were being leveled. Roads were being punched through the virginal landscape. The majesty of nature was being despoiled. There was little concern for the consequences of the economic process being directed from a place thousands of miles away—Moscow. Arctic environments are fragile, much more so than are environments at mid-latitudes and in the tropics. Polar regions take enormous amounts of time to recover from degradation, if they recover at all. Despoiling Siberia and the tundra of the north will result in pollution of the soil and water, unexpected floods, the loss of many species of life—and diminishing returns on the investment.

We know that many, if not most, Soviet decisions to get at their resources were not particularly good ones. Ravaging the landscape is an act driven by ideology, stupidity, or mindless bureaucracy rather than by economic need. Stalin's ethos was that man could and therefore should conquer and master nature. The price being paid for such belligerent behavior is incalculable for today's population and their descendants.

The planet should be viewed as a single ecosystem connecting myriad places. The global environment depends on millions of spatial processes. If we want a harmonious future, if we want place-harmony, we must understand that geography is a dynamic, complex set of processes and that the physical realm must be looked at from a spatial perspective, because all places are connected to one another.

Are we challenging Ecclesiastes's prophecy: "One generation passeth away, and another generation cometh; but the earth abideth forever"? It may abide, but the danger is that it will abide without life.

POPULATION CONTROL IN MEXICO

Mexico is better off than it has been in memory and certainly since this drawing was first published in 1990—birds were falling dead from the poisoned sky. Progressive steps are being taken to alleviate herculean problems of poverty and a burgeoning population and some of the world's worst air pollution. Family planning, resisted so long and so vehemently, is taking root, thanks to the steady encouragement of the government. There is a serious problem of people maldistribution: nearly a quarter of the country's population, or around 20 million people, are piled up in one place, the quake-prone capital, Mexico City, the world's most polluted and populous metropolis. (It has always been the site of one of the planet's biggest cities. It was the Aztecs' Tenochtitlán, where Cortes's early 16th-century conquest was transformed into the center of viceroyalty of New Spain.) The city, which sits in a 7,800-foot-high valley hammered by winter-air inversion, has 3 million vehicles, tens of thousands of factories, and an ever-increasing population of peasants crowding into shantyvilles completely devoid of hygienic facilities. DANZIGER IN THE CHRISTIAN SCIENCE MONITOR © 1990 TCSPS

I am not a card-carrying pessimist. I do not think the environmental damage we are doing will wipe us off the face of the earth. But I do think our lives will be less enjoyable and more dangerous if we don't take a rational stand against the perils.

Worldwatch Institute believes that the world has about four decades to achieve an environmentally sustainable economy or it will descend into a long economic and physical decline. "If we have not succeeded by the year 2030," says the Washington-based research group, "environmental deterioration and economic decline will be feeding on one another, causing social and economic structures to disintegrate. A global population of 9 billion by the year 2020, which the United Nations has projected, would not permit a sustainable world economy."

Despite our understanding of the environment, we still don't understand it well enough. Environmental problems will continue, creating fear, danger, damage, and death. We—individuals, governments, corporations—must try to anticipate the consequences of our actions and minimize their impact. It is a tough assignment. It is not easy being green. Vigilance, intelligence, and creativity are the preventatives to keep Chicken Little's sky from falling.

Can we apply humankind's intelligence in harmony with the environment? In 1992, a United Nations conference in Brazil—an appropriate place—addresses environmental issues and the development of a kind of constitutional convention for the world environment. What will we make of the occasion?

We could adopt the Native American wisdom that we do not inherit the Earth from our parents. We borrow it from our children.

WEATHER OR NOT
CLIMATE

■■■■■■■■■■■■■■■■■■■■■■

*The world is indeed only a small tide pool; disturb
one part and the rest is threatened.*
—GREGORY BATESON

Yuma, Arizona, has at least 300 days of sunshine every year.
That's climate.

It snowed today at Malibu Beach, California. That's weather.

The terms *climate* and *weather* are used interchangeably, as if
climate and weather were synonymous. Many of us say "weather"
when we are really referring to "climate." It is only in the vicinity
of the equator, where meteorological conditions are practically the
same day in and day out, that climate and weather may be construed
to be nearly the same thing.

Seasonal climate and daily weather are spatial processes that
determine and alter, sometimes dramatically, a place's contents and
connections, temperatures and moisture, influencing how people go
about their lives—they are geographic processes. Many daily weather
predictions are based on conditions in another place, from which the
next conditions are expected to originate.

Weather is defined as the state of the atmosphere at a given time
or over a relatively short period, as described by various environmental
processes and conditions. These include atmospheric pressure, temperature, humidity, moisture, rainfall, clouds, and wind speed and
direction. "It's hot today." "It feels like rain."

Climate is defined as the generally prevailing weather condition of a place or region throughout longer periods of time: seasons, the average weather over a long period. It is governed by a place's latitude, or location, relative to continents and oceans, and local topographic conditions, and other factors. It depends on a number of environmental events, such as air movements, jet streams, ocean currents, and land-water heating differentials, to name a few. Nepal, in the great Himalayan ranges, and Florida are on the same line of latitude, but their climate—and their weather!—couldn't be more different. Western Europe is on the same latitude as much of central Canada; Dublin is on the latitude of Lake Winnipeg; Copenhagen is in alignment with Goose Bay, Labrador. The effects of water on a place's windward side temporizes the weather, as can be witnessed in Europe. Still, all weather processes are connected by movements of many elements through the atmosphere.

The energy of the planet derives from the Sun, but the Earth receives solar power unequally over its space. The equatorial zone gets more energy than the northern and southern regions, and, if there were no atmospheric actions redistributing it, the Poles would get colder and the tropics hotter. The weather and the climate of the planet as a whole is influenced positively by large-scale flows of warm air moving from the equator toward the Poles and by large-scale movements of cold air streaming southward toward the equator—an exchange between regions. These movements tend to maintain a balance between surplus heat at low latitudes and deficit cold at high latitudes. A combination of systems—highs and lows—is generated by Earth's rotation, zones of pressure, and wind belts. Northern latitudes are in the belt of westerlies: winds blow primarily from west to east, moving moisture and heat and cold from west to east. In the lower latitudes, the winds are northeasterly, the trades blowing mainly from northeast to southwest. At year's end, when these temperature contrasts are much greater, atmospheric circulation is much brisker and the air is more likely to flow north and south as it navigates its way around high-pressure zones.

The rain clouds that form over the northern Pacific Ocean are deflected from reaching the western United States's Great Basin by the Sierra Nevada and Cascade mountain ranges. The Russian port of Murmansk, the world's largest city north of the Arctic Circle, is essentially ice free because its waters are warmed by the merged flows of the Gulf Stream and the North Atlantic Drift Current. These waters also influence the climate of western Europe, including western Norway, whose high mountain range prevents the air from crossing into Sweden.

The mean daily temperature in Iceland is nine degrees higher than might be expected at that latitude, thanks to the relatively temperate climate maintained by a branch of the Gulf Stream that hits the southwest coast. A mountain in Hawaii is inundated with 450 inches of rain every year. About 20 miles away, the seashore gets barely 20 inches per annum. The rain forest in Washington State receives almost 140 inches of rain every year. Forty miles away, over the Olympic Mountains, farmers must irrigate to keep their farms alive. The devastating storms that sock western Europe several times a winter are caused when an unusually strong jet stream swings out of the northeastern United States and rages across the Atlantic. The stream is related to a condition of very low pressure over northeastern

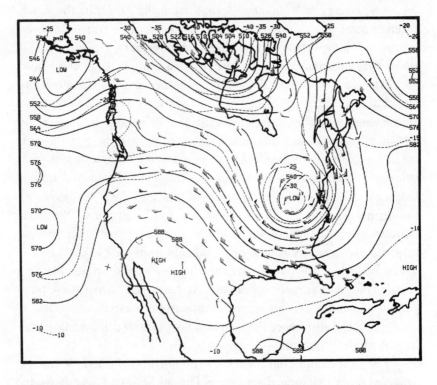

Weather maps track the environment in motion—spatial interactions from place to place.
NATIONAL OCEANIC AND ATMOSPHERIC ADMINISTRATION

Canada, Greenland, and the North Atlantic and unusually high pressure to the south stretching across the Atlantic from the southeastern United States to the Mediterranean. Storms collect extra energy as they cross the ocean, and reach maximum strength as they approach the continent.

Climate has become front-page news. World temperatures are 0.6° F. warmer than they were a century ago. Some scientists believe that global warming, due to the greenhouse effect, has begun. If the warming continues, glaciers will gradually melt and ocean levels will slowly rise. Such a rise would have different results in different places in the world. Sea level rises, like global warming, will affect different places on the planet differently. A rise of three feet would displace tens of millions of people in low-lying Bangladesh who are already being crushed by periodic cyclones and monsoons; they have no higher ground to escape to. A rise of a mere 20 inches would flood much of the Nile delta, where 50 million Egyptians reside. The Indian Ocean would drown the archipelago of 1,200 coral Maldive Islands. At least the process will be slow, allowing time to adjust.

The pattern of the seasonable distribution of temperature and aggregate precipitation is frequently more important than annual averages. Energy regulates the length of growing seasons. Soils that receive their total precipitation at one time may not be so nearly as well off as an area where the precipitation ("precipitation" here means all forms of moisture from rain to snow) is more evenly distributed over time or at the optimal growing time. A place with a scant annual precipitation of 20 inches that falls gently at the appropriate time of the year has the advantage over a place with twice the rainfall that comes in torrents during harvest time. Twenty inches of rainfall a year can produce semiarid pockets in a high-humidity country like India but give Ireland the eternal green of an "emerald isle." The rate of evaporation or drying up of moisture is a critical process. Variations in evaporation rates determine how effective precipitation is in various places.

Climates may be described as maritime or continental. Generally speaking, similar climates tend to occur in similar latitudes. (There are, however, prodigious exceptions.) The northwestern United States and northwestern Europe have roughly the same moist, "maritime" climate because both regions are at similar latitudes and to the east of oceans. Central Asia and Kansas are on similar latitudes. They are both dry. They have cold winters and hot summers: their climates are "continental," because they are too far inland to feel the impact of large bodies of water.

If either the Atlantic Ocean or the Pacific Ocean were not where they are, much more of the Earth would have a drier climate with hot summers and extremely cold winters, quite like the United States upper Midwest or Canadian prairie provinces. If the Atlantic Ocean were dry, Europe would have a climate more like Siberia's; Africa, eastern North America, and eastern South America would be much drier than they are today, and the climate in the Pacific rim would not be much altered from what it is today. But if the Pacific were dry, there would be severe temperature extremes around the rim.

Few countries have a greater variety of climates than the former Soviet Union. Winter temperatures over large tracts of Russia are consistently subzero and even more severe in resource-rich northern Siberia, while resorts along the Black Sea in the south enjoy near-subtropical warmth. The prevailing climate in what was the world's largest country is a continental, humid, cold one. It blankets the vast lands north of the 50th parallel—that is, about two thirds of the former empire.

Weather systems interact nonstop across space. This is why geographers, in common with other scientists, find great complexity in understanding global climate, much less the local, or micro-, climates that are part of larger operating systems.

Comprehending the spatial events between locations would be a good first step. Climatologists have found it possible to construct crude models of global climate. Every region, however, is unique in terms of climatic factors; each must be modeled separately. But regions cannot be studied in isolation from one another. Disruption of Pacific waters feed Asian monsoons. In 1985, a complex chain of events forced the jet stream farther north than usual, leaving fields in the northern United States Plains states dry. The models must be integrated in order to discover how one region affects the regions around it. All the regions then would have to be assembled, like a colossal jigsaw puzzle, to get a handle on the global picture. It is a feat of enormous complexity.

The measure of barometric pressure from place to place is a key to predicting the weather. Week-long forecasts in the late 1980s were about as good as three-day forecasts were a decade earlier.

Television weather-forecasters are so often hedging or mistaken in their predictions because the processes that drive weather fronts are unpredictable and our knowledge of why and how they change is still imperfect. There is no way they can forecast the weather accurately for a week ahead. (Being a weather forecaster on television, says the comedian Billy Crystal, means you can be wrong all the time and never lose your job.) Observable changes are known to be cyclical, but we don't understand the cycle patterns.

When there is a climatic change, the change is not the same everywhere. Is Japan going to become hotter and drier? Will the northern coastline of Australia become more humid? Florida and Iceland once had comparable climates—Florida today–like weather.

Climatologists declared the year 1990 to have been warmer than any other year since scientists began measuring the planet's surface temperature. The seven warmest years within the past century have all occurred since 1980. Some scientists believe that global warming caused by an increase of heat-trapping gases has begun—the green-house effect. These gases—principally carbon dioxide, chlorofluo-rocarbons, and methane—increase mainly because of the activities of people. This warming, as we are always being reminded, can cause sharp changes in climate, agriculture, sea levels, and human behav-ior.

England has the most variable weather in the world. Early-morning sunshine can abruptly give way to a rain shower that is followed by more sunshine. There is the occasional violent winter storm or a dry, warm summer. England's variable weather, in me-teorological terms, is the result of the combination of the Atlantic's Gulf Stream, the uneven land configurations of northwest Europe, and the atmospheric depressions sweeping in from North America. "It's all these depressions that come across and zap the British Isles," is how an English weatherman puts it. "If we could get rid of America, we would have sunny weather all the time."

Currents play an important role in the exchange of heat between high and low latitudes—a remarkable set of spatial interactions. Tem-perature controls the general atmospheric circulation and, consequently, the distribution of precipitation. Temperature is influenced by the move-

WHY IN THE WORLD · 190

ment of air and the condensation of water vapor. But it is solar radiation that provides the *initial* force in the distribution of temperatures.

There is a growing consensus about weather. Scientists are finding that those mysterious areas on the solar surface that are cooler and darker than the surrounding area may coincide with shifts in wind patterns that could help determine the tracks of storms across North America and storms across the North Atlantic. Such shifts shape the weather from Japan to Canada, from Europe to the Middle East. Sunspots vary in a cycle averaging about 11 years. At or near the 11-year peak, the southeastern United States usually has a colder-than-normal winter. Presumably, we will come to understand the important connection between our spatial processes and our sun.

People often ask me to explain why it can be too cold to snow. The colder the air is, the less water vapor it can hold. When the temperature drops to 40 degrees below zero, it is unlikely there will be a snowfall; if there is, it will be a light one. In the Arctic and the Antarctic, where it becomes inhumanly cold, great amounts of snow do not fall with any regularity. Rather, the little water vapor in the air contributes often to the layers of snow that accumulate from year to year. The snow and the ice that pervade the polar regions formed when the climate was warmer, much warmer, and any that forms tends to stay a while.

The North Pole is not as dry as the South Pole because it is at sea level—Antarctica is the world's highest continent—and the Arctic Ocean's ice pack is only a few meters thick. The Arctic is mostly ocean. The South Pole, which is mostly barren desert, has an altitude of about 10,000 feet. It is quite removed from open water, and little moisture reaches it, no more than a few inches a year.

The National Aeronautics and Space Administration has called for an Earth-monitoring spaceship to be developed jointly by the United States, Europe, and Japan. The craft would be equipped with complex instruments to measure such quantities as the rainfall in the tropics, magnetic fields and gravity fields, and the input and outflow of solar radiation in the equatorial region. The instruments would have the power to detect biological changes in small lakes and the stress in global vegetation and the pollutants in the atmosphere. They could measure the depth of glaciers and polar ice sheets to determine how the water vapor of the oceans influences complex environmental processes. Once upon a time, weather systems were predicted on the basis of reports from ships encountering storms at sea. Before even those days, the best weather predictor was sticking a wet finger in the air. In a relative sense, we have come a long way.

Cold air masses over the Antarctic continent sometimes discharge toward the edge of the continent and generate intense low-pressure systems or storms that circle the nearby waters. Strong westerly winds generated north of these storms and south of the southern hemisphere's subtropical belt are called the "roaring forties," a description coined by 19th-century mariners who encountered these winds near the fortieth latitude.

The federal government, concerned in 1984 with dramatic shifts in the weather over the previous decade, launched a new weather satellite. In addition to studying such phenomena as the greenhouse effect and the depletion of the ozone layer, the satellite has become the chief source of data for the National Weather Service. Circling the globe every 102 minutes, the satellite measures temperatures and humidity at all levels. It beams these readings to some 2,000 weather stations operated worldwide by the Weather Service. The Service no longer need rely on less than up-to-the-minute readings from conventional radar systems on land, air, and sea.

Cameras in space have expanded and changed geographic knowledge. They have created new ways of looking at the Earth. Their images clearly identify different zones, measure atmospheric processes, and provide studies of phenomena such as dust storms, volcanic eruptions, the spread of pollutants, large-scale migrations, and "secret" missile silos. They can monitor the state of the natural environment and changes induced by human and natural forces. They help to forecast harvests and droughts with reasonable accuracy.

The century-old United States National Weather Service, jeopardized by deteriorating equipment, will invest $3 billion during the mid-1990s to upgrade its instruments, especially its radar systems, so that meteorologists can better compile basic information and pinpoint the course of dangerous storms.

For a long time, pilots, farmers, and ships have been employing private weather services. Independent meteorologists and climatologists are credited with having an influence on commodity prices, the outcome of lawsuits, and even on government policy on global warming. They advise on where, or whether, factories should be built. Some are called upon to testify on the causes of airplane crashes.

Private weathercasting is certainly one "wave" of the future. The National Weather Service will eventually be limited to general forecasts and emergency warnings; it would like to avoid such questions as whether the temperature and humidity are going to be propitious for putting in the tomato plants a week from Friday. Such predicting is best left to *The Farmer's Almanac*.

About 30 percent of the clients for private weather services are from utilities, somewhat fewer from transportation and from construction. Utilities need advice on environmental issues and early warnings of rough weather that could bring down transmission lines. Construction crews need sufficient notice of storms to come down from lofty building projects.

In most of rural southeast Asia, monsoon means the renewal of life. The monsoon season is celebrated with joy. Monsoons are often confused with cyclones and other destructive winds. Moisture-laden air masses characterize the monsoon for climatologists and meteorologists. The storms may bear rain with the force of a hurricane, but the moisture brings life to dry fields.

The monsoon is not a single event but the beginning of a new weather system, one that brings life-generating moisture to parched areas. Many regard forecasting of monsoons as the world's most important weather prediction. The monsoon is often related even to military actions.

Scientists have created a model of "nuclear winter." An exchange of atomic and hydrogen bombs in the northern hemisphere would set off fires that would destroy cities, forests, and petrochemical facilities, burn for weeks, and sweep up vast amounts of smoke and dust that would encircle the globe, blotting out the sun and causing temperatures to plunge worldwide. A nuclear winter would be the most cataclysmic physiographic event since the "old world" of Panagea was fractured millions of years ago and continents went floating off every which way.

The sky may look blue, but can it really be *that* blue when every automobile in the world is discharging its weight in carbon dioxide into the atmosphere year after year? The air around us, however sweet and clean it may seem on that infrequent number-10 day, is being disastrously polluted. The altered chemistry of the atmosphere portends changes in the weather and eventually the climate everywhere, from the Bay of Bengal and Botany Bay to Hudson's Bay.

The tornado has been called nature's worst storm, its mightiest punch other than the earthquake. It is short-lived, but in seconds the violently rotating column of air in contact with the ground can reduce a bustling street to rubble. Strong wind speeds range up to 200 miles an hour, "weak" ones up to 100 miles an hour.

There are patterns of air movement when nature has a high probability of causing damage and death: tornado paths in the United States, hurricane swaths across the Atlantic Ocean and the Caribbean Sea, typhoon and cyclone zones in the Pacific and Indian oceans. The power of nature is such that a hurricane, which is an enormous heat machine feeding on the water's warmth, flipped a twin-engine C-46 into the air and onto the roof of a hangar in the Dominican Republic. An earlier hurricane swamped Galveston Island, Texas, with a 20-foot storm tide and drowned 6,000 people. Hurricane Hugo leveled South Carolina timber. However, let it be said for hurricanes, they also moderate the atmosphere, sweeping heat north from the tropics, and help to balance world temperatures. Violent natural processes are an integral part of a delicate and complex environmental system, and humans in the places most affected by them must learn to adjust and adapt to them. It is a question of place. WORLD WIDE PHOTOS, INC.

Tornadoes usually develop from strong or severe thunderstorms. Major tornado outbreaks can cause extensive damage over a wide-ranging area. On March 18, 1925, a tri-state tornado traveled 219 miles across Missouri, Illinois, and Indiana. It lasted for over 3½ hours and killed 689 people.

Tornadoes used to occur mostly in "tornado alley" in mid-America. Now they are funneling along the Atlantic seaboard as far north as upstate New York and through trailer parks (there is misplaced confidence in the stability of mobile homes, which are quite defenseless against mighty storms) in the southeastern states.

Kansas and its neighboring states experience tornadoes so often in April because of a "conspiracy of nature": the alignment of the Rocky Mountains, the flatness of the Great Plains, and the movements of the air masses at that time of year. It is said that nobody should be surprised about a tornado in Kansas in April. It is like having snow in North Dakota in January. In April 1991, more than 70 twisters—winds of more than 150 miles per hour—smashed through seven Midwest states.

A nine-month period of atmospheric violence over 1982–1983 marked one of the most devastating upheavals of global weather on record. It began as a storm that broke an Australian drought so severe that farmers were mercifully shooting their starving sheep. On Tahiti, which had been scarcely touched in a century by violent South Pacific storms, 25,000 people ran for their lives from six cyclones in five months. Torrential rains crossed the South Atlantic and unleashed an avalanche of mud and rock, which buried hundreds of people in Ecuador. In the United States, levees bulged along the swollen Mississippi River in New Orleans. Rowboats paddled through the main streets of Gulf-coast towns.

The people of the Netherlands will always remember February 1953. One night, fierce storms destroyed the gigantic sea dike in Zeeland, killing 2,000 people and 250,000 farm animals. The hurricane-velocity winds that tore across western Europe three times in the winter of 1990, killing more than 140 people and costing perhaps a billion dollars in damage, may have occurred because of atmospheric forces that were also responsible for the mild winter weather enjoyed throughout most of the United States.

The greatest natural disaster of modern times was a cyclone and tidal wave that sprang out of the Bay of Bengal and snuffed out 600,000 lives in Bangladesh in 1970. Twenty-one years later, a similar concatenation of violent forces killed another 140,000 in that low-lying south Asian country, where the maximum elevation is only 600 feet.

Normally, the South American coasts of Ecuador and Peru enjoy comfortable trade winds of the Pacific. Fishermen find the cool currents paradisiacal. But for a few weeks each year, the trades slacken, the surface waters stagnate and become silent and "as idle as a painted ship upon a painted ocean," the temperature rises, and heavy rains begin to fall. This warming period in the coupling of the ocean and the atmosphere usually occurs around Christmas, so Peruvian fishermen call it El Niño, Spanish for "The Child." When these warming intervals do not subside but go on and on, havoc from the resulting weather can be felt around the world. A new El Niño was acting up as 1992 dawned.

An interesting aside: some researchers suggest that events on the ocean floor, rather than in the atmosphere and at the surface of the ocean, may set off the stupendous shifts in Pacific currents and air pressure that augur the onset of El Niño. For the 285 months preceding September 1987, a University of Hawaii observer noted a remarkable coincidence between the extent of strain released by earthquakes along the East Pacific Rise and the recurrence of El Niño in a five-to-seven year cycle.

A fascinating sidelight of California's extended drought is that the town of Monticello is reemerging from a receding artificial lake. It was a prosperous farming community before the government claimed it for a dam and reservoir in the 1950s. The government didn't leave anything over six inches, even the trees. The uprooted residents were paid off and told to move elsewhere. As the lake dries, artifacts of the old town gradually come to light: building foundations, a swimming pool, fishing poles, and tennis shoes.

Mark Twain once said that everybody talks about the weather, but nobody does anything about it. That's no longer true, as we have seen. But we may never be able to manipulate it to the point suggested by the boast of an Englishman: "You don't like our weather? Stick around for half an hour and we'll give you something else."

Clearly we are in an era of actively affecting our weather and climate. We have seeded clouds for rain with some minor success and built domes over stadiums to keep the weather out and let the boys play. However, and more important, we are spewing filth into the air, rendering it dangerous, emitting chemicals into the atmosphere, threatening the climatic system, and generally interfering in negative ways with the weather and climate. Already, weathercasters warn of smog and ozone levels in cities. Will the forecasters someday tell us that a front of highly radioactive air is moving in from the west or that a toxic cloud is forecast for the morning from a local accident? Humankind should at least try to let weather be weather and average itself into climate.

There's Something in the Wind

Weather phenomena, as we have seen, move over space and interact with topography, biota, and populations. It pays to know which way the wind blows. These are some special winds:

haboob: a violent duststorm or sandstorm, usually in India or Northern Africa, obliterating vision.

buran: a strong northeasterly wind of gale force in Central Asia and Russia. In winter, it becomes a mean blizzard; in summer, it is identified with hot sandstorms.

purga: an intense Arctic snowstorm, usually occurring in flat open country and characterized by severe cold and wind-driven snow.

foehn: a warm, dry wind blowing down the leeward slope of a mountain; best known in the valleys of the northern Alps, where the name originated; it is superstitiously thought to alter behavior, resulting in nervousness, arguments, and even suicides.

samoon: a hot, dry foehn-type wind from the mountains of "Kurdistan" on the Iraq-Iran border which blows across Iran.

nor'western: a hot, dry foehn wind that descends from mountains on South Island, New Zealand; a type of squall on the plains of northern India usually accompanied by violent thunderstorms and heavy rains and hail showers, usually through the spring.

khamsin: a hot, dry, southerly Egyptian wind, usually in the spring, carrying considerable fine particles of sand because it originates in the Sahara.

bora: an occasional, violent, cold north-to-northeast wind that blows over the northern Adriatic Sea from the interior highlands; it occurs when atmospheric pressure is high over central Europe and the Balkans, and low over the Mediterranean. Trieste and the Yugoslav coasts have been slugged by 150-mile-per-hour bora winds: "the devil's wedding . . . the noise makes it seem like a noisy, windy hell."

williwaw: a sudden, violent squall, a sudden gust of cold air usually along mountainous coasts of high latitudes, such as Alaska's.

chinook: the warm, dry, foehn-type of wind that blows down the eastern slopes of the Rocky Mountains in Canada and the United States, Oregon-northward, especially in the winter and the spring. It raises the atmospheric temperature, sometimes up to 40° F. in a quarter of an hour. When the chinook doesn't make annual appearances, it means that the winter will be severe and pastures will not be available.

sirocco: the hot, oppressive, dust-laden southerly wind, usually in the spring, originating in the Sahara and blowing north across North Africa, Sicily, and southerly Italy, picking up moisture as it crosses the Mediterranean, sweeping over Italy as a moist, warm, southeast enervating wind.

brickfielder: a hot, dusty wind blowing south from brickfields near Sydney, Australia, caused by the movement of tropical air ahead of a trough of low pressure.

mistral: a violent, cold, dry, northerly or northwesterly wind sweeping southward from the central plateau of France; funneled through the Rhone valley, it reaches the Rhone delta as an extremely strong wind, cold and dry.

"roaring forties": the latitudes where prevailing westerly winds of the temperate zones—the westerlies—blow in the southern hemisphere with a constancy second only to that of the trade winds, which are circular updrafts over tropical seas, the source of trade winds.

horse latitudes: in contrast to the doldrums and tropical zones of calms and variable winds, these zones are in belts in the neighborhood of 30° N. and 30° S. latitude where air is fresh and clear—regions of high pressure and light baffling winds.

GEOGRAPHY
UNDER THE MICROSCOPE

There is a genus of geographers who concern themselves with matters of health and illness. They seek to understand why certain illnesses are located where they are, why people are more prone to disease or death in one place than another. They try to discover how disease and environment are related. They look into how the changes that societies periodically undergo affect the health of populations in those places.

Medical geographers map the distribution of disease in order to learn where illness occurs and how diseases spread and to observe changes in the patterns. Even when the primary agent is known and can be treated, as with rubella (German measles) and tuberculosis, outbreaks continue to occur. The explanation might be as simple as the lack of information or the availability of vaccines in those places. It may be that immigrants from a "backward" society settle in a developed country without being immunized.

Detection of heart diseases, cancer, strokes, and other degenerative illnesses can be elusive because multiple factors may be at work. Geographers compile and analyze cultural, ecological, climatic, and locational data that may shed light on causes of morbidity and mortality for medical researchers directly involved in isolating causes of unwellness and death.

The former Soviet Union published an atlas showing the spread of oncological diseases in its 15 republics. Maps revealed that each region was characterized by a specific variation of tumors, including cancer, and by a combination of different forms. Such information is of practical importance to health professionals and ultimately to the citizenry as a whole. They point to clues in space for various ailments. In the United States, there has been a tremendous decline in cancer deaths in people under the age of forty-five, but deaths from breast cancer, brain cancer, and kidney cancer have climbed since the late 1960s, particularly among the elderly. Thirty per cent of all cancer deaths are due to a fully preventable cause—smoking.

Maps have been serving the medical geographer well for decades. During the cholera epidemic in London in the 19th century a map of the whereabouts of victims led to the source of the infection: a few water pumps in common use by neighborhood residents. Mapping the sites of sudden blindness in Africa isolated the catastrophe to people living alongside rivers, and led to the discovery that "river blindness" was caused by a parasite carried by tiny black flies. (Insecticides have been used in recent years to eradicate the black flies. The most relief has come from invermectin, a drug that can kill millions of microscopic worms in an infected person and stave off blindness.)

I cannot exaggerate the importance of the spatial dimension—place—in demystifying the path or the incidence of disease and malaise. A form of meningitis began its westward spread from eastern Africa in 1927. It reached Senegal on the west coast 17 years later. The spread was attributed to the east-west transport routes across the continent. In the 1970s, four distinct route patterns were discovered to be the media by which a cholera epidemic moved between east and west Africa.

During the early 1970s, gasoline processed in the United States contained lead additives boosting the octane. When children were studied for lead content in their blood, it was found that those who lived within 500 feet of a main road had dangerously high levels. Location, then, can put people at higher risk for lead poisoning just by their living reasonably close to motor traffic or even to garages and gasoline stations. Location is often a vital factor in explaining health risk and prevalence.

We are not sure why many spatial-geographic differences exist. We don't know, for example, why there are regional patterns of cancer. Identifying spatial patterns of disease incidence and death—mapping—is often a critical step providing clues to the why of patterns.

Disease eradication can be difficult even when the cause is known. Geographers have pinpointed areas of mosquito infestation. Most of these areas are in Third World countries, which are too poor to pay for pesticide spraying. Malaria claims more than a million victims annually and remains, numerically at least, the number-one disease scourge of the world.

For the first time in a century, Latin America has been afflicted with a cholera epidemic. In the first six months of 1991, at least 200,000 cases occurred in at least six countries, 90 percent of the cases in Peru, and spread into Mexico. The disease, which can be fatal, is transmitted through contaminated water supplies and inadequate sewage-treatment facilities, and tends to spread in waves. It is also sweeping through several African countries at a catastrophic pace, killing people at a higher rate than in Latin America. Death rates range up to 30 percent in some African communities. Until the decaying water sewage and related infrastructure are repaired, cholera goes on spreading, scaring away tourists and commerce of all kinds, putting hotels out of business. The prevalence of disease threatens whole economies.

It is a matter of medical fact that the influenza outbreak of 1918 killed more people faster than any other disease in history, occurring as it did during the last year of the Great War. In that year, the pandemic Spanish flu killed twice as many people—over 21 million—as had died in the war. Some 5,100 perished in a two-week period in October in Philadelphia, Pennsylvania, alone. Victims would be feverish one day, dead the next. No country was exempt. Two thirds of the people of Sierra Leone, in Africa, were stricken. The disease played a role in the war effort itself. United States General John "Black Jack" Pershing called for American reinforcements in late September, but the army, aware by then that the flu was a deadly enemy in its own right, canceled the October draft call for more men and quarantined the camps where sick soldiers happened to be assigned. The flu hit at a time when there were mass gatherings and movements. Troops were packed into crowded trains on their way to crowded camps, and the virus was highly contagious.

The earliest plague may well go back to biblical times, to the time of the Philistines in the 12th century B.C. The first recorded epidemic to ravage the earth began in A.D. 541. Disease was spread through the Mediterranean world by rodents for more than two centuries. The trail was lethal. It killed as many as 40 million people and weakened the Byzantine Empire. At its peak it killed 10,000 people a day in Constantinople (Istanbul today). The second known pandemic began in the 14th century, as lucrative trade routes took form across Asia. In 1347, vessels carrying silk and jewels sailed into Sicily with crews dying from a mysterious disease. Fatalities were so high during the next five years—25 million people in Europe alone perished—that the plague became known as the Black Death. Perhaps 50 million people in Europe and the Mediterranean region died before the plague ended after two and a quarter centuries. The viruses carried by the conquistadors to the western hemisphere killed millions of Native Americans with no natural resistance.

Some of the most momentous geographic processes, with enormous consequences, are the spatial diffusions of diseases. The Black Plague started in one place, spread, and has had the most devastating repercussions worldwide. AIDS, indisputably the plague of our time, is constrained somewhat by the fact that it is contracted only by a fluid transfer.

AIDS was first diagnosed a decade ago. As I write, in the late autumn of 1991, about 200,000 Americans have contracted the disease, and 126,000, or more than twice the number of American fatalities in the Vietnam War, have already died. Friends of mine in the World Health Organization estimate that there are at least 700,000 cases of AIDS worldwide and that 8 million to 10 million people are carrying the HIV virus. (Most of those who test HIV positive will be stricken with AIDS.) Another 20 million—about half of them Asians—are likely to be infected within the next decade. By the end of the first decade of the next century, 25 percent of all sub-Saharan people may be dead of AIDS, which they call "the slim" because its victims waste away.

As late as 1988, Soviet authorities were denying the existence of AIDS within their country. But on my most recent trip to Moscow, in July 1991, I found government estimates that by the end of the century the former Soviet empire may have as many as 15 million HIV-infected citizens and 200,000 cases of full-blown AIDS.

The Soviet Union experienced transmission of AIDS principally through its health-care system. The repeated use of contaminated needles in hospitals accounted for much of the epidemic spread. There was an instance in which one needle used on an infected child carried the virus to 58 other children and 9 adults in one town alone. The transfer of a patient to another hospital spread the infection to 27 additional victims.

The blood supply was another source of infection. After proclaiming its safety, officials admit that Soviet-made equipment for detecting the virus was "very unreliable." Also, there was a critical shortage of AIDS-prevention supplies, such as syringes and condoms. Prostitutes, who exist there in plentiful numbers, were a principal purveyor of the virus. They were ordered to register with health authorities.

In Africa, the AIDS epidemic is outpacing preventive measures, by far. In some countries, in Uganda, for example, the AIDS toll is in a class with famine and tribal warfare, leaving behind a huge population of orphans. In many cities the spread is astonishing. In Lusaka, the capital of Zambia, and in Kampala, the capital of Uganda, more than 20 percent of the adults are infected. Ten percent of the population of Abidjan, the capital of the Ivory Coast, have been stricken. Five percent of Nairobi, the capital of Kenya, have been infected. The rate of AIDS infection in Nigeria, Africa's most populous country, is said still to be low. But the plague, once thought to be confined to Africa's eastern cities, has been traveling across the continent in a westwardly direction. It can be expected to reach everywhere all too soon. (In Rwanda, a woman with AIDS becomes pregnant willingly, explaining, "A child can't die until it's born.")

In Asian countries, the incidence of infection is spreading alarmingly. Even India is feeling the impact. At least a half million people are stricken. Until two years ago, India called AIDS "the foreign disease." A lot of cross-country truck drivers pick up an infected prostitute, drive with her awhile, and pick up another. AIDS among heterosexuals is epidemic in Thailand, where brothels are a way of life. It is said that few Thais know that the disease is closing in on them, or that it will kill them. Thailand is a country with less than a fifth the population of the United States, but there is a prediction of more than a million AIDS deaths there by the year 2000. The plight has attracted medical specialists to Bangkok from around the world.

So we see that AIDS, whose rapid spread was once associated with the industrial world, though its origins seem to have been in Africa, is rapidly becoming the scourge of developing countries as well. Wherever man goes, there goes AIDS. In point of fact, an estimated two thirds of HIV infections occur in Second and Third World countries. This proportion, our studies tell us, could rise to three quarters in the next decade.

Early on, medical geographers discovered that AIDS is no respecter of boundaries or borders. Geographers have divided a map of the world into AIDS North and AIDS South to designate the types of spatial spread in northern and southern hemispheres. AIDS North victims consist mainly of homosexuals, intravenous drug users, and prostitutes. Thus, most AIDS cases are spatially concentrated in the center of large urban areas; policies to combat the disease can be targeted readily. AIDS South victims are mostly African heterosexuals. Typically, a man in AIDS South consorts with an infected prostitute, usually with his spouse's approval, then resumes sexual relations with his spouse, to whom the contagion spreads, and it then spreads to her future babies. AIDS South spreads across rural and urban space, and heterosexually; targeting populations at risk there is very difficult.

An AIDS outbreak forced the People's Republic of China to confront the disease. Nearly 150 cases were discovered among peasants in the southwest, near the border with Myanmar. The victims had injected themselves with dirty needles. China denies that it has a drug problem. The truth is that many peasants, like people everywhere, have the impression that drug taking induces pleasant illusions, and this expectation outweighs all caution. (China promptly executes most apprehended drug dealers.)

The most shocking revelation following the overthrow of the despotic Communist regime of Romania, in 1989, was the number of AIDS-infected children in orphanages and in isolated wards and clinics. Doctors from the World Health Organization said that 700 of the children had been infected with the AIDS virus and that 250 of them actually had AIDS. Willful neglect allowed the disease to spread quickly. "I have never seen anything like this anywhere," cried Jacques Lebas, president of the Doctors of the World, an organization in Paris.

Thus far, AIDS seems to have touched Latin America comparatively lightly. (Haiti, a Caribbean country, has been severely stricken.) But commonplace is the tropical parasite chagas, which has similarities to AIDS. Chagas damages the immune system, as does AIDS, and usually becomes fatal after several years. There is no cure for chagas. All the drugs that have been tested are either inefficient or cause cancer. Chagas was once found only among the rural poor. With the large-scale migration to cities, it is becoming an urban disease as well. Where people go, there goes chagas. (It is believed that Charles Darwin may have been stricken with chagas during stopovers in South America during the voyage of the *Beagle*.)

What has unleashed the plague of AIDS? There is the theory that AIDS has been around a long time and that it was confined to some remote regions until travelers discovered the community or communities and carried away the virus. Another hypothesis is that the AIDS virus may have existed for centuries in African monkeys and apes, and that only in recent times has the virus crossed from monkey into man. There might have been a small genetic change in the virus or simply more contact between monkeys and people as human populations encroached on the jungle.

Interestingly, African chimpanzees can be infected with HIV, but they don't develop AIDS. This suggests that chimpanzees have developed protective immunity. When a new disease first hits a population, it often hits hard. Smallpox, for instance, was brought to the New World by European adventurers, and it took an excruciating toll on the Native Americans. (Once a worldwide killer, smallpox is now sequestered in laboratories in Moscow and Atlanta, Georgia.)

Perhaps the most "optimistic" note has been sounded by an erstwhile United States Surgeon General, the dynamic C. Everett Koop, who has said, "We will not be able to master the AIDS virus in this century. The National Institutes of Health predictions have been right on target all along. But epidemics do disappear/and this one will disappear, maybe because we have suddenly found a cure, which I doubt, or maybe because it will just go the way of all epidemics." In fact, human behavior modified positively by education will or can stem the spatial expansion of the plague.

The World Health Organization has estimated that Guinea worm disease afflicts 10 million people annually in tropical Africa, India, and Pakistan, but nowhere more than in Nigeria. Victims become infected by swallowing contaminated water carrying worm larvae. The disease doesn't in itself kill anyone, but it can lead to malnutrition and a greater susceptibility to disease.

We are still striving to learn why SIDS—sudden infant death syndrome, or crib death—occurs in the United States at significantly higher rates in the Northwest, in Montana, Oregon, Nevada, Idaho, Washington, Utah, Alaska, and Colorado. Rates for those states are nearly double the national average. The lowest rates are for the states of Rhode Island, Massachusetts, Connecticut, and New Jersey. Again, the importance and mystery of place.

Medical geographers center on the villains in the various landscapes and suggest ways to cope with them. In the treatment of an affliction like malaria—the worst of all worldwide maladies, it attacks up to 200 million people every year—we must remind authorities that insecticides alone are not the panacea. Reliance must be put upon sanitary engineering and water management The search for vaccines for every deadly disease must proceed. Lassa fever is a disease that attacks around 300,000 people in Central Africa every year. The victims are mostly children and pregnant women. The scourge was first identified only a score of years ago, but promising vaccines are on the horizon.

The United Nations has estimated that about 15 million children under the age of five die every year because of poor sanitation, malnutrition, measles, tetanus, and environmental pollution. Three million more children become seriously disabled. Diarrhea and respiratory infections are responsible for about 4 million deaths annually. The effects of pollutants are especially hard on children.

Actual starvation among small children is a statistic that will elude us for a while more. We do know that the recent catastrophes of famine, war, and natural disasters have taken an extraordinary toll on infants and small children in Africa, Iraq, and Bangladesh. There is rarely adequate health care where and when it is most critically needed.

Today, thanks to jet planes, someone carrying a deadly disease can travel halfway around the globe before noticing any symptoms. Medical detectives, who are called epidemiologists, are usually right on the heels of any contagious outbreak. They collect evidence on the menace and try to cut the chain of transmission.

There is no gainsaying that medical service is highly irregular, depending on location, quality of practitioners—and the ability to pay for it. In the United States, in the empty space of the West, an ill person may have to travel a hundred miles to visit a general practitioner or even further for a specialist. How far, we ask, will a person travel to see a physician? We find the answer has many variables: his or her age and gender, marital status, nearness to friends and neighbors, economic resources, and attitude toward health precautions and illnesses. We also ask how many people, whatever the circumstances, would refuse to see a doctor.

Ease of access can be the crucial factor. The distance a woman has to travel to get an abortion may determine whether she is able to obtain the service when she needs it. Places like North Dakota especially come to mind. In the entire state there is only one abortion clinic, and it is situated in easternmost Fargo. Some women from the western part of the state must travel more than 300 miles each way, incurring the considerable expense of gasoline, meals, and lodging.

Many parts of the world, and within many states in the United States, have too few medical professionals to meet the needs of their population. In the early 1980s, there was one doctor for every 15,000 people in Zaire, one for every hundred thousand in Ethiopia. In the United States, the figure is about one doctor to every 150 people.

Doctors, of course, are free to decide where they will live and practice, just like geographers. Great urban areas like New York and San Francisco are disproportionately supplied with medical facilities and personnel because doctors understandably are attracted to the amenities of cosmopolitan living, which they can well afford because of the number and general affluence of their patients. Rural areas are less alluring, and have a hard time coming by medical attention. We have all passed through country villages with the sign on the lawn of the city hall: "Doctor Wanted. Free Rent."

Inner cities often share with rural areas and even some Third World countries the same lack of medical attention, with disastrous results. After enjoying an increase in life expectancy since the beginning of the century, the inner-city population has suffered a decline in recent years. Some of this can be attributed to poor or nonexistent prenatal care and malnutrition. Drug addiction and homicide figure significantly in the decline, with young male blacks a disproportionately large factor in these sad statistics.

One role of medical geographers has been to help find the best locations for hospitals, clinics, doctors' offices, and emergency centers. I would like to see them hammer away at the problem of distribution. Maps could show the disparity of medical services available from place to place. Access to good health care should be the birthright of every body and everybody. It is gratifying that universal health insurance has become a rallying cry across the country.

Data relating to a number of programs of the New York City Department of Health, the Commission on Disease Intervention, are constantly being developed and analyzed. Available are the number of cases of tuberculosis, measles, hepatitis, and congenital syphilis in each ZIP code. There are borough maps, by ZIP code, that display this information graphically. The department uses the maps and the data frequently when discussing the impact of these diseases on local neighborhoods. (Acquiring data on AIDS cases takes longer to compile because of the frequent lag between diagnosis and the filing of reports.)

The concern of medical geographers is not limited to the planning stage. There can be stultifying effects on health when sites are developed without regard to the peculiarities of the environment. We can work side by side with architects on all stages of the project, from the drafts of the surveyors to the completion of construction. The principal concern always is to provide the most favorable conditions for work, living, and recreation. I was told of geographers who had been brought into the planning of 20 economic-development projects for Siberia with provisions for adequate medical service. The geographers urged the directors to learn about local diseases and insects and ticks and poisonous plants and reptiles and wild animals.

There isn't much that we can do to reduce the mortality rate significantly. Science has not found a way of extending human life much beyond the age of 85, or to insure healthfulness and meaning-

THE WORLD BANK, CARTOGRAPHY SECTION

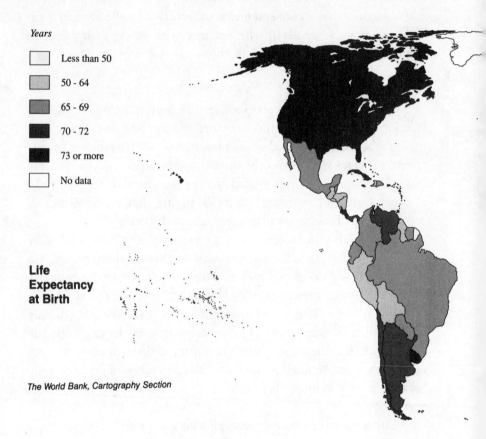

Years

- Less than 50
- 50 - 64
- 65 - 69
- 70 - 72
- 73 or more
- No data

**Life
Expectancy
at Birth**

The World Bank, Cartography Section

fulness to aged people. But some places appear to be more conducive to longer life than others.

Japan has the highest life expectancy at birth. On average, a person born there can expect to live 78 years. On average, a person born in the United States can expect to live 75 years; in the former Soviet Union, 69 years. But a child born in Togo, Ethiopia, or Afghanistan (where nearly 20 percent of the babies die before they are a year old) can expect to live only 41 years. In the United States, life expectancy is enhanced by being born in Oregon, Utah, Colorado,

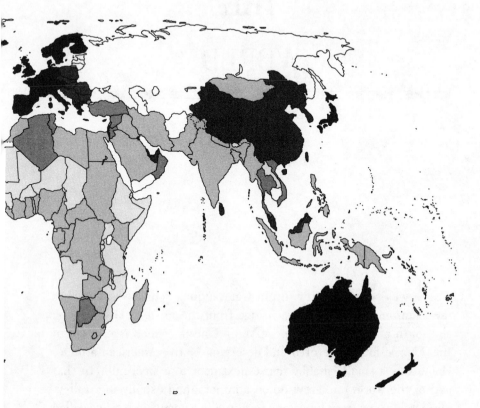

the Dakotas, Minnesota, Nebraska, Kansas, Wisconsin, and Iowa. The lowest life-expectancy rates: Nevada, Mississippi, Louisiana, Alabama, Georgia, the Carolinas, West Virginia, and the District of Columbia. Genes, nutrition, hygienic conditions, and the availability of health care play a role. But we are still left with the mystery of place. When that mystery is divined, we are on the way to making more places celebrated for their promise of better health and longer life.

SAVORING

THE

WORLD

■ ■

I was born and grew up in Catasauqua, Pennsylvania, across the Catasauqua-Hokendaqua Bridge from the town of Hokendaqua and south of Tamaqua and near Mauk Chunk, which is now called Jim Thorpe for the revered and deserving Native American athlete. Home was a small, shabby house next door to a steel mill. In that part of the world one grew up on a mineral and usually in a valley. One could come from limestone (eventually cement) in a valley called Lehigh or from "hard coal" (anthracite) in a valley called Panther. The names of these places evoke memories of childhood and spark waves of nostalgia but little desire to return.

Losing myself in flights of fancy was my sole escape. I became a voracious reader, checking out books by the armful from the local public library, a small, wondrous place still in the same location after all these years. Much of my reading related to the wonders of travel and exploration. I conjured up magical, exotic places in the worlds that lay far beyond the mountains of slag in Catasauqua and far beyond my labors in a drop-forge shop of the steel mill.

I read of Stanley and Livingstone meeting in darkest Africa, T. E. Lawrence challenging the brutal Arabian Desert, Lewis and Clark foraging through the American Northwest. I devoured the practically blind Prescott's accounts of the conquests of Mexico and

My geographic lodestar, Sweden's Sven Hedin, was renowned for his Asia explorations at the turn of the century. He could "hear the melancholy sound of the caravans' bronze bells, that song of the deserts, unchanged through thousands of years." He was a "star trekkie," boldly going where others had never been before, filling in those blank spots on the map labeled "unexplored" with what was there and why it was there, and opening them to the inexorable processes connecting them to the known *places. The late, great physicist Richard Feynman said he wanted to make the trek to Tuva, a purple splotch northwest of Mongolia on a map of Asia, because "any place with a capital called Kyzyl has just got to be interesting."* DR. DAVID HUMMEL, THE UNIVERSITY OF MICHIGAN PRESS, PUBLISHER OF *TO THE HEART OF ASIA* BY GEORGE KISH

Peru. I was a participant in Nordiff and Hall's South Seas adventures and rowed the galleys with Jean Valjean. I joined Sir Richard Burton on his mysterious African wanderings and followed along breathlessly as my companion Richard Halliburton led the way to the remote aerie of Andorra and through the bazaar at Skodra and ultimately to the Taj Mahal awash in the moonlight. I tilted at windmills with Don Quixote, crossing Spain with him, redressing wrongs, rescuing maidens, and finally winning the scepter of Trapizonda.

The 14th-century travels of Marco Polo inflamed my imagination above all else. His description of the palace of the Great Kahn gave me goose bumps. It still does: "The walls and the chambers are all covered with gold and silver and decorated with pictures of dragons and birds and horsemen and various beasts and scenes of battle. The ceiling is particularly adorned, so that there is nothing to be seen anywhere but gold and pictures. The hall is so vast and so wide that a meal might well be served to 6,000 men. . . ." What a place! And I was there.

What a surpassing joy it was to keep company with Mark Twain as he navigated along the Mississippi: "The face of the water, in time, became a wonderful book—a book that was a dead language to the uneducated passenger, but which told its mind to me without reserve, delivering its most cherished secrets as clearly as if it had uttered them with a voice," he wrote in *Life on the Mississippi*. "And it was not a book to be read once and thrown aside, for it had a new story to tell every day. Throughout the long hundreds of miles there never was a page that was void of interest, one that you were meant to skip, thinking you could find higher enjoyment in some other thing. There never was so wonderful a book written by man; never one whose interest was so absorbing, so unflagging, so sparklingly renewed with every re-perusal."

And I discovered Sven Hedin. Sven Hedin, a 19th-century Swede of aristocratic and arrogant demeanor who had the wealth and leisure to indulge his whims. Mostly, his whims spirited him in the direction of Central Asia. He was inordinately fascinated by the high desert and its people and the ubiquitous, ill-tempered, and evil-smelling camel.

It is little exaggeration to say that Sven Hedin determined the direction of my life. Long dead before I was born, he became my spiritual ancestor, my esteemed teacher and guide. In fantasy, I became his devoted traveling companion, following him on expeditions to Tibet and the entrails of Russia known only to the Cossacks, and to Mongolia and across great expanses of inland China.

"Soon the last gleam of light dies away in the west, the curtain falls," Sven Hedin wrote in *Overland to India,* "night builds up its thick walls around us, perspective and distance vanish, the far-distant horizon which lately presented the illusion of a sea has been swallowed up in darkness, and the outlines of the camels are thin and indefinite. . . . But still the air around them is filled with the same never-ending clangs of bells, which follows them through the desert; a sonorous, vibrating, ever-repeated and prolonged peal, melting together into a full ringing tone in my ears, a jubilant chord rising up to the sphere of the clouds and stars, spreading its undulations over the surface of the desert; a glorious melody of caravans and wanderers; the triumphal march of the camels, celebrating the victory of their patience over the long distances of the desert in rhythmic waves of song; a hymn as sublimely uniform as the ceaseless, unwearied march of the majestic animals through the dreary wastes of ancient Iran."

As a geographer and traveler, I have retraced most of Hedin's journeys. But one place has been too politically remote and inaccessible for me to reach. To get to Lop Nor, a visitor from the West has to make his way through Tibet into the autonomous region of Sinkiang-Uigur. Lop Nor, my elusive Olympus, translates approximately as "lost lake," and that is almost literally what it is at times. It has evaporated and become a marshy depression. I still dream of the day when I shall stand where Sven Hedin stood as he made his first map of the basin there.

It may be ironic to speculate on the number of young men whom war has rescued from a hopeless future in a harsh and hopeless place. During the Korean conflict, I joined the Marine Corps and can be thankful that I was one of the fortunate ones who was broadened and liberated by an experience that cost so many others their lives.

Until I entrained for boot camp in South Carolina, I had never been more than 50 miles from my hometown. The journey to Parris Island was more wondrous to me than Dorothy Gale's jaunt along the yellow brick road to the Emerald City. From the train window I saw the landscape change with every passing mile. I knew nothing

The "roof of the world"—Nepal—is probably as close as I'll come to the realization of a dream I've had since childhood: seeing for myself the mysterious, wandering Lop Nor lake, in Sinkiang, a remote, southwest region in the People's Republic of China beyond the distant Himalayas. Stream diversion has emptied the lake. The region, a salty, marshy depression, is used for nuclear tests, and this puts it off-limits to outsiders. As it was, the unique character of Kathmandu in the valley of the Himalayas, where I am standing, and the whole snowcapped Nepalese kingdom were worth the trip, the culture having survived the onslaught of modern processes. The Himalayan chain is almost 1,800 miles long and still the fastest-growing mountain range on the planet. Geologists are concluding that 30,000 feet may be as towering as a mountain can grow. Everest and K2 are 29,028 feet and 28,238 feet tall, respectively.

of spatial variations or spatial flows, but the mosaic of the Atlantic seaboard mesmerized me. Somehow, all those towns, fields, bays, rivers, and landscapes, one telescoping into another, stunned my senses. My love for geography deepened.

My first airplane flight took me from boot camp clear across the country to California. I was shipped out to Korea shortly after my 18th birthday, landing in the thick of the Inchon invasion on the Yellow Sea just south of the South Korean capital of Seoul. For the first time, I began to look at space in the military sense and from a macro scale. Why was I on that beach, and how was it connected to all my past places, all of which had been safer if less interesting?

We fought in especially harsh terrain. We were indeed Mud-marines—grunts. I was surrounded by death and the wounded. I quickly acquired a profound reverence for life. On leave later in the rear, I roamed the villages and the hills behind the lines.

In my 11th month of active duty, I caught mortar shrapnel and was flown to a hospital in Inchon and later was sent to Japan. During rehabilitation, I took every opportunity to explore the floating kingdom and let myself be inundated with its beauties and culture.

My Marine Corps travels were the most illuminating and influential time of my youth. They further stimulated the sensibilities and the awe that would propel me toward a career in geography. I had become my own Sven Hedin. I realized what had driven that Scandinavian adventurer. I, too, had a passion to savor the world and to understand the significant differences between place and place.

I shall treasure until my grave the two summers during my college years that I worked as a guide at the Grand Canyon. I led mules along a five-mile looping trail to the floor of the 280-mile-long canyon. It was a daily geographical fantasy, a lesson in practical transport proving that a straight line is not always the most desirable or safest route between two points. It was also a twofold lesson in climatology. Going from a rim to the floor of the canyon is like moving southward from Canada to Mexico: that is, on a typical day there is a temperature differential, on the average, of 50 degrees. Though the rims are only a mile apart, and perceptions notwithstanding, the difference in elevation is considerable. The much higher North Rim is snowed in during the winter, while the South Rim may be enjoying relatively balmy weather. (The Grand Canyon, surprisingly, wasn't truly discovered until some years after the Civil War; until 1869, to be exact, when Major John Wesley Powell arrived, seeking neither gold nor a route west but to explore a vast unknown region of the United States.)

It was at Penn State, with its excellent department of geography, that I first felt the exhilaration of being challenged intellectually. Having begun to study the Russian language, I chose the Soviet Union as the subject of my dissertation. Its title was "The Russian Peasant Colonization of Kazakhstan from 1896 to 1926." I chose to research a 30-year period during which peasants uprooted themselves amid the poverty and privations of Russian tsardom. No one had studied the Slavic migration from western parts of the empire into Kazakhstan, which became the Soviet Union's third- or second-largest constituent republic, depending on what is measured, area or population.

I was determined that some of my research must be done onsite. A kindly fate made this possible, but it wasn't until I had stepped off the train in Moscow, in 1960, that I really believed it was happening to me. But there were the Kremlin and St. Vasily Cathedral in all their fairy-tale splendor. I really was right in the capital of the Soviet Union. I traveled widely for a difficult but joyous year. I visited Kazakhstan, of course, and explored beautiful and ominous Siberia. The geographical analysis in my dissertation demonstrated how a place can be affected by flows from other places and become transformed as a result into a totally new system of cities, transport systems, demographic types, and agriculture.

After more than a score of rewarding years as a teacher of geography at Ohio State University, I was invited to become director of the United States Office of The Geographer, a research-advisory arm of the Department of State. We prepared myriad charts, maps, and analytical reports. We pinpointed famines in Africa, mapped the spread of AIDS, refugee movements, terrorist activities, and the location of satellites in outer space. Maps were prepared on an almost daily basis for the Secretary of State. If he was setting off on a diplomatic mission, we gave him a set of briefing maps, which enabled him to see at a glance the "lay of the land," so to speak, with analyses of irregularities and potential dangers. I used the Office of The Geographer to promote geography. I extolled geography's virtues, its vital relevance, its unique contribution to an understanding of the world as it really is.

I expect I'll always use whatever facility or forum is at hand to advance the cause of geography. Geography, after all, binds so many factors that shape our lives. *All* our lives. I cannot imagine having passed this way and not known the joy of being a geographer.

SO

YOU WANT

TO BE A

GEOGRAPHER

■ ■ ■ ■ ■ ■ ■ ■ ■ ■ ■ ■ ■ ■ ■ ■ ■ ■

If you are contemplating a career in geography, I can vouchsafe that there has never been a more interesting, a more stimulating time to enter the field.

Today's geographer may find himself wearing any number of hats. Depending on the problem, he could also be something of a functioning sociologist or ecologist or economist or political scientist or historian or climatologist or developer or demographer or financial adviser.

From the 15th to the 19th centuries, geographers were present on voyages of discovery. Their reports were the equivalent of today's front-page stories. With their descriptions of environmental patterns and cultural diversities and natural resources in other lands, they were a boon to commercial interests and governments. One European geographer set down these impressions of America shortly after its "discovery":

"This island is quite big and very flat and with very green trees and much water and a very large lake in the middle and without any mountains, and all of it so green that it is a pleasure to look at. . . . And in between the reef and shore there was depth and harbor for as many ships as there are in the whole of Christendom, and the entrance is very narrow . . . saw a piece of land formed like an island, although it was not one, on which there were six houses."

His view of American space was obviously quite limited.

Geographers in the United States are sought out by both the public and the private sectors. They are valued for their knowledge of regions and environments and for their suggestions in dealing with potential problems. McDonald's and dozens of other major businesses employ geographic counsel. Geographers point out the interrelatedness of phenomena and places, the significance of the spatial dimensions of the problem at hand, and the folly of considering only one aspect of a problem. As my friend Harm J. de Blij emphasizes, geography is a field of synthesis, of understanding relationships.

"You're the only geographer I've ever met," someone accosted me not long ago. "You must belong to a vanishing breed."

I assured the fellow that I was not about to take my place with the pterodactyl. Geographers ply their skills and training as city and county planners; teachers in elementary grades and high schools; college professors and scholars; military planners (General Augusto Pinochet of Chile happens to be a geographer); administrators, department heads, deans, managers; foreign-service officers and diplomats; researchers in think tanks; CEOs of corporations; environmentalists; astronauts and cosmonauts; and cartographers, especially high-tech computer mapmakers now in huge demand and short supply.

Government has always been a major employer of geographers, and at all levels—national, state, and local. During the 1980s, it was estimated that well over a thousand geographers were working for the U.S. government. Besides the relatively few who staffed my Office of The Geographer in the Department of State, they were working for the Defense Mapping Agency, the United States Geological Survey, the Central Intelligence Agency, the Army Corps of Engineers, the National Science Foundation, NASA, and the Smithsonian.

Geographers aid in the planning for land use, resource development, and population projection. They help agencies understand the changing world in such contentious matters as boundaries, terrorism, and refugee flows. They consult on the changing patterns of agriculture and industry, housing and transportation. Maps can be tailored to illuminate every situation, every problem, and even portray projected future patterns (simulations).

All states have agencies that focus on environmental and transportation problems, resource analysis, and planning. These agencies employ geographers, particularly those who are expert in quantitative analysis of spatial-data remote sensing, or cartography. Some states have an Office of the State Geographer.

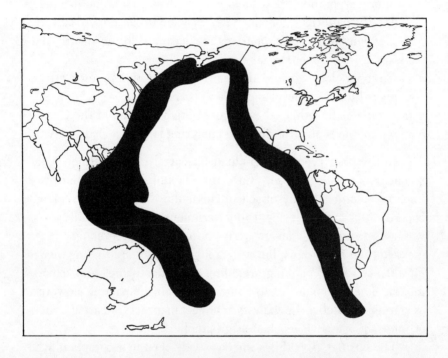

Geographers have always realized that civilization exists by geological consent, subject to change without notice. A prime focus in our studies is the fact that most of the world's 600 active volcanoes loop around the Pacific Ocean in a belt called the Ring of Fire, and a very dramatic fact it is! Huge slabs of rock (or plates) that cradle the world's largest ocean collide and clash with plates that carry the continents encircling the Pacific, unleashing deadly earthquakes and lava eruptions. On August 27, 1883, Krakatau's explosion in Indonesia was more than twice as powerful as the biggest nuclear blast. Charles Darwin more than a century and a half ago described a Chilean earthquake that was the most severe in local memory: "It was something like the movement . . . felt by a person skating over thin ice, which bends under the weight of his body. A bad earthquake at once destroys our oldest associations: the earth, the very emblem of solidity, has moved beneath our feet like a thin crust over a fluid."

University-based geographers may consult on a contractual basis with government or industry on a specific project. These are examples of such involvement:

- Two geographers, in cooperation with the United States Geological Survey, helped to put materials on earthquake hazards into a form comprehensible to the public and employed to develop damage-prevention policies.
- Geographers with the Dayton-Hudson Company and other multi-location corporations have helped to analyze population dynamics and traffic patterns, economic forecasts, and urban-growth patterns that enlighten the selection of optimal locations for new-store complexes.
- Geographers have advised ski-resort operators on long-term effects of air pollution on development and patronage.
- Geographers have helped the United States Bureau of the Census refine methods of gathering and analyzing large-scale survey data.

In the past, geography graduates became teachers either at the secondary or college level. But with the unfortunate, diminished importance of geography in school curricula, they began to seek out opportunities for employment in government and industry. Business was skeptical at first about what contributions geographers could make to their operations; but geography graduates are finding niches for themselves in banking, retailing, manufacturing, and export-import. They are more likely to be carried on the payroll, however, as market researchers or analysts or project managers or cartographers or international advisers than as geographers.

The business world has the door wide open to geographers who have a firm grip on international affairs and fluency in a foreign language. A manufacturer might be planning to open a plant in Turin, for instance, thus creating an opportunity for a young geographer with a facility in the Italian language and a reasonably strong grounding in the history, customs, culture, and current directions of Italy.

How does one become a geographer?

My long-time colleagues agree that it helps, as it does with almost any other career, to have a solid liberal-arts education and to be able to articulate clearly in both speech and writing. The skills of observations, analysis, and logic are a recommendation. Most important is a deep, abiding curiosity.

At Penn State, where I did my postgraduate work, the emphasis in geographic studies was on place and location and movement over space, or spatial analysis—the foundation of modern geography. We were burdened with a fair number of facts, and we had steady recourse to the geographer's standard equipment of maps, mathematics, and statistics.

The approaches to geographical training combine the traditional and the new. Given geography's concern with location, cartographic techniques are basic tools to hone. The curricula may include studies in surveying, library and archival procedures, spatial statistics, aerial photography, remote sensing, qualitative methods, and even artificial intelligence.

An undergraduate who is planning to go on to study geography in graduate school should be advised that choosing the school is very important. The professional foundation a student lays will have a bearing on much of his career. The Association of American Geographers, in Washington, D. C., publishes an annual guide to departments of geography. The guide describes such pertinent data as programs, requirements, and financial backing, and covers geography departments in the United States and Canada.

I would suggest applying to many schools. There is always the chance that some will have a teaching-assistant job open to an applicant who accumulated credentials as an undergraduate. A teaching assistant might find himself put to such tasks as moderating discussion sessions based on a professor's lecture or grading exams or giving a helping hand to undergraduates. A teaching assistant's job should give a graduate student a fair idea if he or she would like a career in the academic world. It should also forgive the debt of full tuition in creditable universities. Graduate students, too, are always eligible for scholarships and fellowships.

There are internships available in various agencies, offices, and commercial companies for students not planning to go on to graduate school.

James Michener, the prolific novelist-journalist, has eloquently celebrated a career in geography. He says that for him geography takes precedence over everything, even history, because in his work he needs to steep himself in the fundamentals that have both governed and limited human development. If he were a young person with a talent for communication who wanted to make himself "indispensable" to society, he would spend considerable time in "mastering" one of the major regions of the world. If that young person were he himself, Mr. Michener says, "I would learn the language, the religions, the customs, the value systems, the history, the nationalism, and above all the geography, and when I was completed I would be in a position to write about that region, and I would be invaluable to my nation, for I would be the bridge of understanding to the alien culture."

SOME
GEOGRAPHICAL
MUSINGS

■ ■ ■ ■ ■ ■ ■ ■ ■ ■ ■ ■ ■ ■ ■ ■

I have been called a practical cogitator, an eternal ruminator, an impossible dreamer, a compulsive notetaker. I have filled notebooks beyond counting during my ramblings along the highways and byways of this ever-fascinating planet. I would like to share with you selective jottings. Their common denominator is geographical inspiration.

"In the beginning," we read in Genesis, "God created the heaven and the earth." It took Him six days to get around to making man. Clearly, in the beginning was geography.

People change their habitats, acquiring new ways of life and new mores. They fix their imprint on the very landscape. Reston, Virginia, a suburb of the nation's capital, was created from a bucolic rural setting; even the lakes there are man-made. Ideas and ideologies change, too, and they can be contagious. After the downfall of the Shah of Iran, Ayatollah Khomeini came to power with his fanatical brand of Islam. This fundamentalism diffused zealously throughout the Middle East and beyond; its influence traveled overseas, giving renewed fervor to fundamentalist sects in many places.

One of the predictable rewards of seeing the world through a geographer's eyes is heightened confidence and pleasure in picking places to live and holiday and in comprehending "the big picture." It's like seeing with a third eye. The smallest palette offers wondrous things to behold. William Blake, in his *Auguries of Innocence,* expressed it so lyrically:

To see a world in a grain of sand
And a heaven in a wild flower.
Hold infinity in the palm of your hand
And eternity in an hour.

All things fundamentally occur in two dimensions and over two palettes, time and space. Everything in the history of the universe and humankind has happened in terms of a time dimension and a space dimension, both of which are dynamic, not static. Some processes are visible, some are not.

I travel so much, so far and wide, I have come to think of my book of maps as being my closest companion. But my favorite getaway is in my mind. It is a piece of space that has no name but where serenity is the prevailing mood. I have always thought the greatest way to escape was to imagine I was somewhere else—in a space created by imagination and, as yet, unmapped and untrammeled by others.

There are patterns of crime. It can be predicted where types of crime will occur, and they can be mapped with some certainty. Criminal acts can be divided into types: there is crime against persons, crime against property, crime across borders. Crime against persons, for instance, tends to be localized in lower socioeconomic areas. In Washington, D.C., almost all crime against persons occurs in black-dominated residential areas. Frustrations are exorcised through attacks on one another. Violent crime is usually between people who know each other. Violence during robbery is a small element in the picture of person-to-person crime. Crime will probably continue growing for the foreseeable future. Part of this is simply a matter of demographic cycles. These cycles are tied in part to how many people are in their teens and 20s, the most crime-prone ages.

The Bible Belt in the United States is a region of the South. Nashville, Tennessee, is the buckle on the belt. I have been told that the list of churches there fills 12 pages in the telephone directory. But the Bible Belt is nothing more than a set of locations with similar characteristics. The boundaries are certain to fluctuate as people move in and out and around, changing ideas and values as television introduces new ways of seeing things. Interacting processes can be mindquakers.

The perspective geographer looks at everything in the physical and human worlds, because he realizes that everything is interconnected, directly or indirectly, and changing at all times. Around the globe is a constantly moving mass of gasses, water vapor, trace elements, ice, pollution, people, ideas, money. This moving mass is driven essentially by the energy of the Sun and humankind.

All spatial distributions change in size and in expression over time. Take, for example, the diffusion of cities in the United States. In the first settling of our country, there were a few colonial centers on the Atlantic seaboard. With time came economic and demographic processes around the country, which in turn led to a movement of people to areas inland from the Atlantic. Thus began the real discovery, the taming and connecting of the land, the urbanization of America.

International terrorism has always been with us. In the early 1800s, Americans were held hostage by the Bey of Algiers. The three-year War of 1812 was brought on in great part by the impressment of American sailors by the British. Nightly television news programs bring into our living rooms the graphic horrors of disaster after diaster: Pan Am 103; the assassinations of Prime Ministers Gandhi; the blowing up of Viscount Louis Mountbatten on his boat; the taking and the murdering of innocents such as Israeli athletes at the Olympics in Munich in 1972; the dynamiting of Marine barracks in Beirut; the ceaseless bloodsheds in such disparate parts of the globe as Northern Ireland, the Middle East, South Africa, and the subcontinent of India. However, one person's terrorism may well be another's liberation struggle. There was the government's terrorism in Beijing's Tiananmen Square, and the terrorism in Medellín, Colombia, the drug capital of the world (and the murder capital of the world, with more than 5,000 murders a year in the record books), and the terrorism of diseases on a rampage, and the terrorism of crime in cities which keeps residents barricaded in their homes at night, sometimes even in the daylight.

The basic cause of international terrorism—the type imposed on innocent people by terrorists to dramatize their cause—is rooted in the loss of space. The driving force of Palestinians is space; of Basques, autonomy over their space; the Tamils of Sri Lanka, the Sikhs, the Kurds, the Armenians, the Tatars all commit violent acts to attract attention or to wreak vengeance for their claims of space. Looking at things from a global perspective is more informative than looking at specific things in isolation.

There is a geographic dimension to terrorism. When I directed The Office of the Geographer, in the United States Department of State, we kept track of terrorists as best we could: the routes they traveled, the sources of their support, the locations of their bases, the places of attack. We were able to map our findings and alert concerned states to the threats. We created maps of terrorism be-

---

Composers have long taken note of geographic names and the character of places for inspiration: "Appalachian Spring," "Grand Canyon Suite," "Tales of the Vienna Woods," "Roses of Picardy," "Moonlight in Vermont," "Take Me Back to Old Virginnie," and "I Left My Heart in San Francisco." The august Smithsonian Institution has declared that music with place names is "eminently marketable and . . . can give wing to the imagination: to sights, sounds, smells, recollections, and unspoken adventures that can be powerfully moving."

KÉK DUNA — *AN DER SCHÖNEN, BLAUEN DONAU*

ginning in 1979, when terrorism against the United States was focused on the country's activities in the Middle East. There is little
or no political terrorism in the United States itself because terrorists
have few support systems there. The U.S. is not a congenial place
for terrorists because of security measures and the openness of the
system.

The landscape of fear is a geographic phenomenon. There are
so many different places where malice prevails: armies of displaced
groups trying to recover vanquished lands, refugees fleeing war or
famine, Chernobyl, and on and on. When we consider transforming
fearful landscapes into peaceful landscapes, we might recall that there
is no more shameful chapter in America's history than the brutali-
zation and removal from their place of Native Americans. In expro-
priating lands that may have set records for Gargantuan plunder, we
turned these peoples into hostile, vengeful enemies. We restricted
them to reservations on remote territories. Only recently have they
begun to raise a collective voice to regain some of their land.

The map of the new United States congressional seats is a spatial
distribution resulting from a set of geographical processes. The most
notable of these processes is migration. More and more people, according to the last census, were electing to move from the northeast
and north-central regions to the South and West, resulting in a
redistribution of population. Florida, for example, grew 33 percent
during the 1980s and gained additional seats in the House of Representatives. Texas increased its population 20 percent to 17 million
people, gaining an additional three House seats. This migration reflected the human yearning for better climates and amenities—and
spatial flows of money (capital) to these destinations, among other
factors.

Given the dynamic content and importance of national and
international activities, it would seem reasonable to ask why there is
no Council of Geographic Advisers to the President or to the Congress. These advisers could predict important geographic changes and

alert the country's policymakers to negative spatial trends as well as to positive trends.

As far back as the Civil War, newspapers started to publish maps routinely, so that a news-hungry public would know where things were. Commercial mapmakers prepared "bird's-eye" views to illustrate the course of the war. Each succeeding war, the two world wars, and Vietnam in particular extended the horizon with scores of new (to us) place names. People could point to a dot on the map to locate the whereabouts of their loved ones. I found myself hoping that all the people who had learned from the maps where things were in the Middle East might also investigate the why of where. What were the commingling of history, location, culture, physical environment, topography, climate, migration, and natural resources? Millions of people wanted to know immediately where things were in the Middle East when Iraq seized Kuwait in the summer of 1990.

Most of the maps I saw in the newsmagazines and newspapers were excellent. They helped us understand the boundary disputes that made Iraq risk a global war.

Geographers interested in distance perception sometimes examine irrational human behavior based on disturbed images of the world. Students in the Department of Human Geography at Utrecht were asked to arrange 12 European capitals in sequence according to their distance from Amsterdam. Eighty-eight percent of the students overestimated the distance to Kiev. Ninety-seven percent of the students found Prague and Vienna interchangeable. The shape of Eurasia led students to underestimate north-south distances; Europe tapers from east to west. Professor Cees D. Eysberg suggests that the results of a more elaborate and large-scale study could be relevant for the tourist industry. Tourism would probably increase if false perceptions of distance and other factors were eliminated.

Almost every athletic activity consists of moving something—a person, a ball, a puck, a horse—from one place to another. This means clearing space that is invariably blocked or guarded by obsta-

Finnair's map is a reminder of the association of the land masses in the northern hemisphere that surround the Arctic Ocean, and why so many intercontinental flights take "the great circle route" over the top of the world. A straight line isn't always the shortest distance between two points. About three quarters of the land above the Arctic Circle is the former Soviet Union. It is said that the most surprising geopolitical idea in the last half century is the notion that the region surrounding the North Pole is the potential core of a future intercontinental community: a global "Arctic Mediterranean." FINNAIR

cles. Understanding the spatial/geographical basis of a game should increase the odds of winning—or at least the odds of losing intelligently. Racquet games and feats on the playing fields demonstrate how the control of space is an adjunct to victory.

There is even a geography of smells. Certain places have an aroma that becomes an integral part of their profile; for me, there will always be a Baltimore smell. For a colleague, there will always be the smell of the manicured cemetery across the road from his home in Champlain, New York. Sri Lanka is known for the fragrance of palm oil, Jakarta for the pervasive aroma of cloves. The smells of cheap gasoline and inferior cigarettes will forever recall the Soviet Union. You can often tell where in the world you are simply by breathing. This is particularly true of the Third World and in any place where primary crops are processed.

Not long ago, 40 or so of the poorest nations in the world met in the hope they could devise ways to capture the attention of the richer countries with their message that a half billion people are desperately in need of help. In these countries, the annual per capita income averages around $200. One can appreciate their desperation. The Cold War is over. There is at least a lull in the superpower struggle in which their favor was sought for political propaganda. It will be a measure of active humanitarianism if the well-off countries can increase their aid to the impoverished peoples elsewhere in the world. It would be a propitious time for a warm rather than a cold peace. Japan is today the most generous with the helping hand for foreign aid. An unseemly amount of foreign aid is given for political purposes rather than for humanitarian goals or to alleviate suffering. The geography of foreign aid by the rich countries is an interesting way of giving to former colonies, supporting allies, and courting favor.

Religions were the first major source of identity. In time, nationalism was often superimposed on religion, but it never replaced or buried it. Some religions are more strongly and widely embraced than ever. The strength of religion—and the tenacity of memory—impressed me in something I read recently. The story was about the impending 500th anniversary of Christopher Columbus's "discovery" of the world across the Ocean Sea. King Ferdinand and Queen Isabella, his patrons, had decreed that all non-Catholics convert to Catholicism or face banishment from Spain. Many Jews became converts and some of them migrated to what became the Spanish-speaking southwestern states of the United States, where many of their descendants still live. Nominally, and to all outward appearances, they have been Roman Catholics for a half millennium. Yet, on Chanukah, they light candles, and they observe other holy days as if they were still practicing Jews. Memories transplanted from one place to another—resisting the erosion of time.

The most wondrous and beautiful place I know? It may sound trite, but I say without hesitation it is the Grand Canyon. President Theodore Roosevelt properly proclaimed, "Keep it for your children and your children's children, and for all who come after you as the one great sight that every American should see." It is a very special place for all humankind and should be so treated. Yet it, too, is under siege with pollution, tourists, and automobiles and buses—the connection of people and places basking in God's beauty.

Seventy years ago, Sir John Arthur Thompson, a Scottish biologist, observed exuberantly: "Living creatures press up against all barriers, they will use every possible niche all the world over. . . . We see life persistent and intrusive—spreading everywhere, insinuating itself—adapting itself, resisting everything, defying everything, surviving everything." I have often wondered what Sir John would say were he with us now to witness living creatures becoming endangered or extinct day by day. In the end, it may be that the unvanquishable cockroaches will indeed be the creatures to inherit Earth.

The United States has celebrated the reopening of a refurbished Ellis Island. Some 17 million immigrants arrived there between its opening in 1892 and its closing in 1954. These immigrants did much to "make" America, and to make it truly a melting pot or, perhaps more accurately today, a salad bowl. I am not certain that the groups now crowding into the larger cities, particularly New York City, are ever going to "melt" into one another and achieve harmony and the sharing of goals. Perhaps "melting" was never appropriate. We can at least be respectful to all peoples and cultures.

I once visited the grave of Ghengis Khan, in Inner Mongolia, or what was said to be his grave. His Mongol nation was one of the greatest empires in world history. His hordes invaded China and Kiev in the empire of Russia. He conquered lands as far west as northern Italy, and spread the Mongol sovereign reign over thousands of miles of space. Imagine the administrative skills it must have taken to hold such a sprawling, disparate set of places together. This was centuries before the invention of the telegraph and the telephone, much less access to express mail or fax machines.

In a perfect world, sovereign states would share resources, transfer capital, allow movements among all places without restriction, relieve famine areas, and aid one another in any way indicated. We live in an imperfect world—a politicized world covered with human-imposed barriers. Such transactions—the spatial processes that in a completely just world would help to even out inequality and tremendous disproportions—are not occurring. In fact, the situation may be getting worse. Third World countries have for many years been asking the United Nations and other forums for a redistribution of wealth. They see the world as becoming more unequal. The poor are getting poorer and the rich are getting richer. And as we all know, the poor get kids.

Our newspapers could provide another service by publishing a regular column on geography. With all the space devoted to astrology and psychobabble, there must be a few inches for a feature that could be called "Knowing Your Place." It could be locally oriented and offer information about the community and the changes, however subtle but real, occurring before one's very eyes.

In terms of individuals, there are nomads and opportunists aplenty. Many will uproot themselves with any encouraging sign in the wind. But in greater numbers there are those of us who don't want to move. It isn't necessarily inertia or sentiment. There are many material and psychic advantages to staying in one place over a

period of time. We establish continuing relationshps and a good rep-
utation, and we feel safer and more secure on familiar ground. There
is the security of tenure. The blue-collar worker is especially place-
bound; he resists moving elsewhere even after his job has been liq-

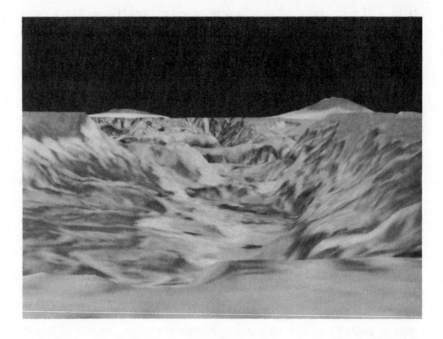

*Geography, as the art and science of place, is constantly changing. In fact, places give way
to each other without real boundaries. Even the geography of the eight other known planets
in the solar system is constantly changing but always in harmony with one another in the
universal system. Mars from three miles out: the sinuous channels were probably carved
by ancient rivers, the sand dunes are comparable in dimensions to the largest dune fields
on Earth. Dust storms carry fine particles some tens of miles out from the Martian surface.
The red planet's Nix Olympica is the largest known volcano in the universe—it is 15
miles high and more than 300 miles across at its base. Venus, which some scientists believe
was Earth's twin at their births 4.6 billion years ago, but went bad, is hotter than a self-
cleaning oven and enveloped by poisonous clouds of sulfuric acid. The planet is dominated
by volcanism on a global scale. Its unique features include more than 600 huge craters,
long rift valleys, and pancake-shaped domes about a half mile high and 10 miles across,
presumably formed by extremely viscous lava, which stretch out in a line remarkably straight
for nature. A channel is longer than any known elsewhere in the universe, a 4,200-mile-
long feature, which is longer than even Earth's longest river, the Nile. Venus's year is
shorter than its day. A temperature of minus 391° F. makes Triton, a moon of Neptune,
the coldest body in the solar system; its water-ice crust is granite-hard. Neptune is currently
the farthest planet from the Sun, its orbit taking it beyond Pluto's until March 1999.* JPL

uidated. Something else will turn up, he'll say, whether he believes
it or not, and he knows little of anything about the opportunities in
other places and cannot afford to go search. "We live here," German
Jews protested while they still had a chance to escape the Nazi
extermination camps. "This has been our home for centuries." People
who "know the risks"—people whose homes have been demolished
in a mudslide or an earthquake say, "Just the same, we want to
rebuild right here." And they do. Thus, the world has people who
bond closely with "their place" and others who prefer trying new
places.

The dreadful events in the Middle East have been an education.
We have learned that that part of the world is by no means just one
vast desert occupied by camels and nomads and punctuated by a few
new cities made fabulously rich by oil fields—the coincidence of the
location of energy. We were reminded that many of the wonders of
earliest civilization were created there and that the architectural
splendors were rooted in deepest history. One can only conjecture
grimly if and when the mayhem in those blighted lands of the Tigris
and Euphrates rivers will be replaced by lasting peace and harmony.
I think especially of Jerusalem, which is sacred to several religions.
It has been a crossroads of caravans and armies throughout history,
and should be spared forever more to serve symbolically as the capital
of the new One World sensibility. (The ancients did not know about
oil literally underfoot. I wonder what we don't know.)

By the miracle of television, 500 million people have watched
the Academy Awards ceremonies live, and as many as a billion the
Olympics. An unfolding disaster is immediately broadcast live to the
farthest corners of the globe. Instant globalization and communication, including war, do not create homogeneity. Quite the opposite.
It sharpens diversity and the awareness of dissimilar environments,
and it stimulates the demand for the kind of education that insures
freedom of self-expression.

It is obvious why I keep this prediction at hand and broadcast it often: "By 2000, it was recognized that modern geography is the integrated view of man and his planet," Louis Branscomb wrote in an article in *American Scientist* magazine in 1936, "the bringing together of ecology, the study of human habitats, geomorphology, social anthropology, and economics—in short, all the tools necessary to understand how human beings should view their planetary home. Once again, geography became a popular course of study in school, particularly since students were no longer required to memorize state capitals and map features."

The announcement that Atlanta would be the site of the 1996 Olympic Games took many people by surprise. Atlanta? *Atlanta, Georgia?* A provincial Southern capital? I was reminded of many discussions I have had with students and friends about the blurring of boundaries and the changing character of places. Some of these discussions focused on the American North and South. Gone is the sleepy, racist, stereotypical South, which now often seems more progressive and flexible than large pockets in the decaying northern rustbelts. For the first time in decades, the population in the South has been increasing in part because of reverse migrations. The shifts

---

Ocean bottoms are neither homogeneous nor flat, as soundings in fathoms of the North Atlantic reveal. Around Rockall Bank, for instance, adjacent mountain peaks range from 203 fathoms (1,218 feet) to 1,218 fathoms (7,608 feet) below the surface. Earth's longest mountain range sits on the bottom of the Atlantic: the volcanic, tortuous, craggy Mid-Atlantic Ridge, still unknown a century ago, stretches 12,000 miles along the center seam of the spreading floor from Iceland almost to Antarctica—a system greater than the Rockies, the Andes, and the Himalayas combined. Built by material emitted from the molten interior, the chain's highest peak is Pico Island, in the central Azores; it thrusts 27,000 feet skyward from the floor of the ocean, 7,711 feet above water. Seven underwater sites have been designated world wonders: the waters around the Republic of Belau (Palau) in Micronesia, Russia's Lake Baykal, Ras Muhammad reef in the northern Red Sea, the Galápagos Islands, the Belize barrier reef, the northern portion of Australia's Great Barrier Reef, and the deep ocean vents found in Pacific, Atlantic, and Indian waters. New ocean floor is created continually. The deep-sea bed is regarded as a common resource for humankind. DEFENSE MAPPING AGENCY, HYDROGRAPHIC/TOPOGRAPHIC CENTER

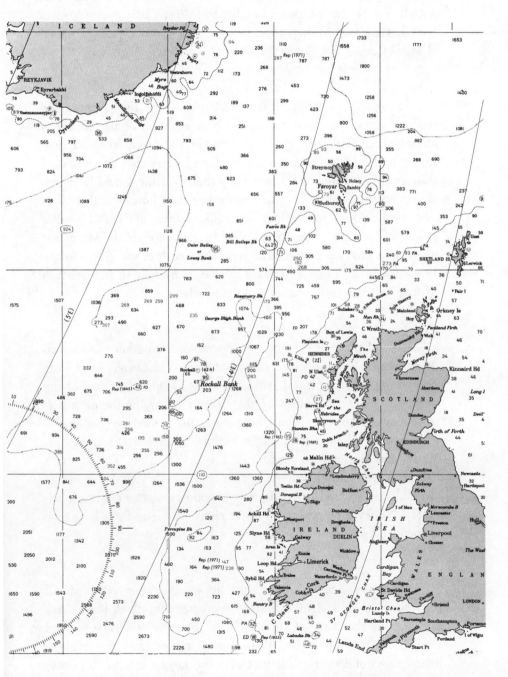

in America are evidence of the dynamic of place and the importance of spatial processes. The Northeast is in decline—in some sections, not all, the Southeast grows rapidly; the West Coast is a magnet of growth; the upper Midwest declines. The U.S. is in a state of regional change of enormous proportions.

Geography students learn the dynamics of the political and economic processes that prompt migration. Who of us, with any acquaintance with the 1930s, can forget the tragic skein of dust storms and depleted crops and bank foreclosures that drove tens of thousands of Oklahoma farmers from their worthless land in search of survival as fruit pickers in California? And how well their epic story was told in John Steinbeck's *The Grapes of Wrath*. Similar geographic peregrinations are recalled in the history of the Mormons, Boers, and the American blacks from the South to the North in the 1930s and '40s. These processes can be seen today in sub-Saharan Africa, Vietnam boat people, and many others.

One of the best examples of geographical ignorance in modern times was the Soviet invasion of Afghanistan. The Soviets knew little if anything of the cultural, ethnic, and economic geography of the country. They were not aware they were challenging the most het-erogeneous, clannish, tribally disparate, un-unified peoples in the whole world. Afghanistan is an incredible, geographical hodgepodge. No one has been able to conquer it. The British tried. The Russians had tried even earlier. They went on thinking they could exploit an internal conflict and bring the whole country together to their ad-vantage. They'll be paying for their eight-year folly for a long time to come. *Sovietnam!*

Capitals are capital places, usually. Capitals serve citizens, and their location is a significant factor in their ability to serve. Ask someone for the name of the capital of the state of Delaware, and the answer is likely to be "Wilmington" rather than the smallish inland town of Dover. Wilmington is the largest city in the state, though its population is only about 70,000. It is the only city in "The

First State" known to most Americans. Why is upstate Albany, with hardly more than 100,000 people, and not New York City, the capital of the state of New York? Or downstate Springfield, only a thirtieth the size of Chicago, the capital of Illinois? Centrality, rather than population concentration, has usually been the main factor in the selection of capitals.

Brazil moved its capital from beautiful, coastal Rio de Janeiro to Brasilia, a manufactured city designed to encourage the exploration of the Amazon region and slow the growth of Rio. Australia has a string of seaside metropolises, but it went inland to create its capital at Canberra. Argentina, not too seriously, speaks of moving its capital from magnificent Buenos Aires northward to Viedma. A dozen countries, most of them in the Third World, are talking about naming or even building totally new capital cities.

Is there a chance that Washington, D.C., would be superceded by a place more central in the continental United States? Say to Kansas City or Omaha? Quite certainly not. When the Constitution was ratified in 1788, the capital of the United States was New York City. It presently moved to Philadelphia for 10 years. A political compromise settled the capital permanently at the approximate midpoint in the original 13 Atlantic-seaboard states. Waves of immigrants pressed westward over the years because of greater opportunities and the chance to acquire ownership of land. Today, the geographic center of the population of the United States is in Missouri. But there have been no cries for a more centralized capital. Washington has become a mecca for entrenched social and political power. It is a world capital. With communication systems as sophisticated as they are these days, capitals need not be accessible places. In the United States, given its wealth and communication systems, Washington serves very nicely.

It is interesting to note how place names have enriched the English language. Names of places have moved into the language with meanings derived from their places. Donnybrook, Ireland, has an annual fair that was notorious for brawls. Donnybrook came to mean wild, free-for-all fights. The definition has broadened to mean "rowdy contention carried on between rival forces," such as on the floors of legislatures. . . . The course of the Meander River in Asia

Minor twists and turns for about 250 miles before flowing into the Aegean Sea. Meander has come to be synonymous with meandering, taking a winding course. . . . The scene of Napoleon's final defeat in central Belgium in 1815 has long passed into everyday English; we refer to meeting our "waterloo" when suffering a decisive setback. . . . China's chief seaport and largest city has long since entered the vernacular. When sailors were kidnapped for duty on vessels to the Orient—a practice associated with drugging or coercion—they were said to have been shanghaied. The broader meaning today is to be forced into doing something. . . . Among contributions from the Greeks of antiquity are Sparta and Marathon. Sparta was a city in southern Greece that produced citizens known for their strict self-discipline, frugality, and simplicity; we speak of someone who is Spartan in his or her habits. Marathon was the site of a victory of Greeks over Persians in 490 B.C., the news of which was carried to Athens by a long-distance runner. Marathon means any race of competition requiring endurance. . . . Bikini conjures up the briefest of women's swimwear. But there is that other Bikini, the atomic-bombing-test atoll in the Marshall Islands in the western Pacific that inspired the name of the bathing suit. Webster's Third finds a comparison with the "effects of an atomic bomb." . . . Then there's madras and tequila and sherry and fez and geyser and paisley and hamburger and frankfurter and cologne and bialy. . . . And so place has enriched our vocabulary and provided generic descriptions of our activities and products.

In the beginning was geography: infinitely dense space, then time, continually in motion, changing and interconnected—the origin of the universe. The Sun and eight of the nine known planets in the solar system were captured electronically by the United States interstellar-space vehicle Voyager I *3.75 billion miles from Earth. There are more than 200 billion ordinary stars in the Milky Way galaxy—200 billion suns. "Wake!" wrote Omar Khayyám: "For the Sun who scatter'd into flight / The Stars before him from the Field of night, / Drives Night along with them from Heav'n and strikes / The Sultan's Turret with a Shaft of Light."* NASA

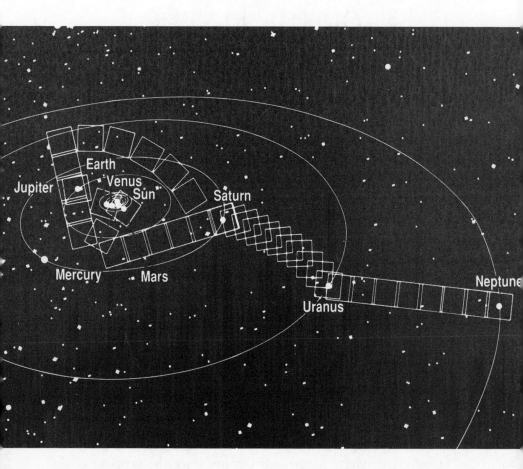

INTIMATIONS
OF THE FUTURE:
PROFILES OF 173
COUNTRIES—AND COUNTING

Geographers define a sovereign state, or, more commonly, a country, as a bounded territory with a population, an economy, a system of circulation, a government that can defend the space and the population, and recognition by a majority of the world's other sovereign states. A nation, on the other hand, is a people who share a common historical/cultural characteristic but do not control their space. Many countries are home to one or more nations. Nations that go on to incorporate and control territory begin to assume the profile of sovereign states. In the half century since the breakout of the Second World War and the breakup of colonial empires, the number of sovereign states more than doubled, from 70 to 173. The newest ones include Namibia, which gained independence from South Africa

---

EARTH: The fifth-largest planet in the solar system and the only planet definitely known to support life. Seventy percent water and thirty percent land, Earth has seven continents, in order of size: Asia (the largest); Africa; North America; South America; Antarctica ("the peopleless continent"); Europe; and Australia. Ten ecosystems—mountain, desert, rain forest, savanna, steepe, prairie, broadleaf forest, needleleaf forest, tundra, and ice cap—are scattered across the 13 land masses. These regions, like all regions, continually change, of course, because of spatial processes like wind currents and migrations. Since Hitler's demise and Nagasaki in 1945, there have been 140 wars, resulting in 8 million dead.

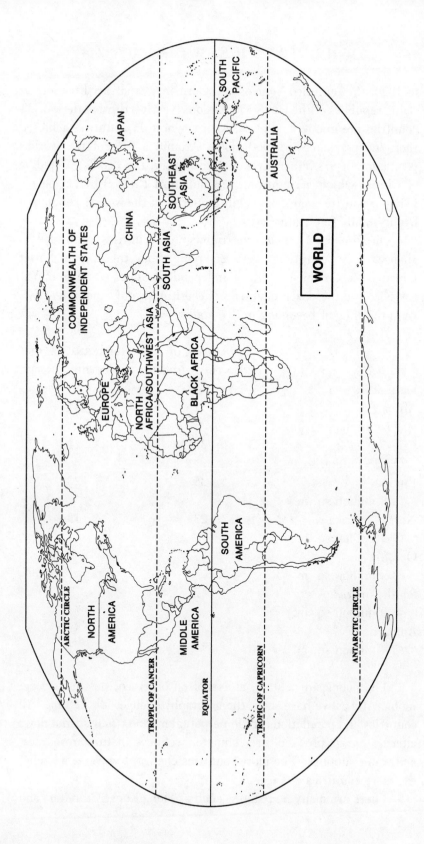

WORLD

ARCTIC CIRCLE

NORTH
AMERICA

TROPIC OF CANCER

MIDDLE
AMERICA

EQUATOR

SOUTH
AMERICA

TROPIC OF CAPRICORN

ANTARCTIC CIRCLE

COMMONWEALTH OF
INDEPENDENT STATES

JAPAN

CHINA

EUROPE

NORTH
AFRICA/SOUTHWEST ASIA

SOUTHEAST
ASIA

SOUTH ASIA

BLACK AFRICA

SOUTH
PACIFIC

AUSTRALIA

in 1990, the reunified Yemens and Germanys, and the three former
Baltic republics of the late Soviet Union. Nearly a third of the world's
countries are in one land mass, Africa, where 17 countries achieved
their independence in 1960 alone. Around a fifth of the world's
population of 5.3 billion lives in one country, the People's Republic
of China, which accounts for 1.3 billion of the 3.9 billion people
living in non-free countries. Only a quarter of the world's population
live in totally free countries.

On January 1, 1992, the number of countries continued to
change. New countries were emerging from the former Soviet
Union—the Commonwealth of Independent States—and from Yugoslavia, and the dozen countries of the European Community (EC)
were thinking of becoming united states.

LAND MASS	NUMBER OF COUNTRIES IN THE LAND MASS	PERCENTAGE OF COUNTRIES IN THE LAND MASS
Africa	52	31
(11.7 million sq. m.)		
Asia	38	21
(17.25 million sq. m.)		
Europe	35	20
(3.8 million sq. m.)		
North America	23	13
(9.4 million sq. m.)		
Oceania	13	8
(3.3 million sq. m.)		
South America	12	7
(6.9 million sq. m.)		
Antarctica	0	0
(5.4 million sq. m.)		

The policies of a state, as Napoleon observed, lies in its geography. And as we have seen, the geography changes all the time. All
countries and populations are in perpetual motion, their spatial flows
altering topography, culture, climate, wellness, natural resources,
and water supplies. The more countries change, to reverse a cliché,
the more countries change.

There are many non-states, such as Hong Kong, Gibraltar, and

Taiwan. They are "leftover" spaces whose sovereignty rests with other countries, or is unclear, at least temporarily.

With time, countries can change beyond recognition. An understanding of their spatial composition, dynamism, and interaction internally and with other sovereign states is information that may be critical—providing intimations of the future.   2/1/92

EUROPE: the world's second-smallest continent has been blessed by location and an almost infinite range of natural environments. Easy access to the Atlantic Ocean and to the North, Aegean, Baltic, Adriatic, and Mediterranean seas spawned global empires, populated the western hemisphere, and extended its cultures worldwide. Revolutions and wars plagued an early unruly continent; harmony and cooperation are now being tried in the West: the European Community—EC—seeking a truly common, seamless market— promises to enhance economic integration and potency, political unification, and permanent peace—the United States of Europe, a key stage in the planned coalescence of space, superseding fragments. (Ironically, Eastern Europe appears to be creating more borders as Western Europe eliminates its borders.) The North Atlantic to the west is the "inner sea" of the Euro-American world and the countries in its geological orbit. Strata on the northwest coast of Scotland match Newfoundland strata 2,500 miles to the west.

ALBANIA: known to its tyrannized citizens as the Land of Eagles, to outsiders as the Land of Nod, and to the world at large as the first constitutionally atheistic state, Albania has been a dirt-poor, brokendown Iron Curtain country, the most destitute of the states that came under Soviet domination after the Second World War—Europe's darkest country. Contact with other countries was minimal. The Communists, who took power toward the end of the war, had been the strongest of the partisans resisting the Nazi conquest. Their hardline, monolithic rule didn't have an ally after 1978, when China cut off aid.

The southeast European country, which is slightly larger than Maryland, is on the Balkan peninsula between Greece and Yugoslavia, and 55 miles east of Italy across the Adriatic Sea. Except for a fertile coastal lowland, Albania is mountainous and rugged, 210 miles long from north to south and 90 miles at its widest.

Prior to Stalinism, with its cast-iron ban on dissent, Albania had endured centuries of harsh Turkish rule. After the First World War, United States President Woodrow Wilson was the most powerful defender of Albania's territorial integrity in international councils as

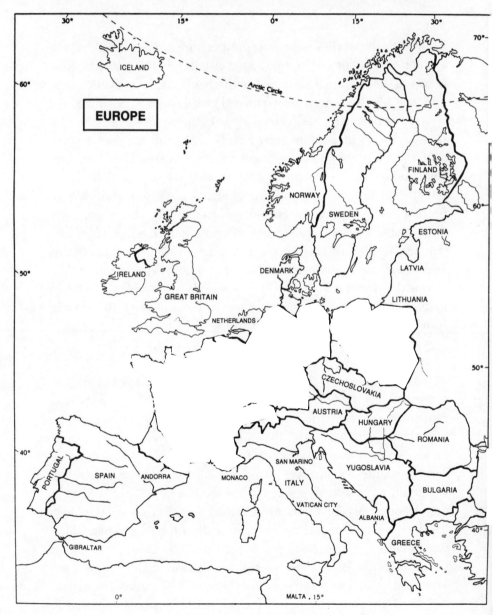

Italy, Yugoslavia, and Greece vied to bite off chunks of the country. Albania became a republic in 1925, was taken over by Italy in 1939, and occupied by Germany in 1943. In 1950, the United States was actively engaged in a plot to overthrow the government with paramilitary forces. (In 1991, Washington ended 52 years of nonrecognition.) Albania, which was called Illyria when it was part of the Roman Empire, broke with Moscow in 1961 and switched its allegiance to both the People's Republic of China and the Democratic People's Republic of (North) Korea.

The "Rip Van Winkle of countries" has been virtually car-less. A person could be jailed for sporting a beard or wearing jeans. Housewives waited in line eight hours for their rationed quart of milk. Cities were fortified with oval steel and cement pillboxes mounted with gun turrets. There were no restaurants or cafés in the capital. Transportation and communications were primitive. In this last bastion of orthodox Communism in Europe, a peaceful liberalization is under way, influenced by spatial diffusion of democratizing processes across eastern Europe. Albanians may now apply for passports. Churches and mosques, sealed under a rigorous state policy for 24 years, are open to worshipers. Free-market principles have been introduced.

Albania has considerable resources. It is self-sufficient in grain, it sells electricity to neighboring countries, and there are oil, copper, and coal deposits. In a population of 3.5 million, the ethnic Greek minority totals 400,000. The average monthly income is $60. Until recently, Albania boasted that it had no foreign debt, but there had been a ban on foreign capital. Hopeless Albanians by the tens of thousands fled across the Adriatic to Italy in the summer of 1991, only to be turned back by Italian authorities—the faltering economy of the long boot-shaped peninsula could not absorb the mammoth migration.

ANDORRA is a political-geographic anomaly. The postage-stamp-size coprincipality in the eastern Pyrenees has had two heads of state since 1278: the president of France and the Roman Catholic bishop of Urgel in Spain. Though under their joint suzerainty, the state is in fact ruled by a council. There is no formal constitution, no income tax, no army. Abortion and divorce are prohibited. The Catalan-speaking Andorrans are a minority in their own state: less than 20 percent of the population of 50,000 are natives; the other residents are immigrants and illegals who seek tax relief and/or freedom from close scrutiny. A generation of American readers learned of Andorra's very existence in the romantic travel adventures of Richard Halliburton. There it was, this exotic little place about a fifth the size of Rhode Island, valleys enclosed by steep mountains, going about its affairs oblivious and unknown to most of the outside world. Today, the poor little rich land is a kind of Hong Kong, duty-free bargains attracting millions of tourists and shoppers. Farming has been abandoned for commerce and finance (and smuggling). Little of the soil is arable, but the upland pastures are conducive to the raising of cattle and sheep. The literacy rate is 100 percent.

AUSTRIA has hills alive with the sound of music and schussing and couches lined with psychiatric patients. Vienna, the capital, has been home to some of the world's foremost composers (e.g., Beethoven and Mozart in the city's romantic past) and it was the birthplace of psychoanalysis (Breuer and Freud). In 1938, the parliamentary democracy agreed enthusiastically to becoming incorporated into the German Reich. (Hitler had been born in Braunau, in Upper Austria.) After the Second World War, the mountainous, landlocked south-central European country (it is slightly smaller than Maine) was occupied for a decade by four allied powers. A little over 18 percent of the land is arable, most productively in the Danube River basin in the east. State enterprises control a large part of the economy. Tourism, in part because of the Alpine skiing, is a major industry. Ninety-eight percent of the 7.5 million Austrians are German, Roman Catholic, and literate. Per capita income is $12,521. Universal suffrage begins at 19. Life expectancy is high—76.4 years for women, 69.3 years for men—but the suicide rate has been the highest in the world for some time. As in much of western Europe, the population growth rate in Austria is low.

BELGIUM, which is slightly smaller than Maryland, is at the economic center of European enterprise; it lives by trading. A flat northwest European constitutional monarchy on the North Sea, it is heavily industrialized and nearly self-sufficient in food production. In 1947, Belgium, the Netherlands, and Luxembourg formed a customs bond, which in 1960 became the Benelux Economic Union, over time evolving into the European Community (EC). Belgium was an early champion of the Community. The region had been annexed by France in 1797, and in 1815 it was handed over to the Netherlands by the Treaty of Paris. It proclaimed its independence in 1839. Belgium was occupied by German armies in both world wars. Language still splits the state: the north, which is known as Flanders, is dominated by Flemings, who are Flemish-speaking descendants of Germanic Franks; the south, which is known as Walloonia, is dominated by French-speaking Walloons, who are descendants of Celts. The linguistic division marks nearly all political decisions. (Ten percent of the population of 9.9 million is bilingual.) The capital of Brussels is one of the continent's most cosmopolitan cities, and is celebrated for its gastronomic pleasures.

BULGARIA was ultraconformist when it was under the four-decade totalitarian rule of a dictatorship, the oldest Communist police state in eastern Europe. A mountainous country the size of Ohio in the eastern Balkan peninsula on the Black Sea, it is surrounded by Turkey, Greece, Yugoslavia, and Romania. East-west mountains bisect the country. There are rich soils in the river valleys, and the climate is similar to that of the American Midwest. A German ally in the First World War, Bulgaria lost its Aegean coastline after the Treaty of Versailles.

Following the Communists' overthrow of the monarchy after the Second World War, it became an industrial state modeled on the Soviet Union. Industry came to rely on Soviet fuel and matériel, and Moscow in turn bought around 60 percent of Bulgarian exports.

In the population of 9 million are around 2 million ethnic Turks, many of whom are Muslims, and Gypsies. The highest birthrates are in the Muslim regions. Tensions born of centuries-old hatreds persist, especially between Turkish Bulgarians and ethnic Bulgarians. Earlier government policy demanded assimilation. (Assimilation compelled Turks to convert their names to Slavic forms or to leave the country.) The policy has since changed, but enmity with Turkey and Yugoslavia persists. In the wake of the democratization processes that diffused across eastern Europe in 1989, the country was wracked by conflict. Ninety-three percent of property was still state-owned in January 1992 when a fifty-six-year-old philosopher won Bulgaria's first direct presidential election. Privatization was seen as crucially important to revival of the crumbling economy.

CZECHOSLOVAKIA—officially today The Czech and Slovak Federal Republic—was created after the First World War as a homeland for Czechs, Moravians, and Slovaks, minorities in the former Austro-Hungarian empire. Cradled inside the great arch of the Carpathian Mountains, in the landlocked heart of Europe, the country has been renamed to placate a reassertive Slovak nationalism in the separatist-oriented eastern region, a birth largely suppressed during four decades of rickety Communist rule and economics. The democratic Czech republic was established from the ruins of the Hapsburg Empire. Agricultural Slovakia was a nominally independent state during Nazi occupation.

Industrious, prosperous Czechs make up 63 percent of the population of 15.5 million and control roughly the same percentage of territory. The Slovak republic earns about 30 percent of the national income. The country was demoralized when the Soviet-led Warsaw Pact invasion of 1968 crushed hopes for liberalization—the intoxicating "Prague Spring." For two decades, about 140 cities and villages were occupied by the Soviet military. A "velvet revolution" in 1989 toppled totalitarianism and led to the first free elections since 1946. A nearly 100 percent turnout of voters chose a playwright and former political prisoner, Vaclav Havel, as president.

The country is about the size of New York State. Decades of unchecked production from coal mines and industry ravaged the environment; a third of the country is ecologically devastated. Acid rain has destroyed at least 60 percent of the forests in the central and western regions and more than 20 percent in Slovakia. The population size has been static despite government incentives for childbearing. The sale of military equipment is a key source of foreign exchange.

DENMARK, the southernmost of the four Scandinavian countries and once in union with Norway and Sweden, is the Jutland peninsula jutting into the Skagerrak and the Kattegat and bordering on Germany, plus a number of islands in the Baltic Sea, chief among them Zealand, Falster, and Bornholm. It is the size of New Hampshire and Massachusetts combined, low lying and densely cultivated. Overseas territories are the self-governing North Atlantic province (since 1953) of Greenland, the world's largest island (exclusive of Australia), and the windswept Faroe Islands (also self-governing) between Scotland and Iceland.

Denmark excels in dairying and stock raising and has prosperous commercial and fishing fleets, but lacks good harbors. In common with the other four Scandinavian countries, it provides cradle-to-grave social services for its 5.1 million citizenry. Denmark was occupied by German forces during the Second World War and earned the civilized world's admiration for the protection it gave its Jewish citizens. Now a constitutional monarchy, the state was first settled in the 500s by Danes, a Scandinavian branch of Teutons. There is no poverty and there are few examples of extreme wealth; per capita income is $11,312. Most of the people are Danes and follow the Lutheran faith. The literacy rate is 99 percent. Women live to an average of 77.3 years, men 71.5 years. In Copenhagen, the country is blessed with one of the most agreeable capitals in the world.

ESTONIA, which is a third smaller than either of the two other Baltic states, and has the least population (1,573,000) of the three, became an independent sovereign country again in September 1991, fifty-one years after having been incorporated as a constituent republic by the conquering Soviet Union. The incorporation was never recognized by the United States.

Estonia was ruled by Sweden before 1721, then by Russia until 1920. Independence after the First World War led to Estonia's membership in the League of Nations. A dictatorship held sway during the years 1934–1937. A year after the 1940 annexation by the Soviet Union, Nazi forces drove out the Soviet occupiers and Hitler incorporated Estonia into his Ostland province. The country was recaptured by the Soviet army in 1944.

Estonia faces the Baltic Sea on the west and the gulfs of Riga and Finland, which are northern and southern arms of the Baltic. Because of these marine influences, Estonia enjoys a mild climate despite its northerly location. The country borders Latvia on the south and Russia on the east. Estonia is generally flat. Natural resources include shale, peat, limestone, dolomite, marl, sand, and timber. The most important industries are fishing, dairy farming, and shipbuilding. During the Soviet reign, the country made military equipment, and the economy was integrated with the economies of the 14 other republics.

Ethnically and linguistically, Estonians are related to the Finns. Large numbers of Russians migrated to the northeast section of Estonia.

FINLAND has a highly developed social-welfare program that provides national health insurance and free education through the university level. The infant death rate is one of the lowest anywhere. All 5 million Finns can read and write, and the average per-capita income is $11,000. (Unemployment, however, reached a record 7 percent in 1991.) Basically a bilingual population, Finnish and Swedish, the Finns are of Fenno-Ugrian stock, related to the Magyars of Hungary.

The state is big, about the size of New Jersey, New York, and the six New England states combined, or slightly smaller than Montana. It once was ruled by Sweden, once was united with Denmark, once was a grand duchy in the Russian empire. It became an independent state only after the First World War. Finland lost a portion of its territory to the Soviet Union in a peace treaty after the Second World War.

The north European republic is bounded on the north by Norway, on the east by the Soviet Union, on the south by the Gulf of Finland, and on the west by Sweden and the Gulf of Bothnia. In the winter, Finland has no ice-free port. About a tenth of the country is lakes, about 60,000 altogether, many of them interconnected. About 70 percent is forested. Only 8 percent is arable. The natural beauty is a tourist attraction. Northern Finland is the land of the midnight sun, and indigenous nomadic Lapps live within the Arctic Circle. Most Finns reside in the south.

Life expectancy for women is 78.8 years, for men 70.4 years. Finland is a leader in peat technology and competes with Ireland for the largest share of the world market. The economy is based upon dairy and timber products, copper, shipbuilding, and textiles. Finland has not sought membership in the European Community. The capital city of Helsinki is called the White City of the North because of its cleanliness and buildings of native granite.

FRANCE, which sets the highest standards for good eating, is a Texas-size land of small independent farmers blessed with some of the finest soils and vineyards—a third of the land is arable. During the Fifth Republic, President Charles de Gaulle asked, "How can you govern a country with 246 varieties of cheese?"

France stretches from the English Channel in the northwest to the Mediterranean Sea in the south. A wide plain covers most of the country, and it is mountainous in the eastern region bordering Switzerland. Mont Blanc, the continent's tallest mountain, is 15,771 feet. France and Spain are building a toll-free, 5.3-mile-long road tunnel through the Pyrenees Mountains under the Col du Somport. France is also completing with the United Kingdom the "chunnel" under the English Channel. Both projects will extend France's geographic connectivity.

France, which can trace its history to Caesar's conquest of Gaul in the first century B.C., fought a revolutionary war—some years after the American colonies did—to put an end to royal absolutism and gross inequalities among the population. The culture has attracted exiles from everywhere, especially between the two world wars when writers and artists ("the lost generation") were finding their spiritual home in Paris, "the city of light." (Seventeen thousand years ago, paintings and engravings of bison, stag, and reindeer were already decorating the Lascaux Cave.) Already weakened by a massive loss of life in the trench warfare of the First World War, 12 years earlier, France capitulated quickly to German armies in May 1940, then was liberated by Allied troops in 1944. Postwar recovery was slow, and colonial wars were lost in southeast Asia and Algeria.

The country has been undergoing "la malaise française," a sort of national liver complaint, as it questions its identity and answers the charge that it suffers "progressive Americanization." In serious decline since 1970, French industry has had trouble adjusting to much-higher oil prices and the modernized world economy. A unified Germany's emergence as a power center tests France's historic role on the international scene. (Before the Persian Gulf war, in 1991, its closest ally in the Middle East was Iraq.) France still has overseas departments and territories: undeveloped, tropical, rain-forested French Guiana on the northeast coast of South America; the impoverished Mediterranean island of Corsica ceded by Genoa in 1768; volcanic, coral French Polynesia—tropical paradise islands—in the South Pacific. France is 99 percent literate and 90 percent Roman Catholic.

GERMANY is a behemoth again, thanks to East Germany's successful uprising for political and economic freedoms, the collapse of the Berlin Wall, and unification after 45 years of division as West Germany and Soviet-occupied and centrally planned East Germany. The divided city of Berlin was the seismograph of the 45-year Cold War. In 1990, the first free, all-German election in 58 years (or since Hitler's ascension) endorsed internal unity and full sovereignty for all Germans. Germany, which is the size of Virginia and Wyoming combined and has a population of 77.6 million, is the most populous, the wealthiest, and the healthiest democracy in Europe, eager to play a pivotal role in the European Community. It is a NATO member. Rapid resurrection of West Germany after ruination in the Second World War bordered on the miraculous.

Germany extends from the Baltic and North Seas in the north to boundaries with Austria and Switzerland in the south. Other immediate neighbors are Denmark, the Netherlands, Luxembourg, France, Poland, and Czechoslovakia. The Ruhr, in the West, is the powerhouse of the industrial economy. About 30 percent of the land is arable. A daunting challenge for years to come will be rebuilding and integrating the shattered, formerly socialist economy in the east; heavily polluted, ecologically ravaged, and industrially traumatized, the region is a gigantic scrap yard. Businesses there released a million workers after unification—joblessness had been nonexistent—and subsidies for housing, fuel, and public transport expired. Many believe that when the capital city moves from Bonn to Berlin, a revitalization of the east is assured.

GREAT BRITAIN, comprising England, Scotland, and Wales, is Europe's largest island. It is separated from the northwest coast of the continent by the English Channel and the North Sea. Shakespeare described his native England as "this royal throne of kings . . . This earth of majesty . . . This fortress built by Nature for herself . . . This precious stone set in the silver sea . . . This blessed plot . . ." The United Kingdom, with a population of 57 million, is Great Britain plus the strife-torn provinces of Northern Ireland, which is separated from Britain to the east by the Irish Sea. In recent decades, Irish Republican Army guerrillas have terrorized Britain to bring about the end of its involvement in Northern Ireland; the breakdown of law and order there in 1968–1969 led to direct rule.

Overseas dependencies once included 13 colonies in North America and territory in the Far East (including the subcontinent of India), the South Pacific (including the continent of Australia, which was founded by Britain as a penal colony, and New Zealand), and the West Indies. At one time, the sun never set on the British Empire, through which the English language spread and became preeminent.

Britain has limited natural resources—large deposits of coal and, now, extensive oil and gas fields in the North Sea—but the industrial revolution and access to the channel, the sea, and the ocean led to domination of world trade. By the rule of natural size, Britain is "entitled" to 3 or 4 percent of the world's wealth and power; at the height of the empire, it enjoyed around 25 percent. After industrializaton, Britain became the first urban state. Opening of the $13.44 billion three cross-channel tunnel tubes in mid-1993 will physically link "this sceptred isle" with the continent, bringing closer involvement in the European Community. London–Paris by train will take three hours.

The climate is mild due to the Gulf Stream; it rains a good deal. Britain has rolling land: uplands, lowlands, highlands. The highest peak is Ben Nevis, in Scotland, at 4,406 feet. The waters of Scotland's fjordlike sea lochs are pristine. Welsh, that eloquent offspring of the official Celtic language of King Arthur's legendary court, is spoken by less than a fifth of the people in the principality of Wales. Forty-six million of Britain's 60 million acres are farmed. The population is totally literate. London, the capital of the constitutional monarchy, is a world-class city, an international financial center with 7 million residents.

Turning back Hitler's air power in the Second World War was the country's finest hour. "The man of the century" was Sir Winston Churchill, the indefatigable prime minister and war leader. Between 1957 and 1970, the monarchy relinquished control over most of its overseas dependencies. The country maintains affiliations with the Channel Islands, the Isle of Man, Gibraltar, the British West Indies, Bermuda, the Falkland Islands, St. Helena, Ascension, Hong Kong (until 1997), and Pitcairn Island. The British Commonwealth of Nations is an association of 48 sovereign states, formerly British colonies.

GREECE is the mountainous southern tip of the Balkan Peninsula on southeast Europe's flank, and some 2,000 islands in the Ionian and Adriatic seas. Only a quarter of the land can be cultivated. Lemons and olives are the chief exports. Within the European Community, Greece has become the black sheep. It has been "leading" in inflation, lagging in growth, and trailing in income. Civilization in the Hellenic Republic goes back to the early Bronze Age; Greeks played a central role in the development of European, African, and Asian cultures, producing masterpieces of art and literature and breaking ground in philosophy, political thought, and science. Almost everyone in this Connecticut-sized land of 10 million is literate, which is hardly surprising in a locale with a heritage of respect for knowledge and discovery.

HUNGARY, formerly part of the powerful Austro-European empire, is restructuring its political and economic systems after throwing off all vestiges of its Soviet satellite status. Moscow's removal left Hungary, as it moves toward a market economy, with the highest per-capita foreign debt in eastern Europe. (The Soviet Union stored nuclear devices here until the late 1980s.) Hungary joined the Axis powers in 1941 and was invaded by the Red Army in 1944. Until nearly the end of the Second World War, it was able to provide a haven for its native Jewish population and for Jewish escapees from Nazi-occupied lands. An anti-Communist uprising in 1956 was ruthlessly crushed by Soviet troops. Communist leaders reigned for 32 years.

Considered by many to be the jewel of mitteleuropa, Hungary's dazzling capital of Budapest was world renowned for its restaurants, Gypsy music, and general air of romance. The country was the most liberal of the Warsaw Pact nations.

Much of the landlocked people's republic is blessed with the Great Hungarian Plain east of the Danube. Fifty-seven per cent of the land is arable. The leading agricultural output is meat, poultry, maize, wheat, and grapes for wine. Textile, machinery, metal products, and chemicals are major industries. The transition to a free market—the growing development of private enterprise and the growth of industry—is fueling a population shift from rural to urban living. An estimated 190,000 Hungarians fled the nation during Soviet control. Ninety-two percent of the population of 10.6 million is Magyar, and 98 percent is literate.

ICELAND, the westernmost state of Europe, includes a major volcanic island and several smaller ones. Strategically set at the edge of the Arctic Circle in the North Atlantic, it is about 70 percent uninhabitable. Lying along the tortuous Mid-Atlantic Ridge, Iceland is literally being torn apart by the gradual spreading of oceanic plates. It has more than 100 of the world's 600 active volcanoes; many seem to be interconnected. In the past five centuries, the volcanoes have poured forth a third of the world's output of lava. There are black lava deserts. United States astronauts have trained in a lake-filled caldera. There are about 120 glaciers on the jagged mountains; one of them occupies a twelfth of the whole country. The harnessing of geothermal heat is widespread; many homes are heated with hot water pumped directly from the earth. The cool, temperate oceanic climate is influenced by both the Gulf Stream and Arctic currents.

A former Danish possession, Iceland passed a popular referendum in 1944—and the kingdom became an independent constitutional republic; it is the size of Virginia. Fishing represents 17 percent of the gross national product and 70 percent of export earnings. The country has no timber, and agriculture is limited (principally hay for fodder), but there is extensive grazing land for sheep, horses, and cattle. Norse settlers came ashore in the ninth century, and Norse sagas have inspired high educational levels. All 250,000 Icelanders are literate, and they are reputed to be among the most widely read people in the world. More than half of the population lives in the capital city of Reykjavík and its suburbs. The Althing is the world's oldest surviving parliament; it first met in A.D. 930. Iceland boasts an advanced level of social-security legislation.

IRELAND (Eire) lies in the Atlantic Ocean, in northwest Europe, separated from Great Britain to the east by the Irish Sea, the North Channel, and St. George's Channel. The republic was born out of violent, bloodletting guerrilla and civil wars. It occupies about five sixths of the Emerald Isle, which is slightly larger than West Virginia, and is predominantly Roman Catholic. To the north are six counties, overwhelmingly Protestant, which are part of the United Kingdom and are known as Northern Ireland. Protestants and Catholics in Northern Ireland have been in mortal conflict for more than two decades, fighting and killing in the streets without much hope of a lasting peace. Terrorism reigns. The Catholics are intent upon cutting the Protestants' ties to the British crown and reuniting Northern Ireland with the rest of the republic.

A rim of mountains encircles a central plain whose grasses are made greener by the temperate maritime climate and rainfall averaging 80 inches a year. There are many lakes and rivers. Bog covers about 16 percent of the land. Commercially harvested peat, or turf, generates 21 percent of the electricity. Fourteen percent of the land is arable; the economy is mainly agricultural, with dairy produce and flax among the chief products. Enough food is produced for both domestic consumption and export. Ireland's fine linen, laces, china (Balleek), glassware (Waterford), and ale and spirits are world famous. In recent years, many foreigners have bought property here for a first or second home and for raising horses. There are not enough employment opportunities for the population of 3.5 million.

In the 1840s, migration began in desperation when the potato crop—the dietary staple—was blighted from both a fungus growth and the harshest and longest winter in memory. Migration continues but with less intensity and urgency. About 100 Irish leave the Old Sod every day, most of them heading for Great Britain, the United States, Canada, or Australia.

ITALY, which had 49 peaceful changes of government between the close of the Second World War and March 1991, has become the sixth-strongest economic power in the world, though the generally rugged and mountainous country doesn't have oil or coal resources. The commercial hub in the north prospers from booming industrialization centered in Milan, Turin, and Genoa. Automobiles, chemicals, machinery, textiles, and luxury goods are the leading industries. With its talent for design, Milan is recognized as the world capital of fashion, equal to, if not surpassing, Paris and New York. Industry has had an impact on the impoverished, hardscrabble, agrarian south. Fruit, wine, and olive oil are derived from the rocky but patiently worked earth there. Almost the entire population of 57.6 million is Roman Catholic. Probably no sovereign country has given the world more enduring art in the areas of painting, sculpture, architecture, opera, and cinema.

Italy is a 100-to-150-mile-wide boot-shaped peninsula, about the size of Florida and Georgia combined, extending 760 miles southward from the Alps into the Mediterranean Sea. Offshore appendages are Sicily (the largest island in the Mediterranean and separated from the toe of the boot by the narrow Strait of Messina) and Sardinia, 115 miles west of the mainland.

Italy entered the Second World War in 1940 as one of the Axis powers and surrendered to the Allies in September 1943. It has the largest nonruling Communist party in the world. Despite general prosperity, there is continuing government malfeasance and, currently, high unemployment. Only about half of all telephone calls get through, and there are periodic water shortages. There are occasional geodramas. Since the burial of 2,000 Pompeiians and 160 acres of Mount Vesuvius in A.D. 79, Italy has trembled in the shadows of active volcanoes. (In the summer of 1991, several more bodies were discovered in the ashes of Pompeii.) Europe's highest volcano, the cone-shaped "forge of Vulcan," Mount Etna (10,900 feet), has erupted all over Sicily 140 times. In 1908, an earthquake killed 83,000 Italians. In 1980, Naples was paralyzed and demoralized by a quake. In 12 months spanning 1983–1984, there were more than 4,000 jarring quakes near Naples.

LATVIA, whose patriots' proclamation of independence from Russia after the First World War held for 22 years, gained its freedom from the Soviet Union in September 1991, and, like two other Baltic states, Estonia to the north and Lithuania to the south, joined the United Nations. Russia had seized the capital city of Riga in 1710 and the rest of the country, which is the size of West Virginia, by 1721. Free from Russia since 1918, Latvia was invaded by Soviet troops in June 1940, ten months after the Hitler-Stalin Pact of August 1939, and was incorporated into the Soviet Union as a constituent republic, an annexation never recognized by the United States. Latvia was conquered by German troops in 1941, then reconquered by Soviet troops in 1944. Latvia is largely a fertile lowland with extensive dairying and stock raising. Resources include timber and peat. It is the most industrialized of the Baltic states, with its textile, chemical, and electronic industries. The Baltic Sea is to the west, Russia to the east. The population of 2,681,000 is largely Lutheran; Latgale is a Roman Catholic region. The earliest Latvians arrived in the present lands eleven centuries ago.

LIECHTENSTEIN is the most intensely industrialized area on the planet and one of its smallest countries—625 square miles—and it has one of the highest per capita incomes in the world: $26,296. The sole intact sovereign state of the Holy Roman Empire, it is an Alpine, hereditary, constitutional monarchy landlocked at the crossroads of central Europe—tucked between Switzerland and the Rhine River on the west and Austria on the east. Liechtenstein was not invaded by Hitler. Taxes are rock bottom, and many foreign businesses are headquartered here. The country also leads the world in per-capita exports. One company, for example, makes tens of millions of artificial teeth. Postage stamps provide a tenth of the government's revenue. There has been a long open-border economic alliance with Switzerland, which administers the postal service. The Swiss franc is the official currency.

Sixty percent of the land is mountainous, and there are no natural resources other than beauty. The population of 28,000 is 95 percent Alemannic and 5 percent Italian and is concentrated on the fertile plains near the Rhine River. The minuscule country has a large contingent of foreign labor and a literacy rate of 100 percent. It has been independent since 1806 and the breakup of the Holy Roman Empire. The official language is German. Universal adult suffrage has existed since 1986. Liechtenstein was named for the family that bought the land about 1700.

LITHUANIA, the southern Baltic state, is, like neighboring Latvia, about the size of West Virginia. It borders the Baltic Sea for only about 15 miles. The land is flat, and the economy is agricultural. Lithuania merged with Poland in 1569 and was annexed by Russia when Poland was partitioned in the 1700s. It declared independence as the occupying German army collapsed near the end of the First World War. In the 1930s, the country had a booming agricultural trade with western Europe. In 1939, the Soviet Union occupied the capital city of Vilnius, which had been annexed by Poland in 1923 with the approval of the League of Nations, and returned it to Lithuania. But a year later, Stalin's troops took possession of the whole country, an absorption as a republic never recognized by the United States. Hitler's military occupied the country during the period 1941– 1944 and exterminated almost all of the considerable Jewish minority. In Snieckus, in the 1970s, the Soviet Union built a nuclear power plant to the same specifications as the one in Chernobyl that all but melted down in the mid-1980s. Most of the 3,690,000 Lithuanians have very ancient roots, and they are largely Roman Catholic.

LUXEMBOURG is a banking haven, a prosperous constitutional monarchy landlocked in western Europe. Smaller than Rhode Island, the 999-square-mile picturesque grand duchy snuggles in a crevice amidst France, Belgium, and Germany. The Ardennes Mountains are in the north, an open plateau in the south. The capital city, also called Luxembourg, has been dubbed the "Gibraltar of the North." Caesar's armies marched through the country, which in this century was conquered by Germany in both world wars. The 380,000 Luxembourgers, who are 81 percent urban, are a mixture of Celtic, French, and German. Ninety-seven percent are Roman Catholic, and literacy is 100 percent.

MALTA, principally the three inhabited, hilly, indented mid-Mediterranean limestone islands of an archipelago, was the most bombed site in the Second World War, because the British crown colony was strategically situated between southern Europe (Malta is 58 miles south of Sicily) and northern Africa (180 miles to the south of Malta). There were around 1,200 Axis air raids on the Allied naval and air bases there. The instinct for survival overcame the geography. U.S. President Franklin D. Roosevelt called Malta "one tiny bright flame in the darkness." The population was decorated with the George Cross for heroism and devotion by Britain's King George VI. For centuries the crossroads of war, Malta had been annexed by the British in 1814 and gained full independence exactly a century and a half later. Its first-class harbors became a haven for NATO warships.

Malta now maintains a policy of steadfast neutrality and prohibits foreign military bases. (In 1986, it tipped off Libya that United States bombers were heading toward North Africa.) The democratic republic, wrapped in its cloak of history, is one of the world's most densely populated places, with 371,000 people shoehorned into an area merely twice the size of Washington, D. C. There are no mountains or rivers. There is virtually no rain in the breath-snapping heat of the summer. Most food must be imported. Only a third of this tiny rock of history and romance is arable. Mediterranean salt water is made potable. On a population percentage basis, Malta is nearly as Catholic as Vatican City.

MONACO, a constitutional monarchy on the southeast Mediterranean coast of France, is the world's second-smallest independent state: its 0.6 square miles are about the size of New York City's Central Park. With a population of 29,000, Monaco has the highest density of people of any country in the world. Tax benefits and the Monte Carlo gambling casino are principal attractions. Founded as a colony of Genoa in 1215, Monaco has been ruled since the 14th century by the Grimaldi family.

NETHERLANDS, a quarter of which lies below the level of the North Sea, has the image of being a tiny country, though it is slightly larger than Connecticut, Massachusetts, and Rhode Island combined.

It has one of the most densely settled populations in the world, but the standard of living remains high. The per capita income of the 14.8 million Dutch averages over $13,000, and the literacy rate is 99 percent. Women have a life expectancy of 80 years; men 73 years. About half of the northwest European kingdom has been reclaimed from the sea through a rampart of dikes, dams, and dunes stretching 1,200 miles. Were it not for the barriers, half of the Netherlands would be flooded twice a day.

The first polders, or reclaimed lands, date from about A.D. 100. In 1940, the Dutch military breached some dikes to try to stem Hitler's blitzkrieg. The Netherlands was a colonial power, the seventeenth century its golden age. Through the Dutch East India Company and the Dutch West India Company, it ruled a vast empire and had a stranglehold on most of the world's trading. This also was the era of exemplary Dutch painters (e.g., Rembrandt and Vermeer). In 1953, a North Sea storm devastated much of southern Netherlands.

Today's prosperity is nurtured by flowers, electronics, oil, and agricultural surpluses. Rotterdam, which was demolished by Nazi bombs during the Second World War, and its Eurooport facility handle more cargo than any other seaport in the world. Refugees, particularly from former colonies, have been drawn by the country's religious tolerance. The constitutional capital of Amsterdam, where a quarter of the population is on welfare or social security, has been called the Venice of the North because of its canal system. (The country has 3,478 miles of canals.) The seat of government is The Hague, where the World Court also sits.

NORWAY is justly famed as the land of the midnight sun. The northernmost country in Europe, it occupies the western part of the Scandinavian peninsula and borders on Sweden to the east and on Finland and Russia to the Arctic North. Its Atlantic coastline is 1,500 miles long. The climate is mild because of the North Atlantic Drift. Norway's position at North Cape, off the northern coast, and the northernmost point of Europe, is strategically vital but highly vulnerable: Barents Sea, between the cape and the Svalbard Islands, is the only accessible passage to open waters for Russian ships.

Proximity to the Atlantic and to several seas has spurred an economy based on shipping (Norway has the world's third-largest merchant fleet) and fishing. Only three percent of Norway is under cultivation. The major natural resources are oil, which has improved the country's economic position considerably, and timber (forests cover a quarter of the land). The standard of living is high (per-capita income is $13,790), and there is 100 percent literacy. Norway, like the other Scandinavian countries, enjoys progressive social legislation. The population of 4.2 million is homogeneous and 94 percent Lutheran. The hereditary constitutional monarchy is on a friendly footing with all states. Democratic, beloved King Olaf died in 1990 and was succeeded by his son, Harold. German armies occupied the country during the Second World War. Norway prohibits foreign military bases and nuclear arms on its soil. Life expectancy for women is 79.5 years; for men 72.7.

POLAND is striving to create a free-market economy as it discards and transforms the stultifying, centrally planned economy that was imposed by Moscow at the end of the Second World War. Soviet troops, an occupying force since the Red Army liberated the country from Nazi control during the last months of the war, were to be stationed here into 1993. Awakened expectations have slammed into the grim legacy of the 45-year rule by the monolithic Communist Party. A spiraling inflation that reached several thousand percent annually led to near-destruction of the currency, a thriving black market, and the smuggling of consumer goods. Poles could no longer afford to own or drive an automobile. Citizens must still wait more than a decade to get a home phone. The foreign debt has been prodigious.

Stalin once remarked that Communism fit Poland as well as a saddle fits a cow. The country experienced centuries of struggle with dismemberment and domination. In 1939, the first shots of the Second World War were fired at Poles, and the blitzkrieged state was divided less than a month later by the invading German and Soviet armies. The largest state wholly in eastern Europe, and the sixth largest on the continent, it has been at the crossroads of change since 1980. Poland was the first Eastern-bloc country to form a non-Communist, multiparty parliament. The presidential election of the charismatic Solidarity trade-union chief Lech Walesa marked Poland's first democratic election in more than six decades. (Under martial law, Walesa had been imprisoned for nearly a year.)

Poland is on the Baltic Sea and shares a 540-mile frontier with Russia. It is a wheat farmer, a coal producer. There are reserves of 35 million metric tons of coal. The soil is generally poor and the growing season short. The northern and central regions are mostly lowland. The Carpathian Mountains in the south rise to 8,200 feet. Roman Catholicism is a unifying force for the homogeneous population of 38.4 million; 95 percent of the population is Roman Catholic. Poland celebrated its millennium as a Christian state in 1966, and the 200th anniversary of Europe's first written constitution, reforming the gentry-ruled country, in 1991. Poland has always been a despotic state, lacking even a modicum of civil rights or tolerance toward minorities. Endemic anti-Semitism continues. Three million Jews lived in Poland before the Second World War; today, only about 5,000 live there. About 16 percent of all Poles died during the war.

PORTUGAL, the poorest country to join the European Community, has been an independent state for nine millennia. With a 500-mile-long northwest-west-south stretch on the Atlantic Ocean, it was once the world's foremost commercial state. Henry the Navigator in the fifteenth century laid the foundation of the empire, sending explorers in all directions. Portugal dominated the East Indian and African slave trade and established colonies in Africa, South America, and Brazil. Neglect of agricultural opportunities and depopulation through colonization and expulsion of its Jews led to a rapid decline in power. Today, Portugal has a modest economy. It administers Macao (on Chinese territory in the Far East) until 1999 and commands two autonomous regions, the Azore Islands, 740 miles west of Portugal, and the Madeira Islands, 350 miles off the northwest coast of Africa. Including the volcanic Cape Verde Islands, once an overseas province, the islands constitute unparalleled defensive outposts. (In antiquity, the Azores were thought to be the western edge of the world.)

Somewhat smaller than Indiana, Portugal shares the Iberian Peninsula with Spain at the southwest section of Europe. The Tagus River bisects the country northeast/southwest. North of the Tagus are mountains, and the region is cool and rainy. In the south, it is drier, and there are rolling plains and a warmer climate. Thirty-nine percent of the land is arable. Portugal is a leader in cork production. All but 3 percent of the population of 10.5 million is Roman Catholic.

One of the main exports is wine. Portugal established itself as a republic in 1910 and remained neutral during the Second World War. Well over 2 million Portuguese reside as permanent emigrants or temporary workers elsewhere in the hemisphere.

ROMANIA, one of the six former Soviet-bloc countries, is poor, barely emerging from the cruelly repressive regime that ended in 1989 with the jettisoning and murder of dictator Nicolae Ceauşescu. The Ceauşescu government had banned all intellectual pursuits, from cybernetics to Scrabble. The revolt began in Romania's second-largest city, Timisoara, near the Hungarian border in the west. Two million Hungarians live in Romania, and they often come to savage blows with Romanians, as have the Gypsies and the long-settled German minority. To bolster the population, abortions were virtually forbidden and birth control banned.

Industry, once a prominent maker of chemicals, petroleum products, steel, and textiles, is being reconstituted, a market economy established, and a framework defending the rights of citizens established. Another principal task: purifying the largely undrinkable water supply. This southeast European state on the Black Sea gained independence from Turkey in 1878, after the Russo-Turkish War. It is almost as large as New York and Pennsylvania combined and 98 percent literate. The population of 23 million—50 percent of it urban—is a mix of Latin, Thracian, Slavic, and Celtic peoples. The country has no foreign debt, because President Ceauşescu's austerity crusade deprived the population of decent food, heat, and electric light in order to export enough goods to pay off loans. Even during a food shortage, food was exported. One of the most horrifying revelations in the "opening up" of Romania was the number of children perishing in the wards of hospitals from neglect, hunger, and disease (often AIDS). And there is a generation of abandoned children.

SAN MARINO is the oldest state in Europe and the world's smallest republic (24 square miles). Dating back to the 1300s, the Most Serene Republic sits on the slopes of Mount Titano in the rugged Apennines in northeastern Italy. San Marino and Italy have had such close ties since 1862 that there are no customs restrictions between them, and

Italy helps the enclave meet its annual budget. Tourism and the sale of postage stamps boost the economy. The population of 23,000 is 90.5 percent urban and 96 percent literate. The government was communistic in the period 1945–1957. San Marino joined the United Nations in 1992.

SPAIN was legally neutral during the Second World War, but because of its pro-Axis sympathies it was not accepted for membership in the United Nations until 1955. The southwest European constitutional monarchy, in one of the oldest inhabited regions on the continent, was ruled by a dictator for 35 years during the middle years of this century. A half million Spaniards died in the 1930s civil war, as did thousands of foreigners who came to Spain to take up arms with the Loyalists. (The bloody war has been memorialized in Ernest Hemingway's novel *For Whom the Bell Tolls* and in Picasso's *Guernica*.) The insurgents won the war in 1939, and General Francisco Franco set up an oppressive Fascist regime.

Three quarters arid, and mainly a high plateau, Spain occupies most of the Iberian Peninsula and is about the size of Arizona and Utah combined. The country also possesses mountain ranges and river valleys. There are few good harbors, and no large lake. Madrid is the highest (2,133 feet) capital city in Europe, and at 11,407 feet Mulhacén is the continent's highest peak outside the Alps, which cross Spain, and the Caucasus. Point Tarifa is the southernmost spot in Europe.

Industry employs over 30 percent of the population of 40 million, 90 percent of whom are Roman Catholic. The state was once ethnically homogeneous; it is becoming a multiracial society, with as many as a half million illegal immigrants from the Third World; the land border with Portugal is largely unguarded. The economy includes uranium, olive oil, citrus fruits, shoes, mercury, wines, and automobiles. After the long somnolence induced by the Franco dictatorship, Spain awoke with a vengeance and is marching toward equal status with the world's most progressive powers. Militarily, it wants to assume a strategic role from the Canary Islands to the western Mediterranean. Basque extremists continue their campaign for independence, and Madrid continues its campaign for return of Gibraltar from Great Britain. The Spanish colonial empire once reached around the globe.

SWEDEN last fought a war in 1813–1814, against Napoleon; it was rewarded at the Congress of Vienna with union with Norway, which lasted until 1905. Sweden has remained neutral for more than 175 years. Though traditionally nonaligned, it has sought membership in the European Community. The largest of the Scandinavian countries (and larger than California), this watery northern European kingdom is on the Baltic Sea, with Norway its western neighbor and Finland in the far northeast. Wheat growing and dairying prosper on the limited land under cultivation. Farms are part of an agricultural-cooperative movement. The major resources are timber (more than half the country is forested), high-grade ore, and hydroelectric power.

The last century saw industrial progress, liberalization of government, and massive emigration (mainly to the United States). This century also brought the rise of the Social Democrat party and the passage of social legislation transforming Sweden into a welfare state. Taxes perforce are high. Daily absenteeism from work has reached 14 percent. The 8.5 million Swedes have a per-capita income of around $12,000. Life expectancy for women is 79 years; for men, 73 years. The literacy rate approaches 100 percent. Greeks, Turks, and Yugoslavs have found a home here as laborers. The capital city of Stockholm is an island-city of charm and sophistication. The generosity and egalitarianism of the benign constitutional monarchy may be robbing the citizens of their drive and competitiveness.

SWITZERLAND, with a 700-year tradition of independence and isolationism, is a financial center of supreme importance. It once had the highest national income among industrialized countries. It boasts the world's highest per capita GNP—more than $26,000. The mountains are sites for soaring eyes. Some of the most privileged and pampered people in the world call Switzerland home, or a home away from home. Attractions include the scenery and the skiing, the clean air and the clean cities, the peace and the quiet, the relative absence of street crime, and—for the well-heeled and those with a passion for financial secrecy—hospitality and hoteliers nonpareil. The country is also distinguished by a commitment to neutrality, surpassing chocolates and cheeses, and peerless skill in watch making, which accounts in terms of dollar value for 50 percent of all worldwide watch production.

A landlocked central European country on a central plain between the mountain ranges of the Alps and the Juras, it is not large—about the size of Rhode Island, Massachusetts, and Connecticut combined—a 23-canton federation. Revolutionary outbreaks led to occupation by France in 1798. In 1815, the former regime was restored, and perpetual neutrality was guaranteed. Switzerland is not a member of the United Nations, and it has expressed reservations about a scheme to create a free-trade area stretching from the Arctic to the Mediterranean. In an earlier time, the country beckoned as a sanctuary for tubercular patients, as dramatized by the dying Hans Castorp in Thomas Mann's *The Magic Mountain*. There are four official languages: Romansch, Italian, French, and German. Sixty-five percent of the population of 6.6 million is of Germanic origin. Women have a life expectancy of 81.3 years; men 70.3 years.

VATICAN CITY, the headquarters of the Roman Catholic Church and the residence of the Pope, has been since 1927 an independent state, the smallest in the world. Its 108.7 acres are surrounded by Rome, the Italian capital. The enclave includes St. Peter's Church (the world's largest Christian church) and the Vatican Library, founded in the 1400s (the oldest public library in the world). The permanent population of 755 is 100 percent literate. Suffrage is restricted to Roman Catholic cardinals less than 80 years of age.

YUGOSLAVIA (as of December 19, 1991), in one of whose six republics, Serbia, an assassination triggered the First World War, sits largely on the deeply indented 1,741-mile-long Adriatic coast of the Balkan Peninsula and has been inhabited for at least 100,000 years. (More than 700 islands are also part of Yugoslavia.) The Kingdom of the Serbs, Croats, and Slovenes was renamed Yugoslavia 60 years ago. In recent times, the country was a feudal state with an eight-man presidency (which split into two four-member groups) and 20 major ethnic groups. The north and the west are Catholic, the south and the east are Muslim and Eastern Orthodox. Thirty-seven percent of the 23.8 million population are Serbs, 20 percent are Croatians.

During the Second World War, the pro-Nazi Independent State of Croatia, known as the land of the south Slavs, murdered about a million Serbs, who were staunch allies of the Western powers. About the size of Wyoming, the mountainous (70 percent) southeast European country became a republic in the wake of the war; the Communist Party gained power in an armed uprising against German and other occupiers. The Croatian guerrilla leader Marshal Tito eventually broke with Stalin. "The Great Unifier" became a pioneer of economic reform, shaping a socialist-market-oriented economy.

Each of the six republics and two autonomous regions has expressed, usually quite belligerently, separatist preferences. Nonintegration of the ethnic groups exacerbated the chaos. After a decade of rising tensions, Yugoslavia in the 1990s began to disintegrate, sliding to the brink of dissolution and the creation of a new, smaller Serbian-dominated country. Inflation surged and labor productivity and real personal income plummeted. Violence and threats to order and safety swept the country as it searched for new forms of coexistence.

"Yugoslavia" could be erased from the map of the world. The European Community planned to recognize the separatist republics of Croatia and Slovenia as independent countries in mid-January 1992. Serbia, under Yugoslav Army protection, was the source of much of 1991's violence, especially the savaging of Croatian cities.

GIBRALTAR (a non-state) is a minuscule (2¼ square miles) British colony perched on a 2.5-mile-long mountainous peninsula on Spain's southern coast, in southwesternmost Europe. Naval and air stations are based at the Atlantic entrance to the Mediterranean Sea. The Rock of Gibraltar, a limestone-and-shale ridge (one of antiquity's pillars of Hercules), on the eastern end of the Strait of Gibraltar, rises to a height of 1,398 feet. A Jurassic limestone fortress, its caves have yielded important archeological finds. Gibraltar is home to 30,000 British subjects. The Strait of Gibraltar is a 36-mile-long waterway (23 miles at its widest, 8 miles at its narrowest) connecting the Atlantic and the Mediterranean, between Africa and Spain. Gibraltar Bay, an inlet of the Strait, in the southern extremity of Spain, has an enclosed harbor. It has been a possession of the British since its capture in 1704. In 1967, residents voted to remain under British control. Repeated efforts by the Spanish to claim Gibraltar have met with defeat; the dispute is the major interstate problem in Europe. Gibraltar is a free port, and the strait is an international passage.

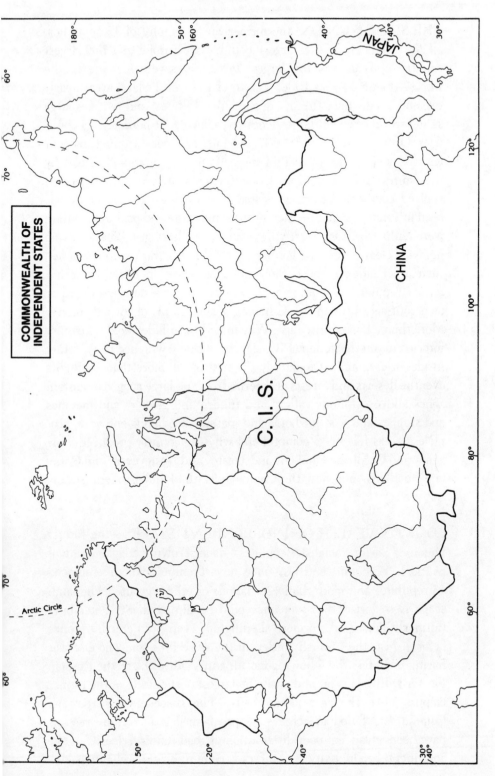

THE SOVIET UNION: the former Soviet empire of 15 republics, all but landlocked, the largest country in Europe *and* the largest country in Asia was the largest territorial sovereign state in the world—11 time zones wide, a sixth of the world's land area, nearly two and a half times the size of the United States, and twice as large as the world's second-largest country, Canada. It bordered 15 countries. Only five percent of the 29,140-mile-long coast fronted on open, ice-free ocean. The turmoil of the early 1990s, prompted in part by the painful transition from Communism to a free-market, decentralized economic union, persisted. For decades, the economy has been mismanaged and inefficient; poor planning and poor distribution were hallmarks. One terrifying result: food shortages. There was no highway clear across the country. The three Baltic republics gained their independence and became sovereign states in 1991. The other former republics and autonomous ethnic units are determining their own political status in the reshaping of the destiny of the old empire. More than a half century ago, Winston Churchill declared, "I cannot forecast to you the action of Russia. It is a riddle wrapped in a mystery inside an enigma." It took only the first of my more than 30 flights over the deserts and forests and permafrost and large rugged mountain zones and precipitous valleys and tundra and glaciers and marshes and steppes and vast lowlands and massive floodlands and vast, peopleless areas for me to comprehend why the country produced Turgenev and Gogol and Chekhov and Tolstoy and Dostoyevsky and Gorky and Pushkin—and now the Commonwealth of Independent States.

COMMONWEALTH OF INDEPENDENT STATES—the former Union of Soviet Socialist Republics—is an evolving conglomerate of political, economic entities, once a vast empire which embraced 15 republics and more than 126 nations or ethnic groups. The landscape is so broad that someone could be dining in St. Petersburg (formerly Petrograd, formerly Leningrad, even St. Petersburg once before), near the Finnish border, at the same time someone else was having breakfast the following morning in Vladivostok in the Russian Far East. Russia is far and away the largest of the states of the former empire, once the seat of authority. The space-age superpower's Communism failed to achieve its proclaimed goals. Many reasons have been advanced: poor distribution of goods and services . . . low and inept productivity . . . inferior communication channels . . .

government-directed terrorism . . . the suppression of human rights
. . . cultural differences . . . the bureaucracy's lack of incentives
and the profligate ways of the entrenched plutocracy. Reform measures assuring freedoms were promised by a "new deal" in the mid-
1980s. The leadership of President Mikhail S. Gorbachev tilted
toward enlightenment: it led to disintegration of the empire, widespread maligning of Gorbachev, his resignation, the unpredicted
dissolution of the Soviet Communist Party, and the ascendancy of
both Russia's president, Boris N. Yeltsin, and the giant former Russian republic itself, the dark, rich locus of Pushkin's "deeds of days
long vanished/traditions of deep antiquity."

Reform was slow because of resistance to glasnost (openness)
and perestroika (restructuring). Reformers and reactionaries engaged
in ferocious power struggles in the face of pressing demands, such
as food for the table and the modernization of antiquated industry.
Prior to the putsch in mid-August 1991, republic after republic and
nations within republics declared their ethnicity, their nationalism,
and pressed for independence and a loose confederation. Emigration,
particularly among Jews, rose to high tide. The three Baltic republics,
which had been illegally incorporated into the union in 1940, openly
challenged Soviet authority for years, finally gaining independence,
and then membership in the United Nations in September 1991.
Ethnic violence and border clashes in the Caucasus and Central Asia
are rooted in poverty, long-repressed government-induced inequities,
and enmity. Ukraine, after Russia, is the most organized and most
economically mature and westernized of the regions, and it is larger
in area than France: the "breadbasket" is the foremost European
producer of grains, cattle, dairy products, and sugar, and its industrial
engine leads the continent in the unearthing of coal, iron ore, manganese, titanium, and natural gas. Ukraine, which has a virulent
record of anti-Semitism in its history of ethnic hatred, bore the brunt
of Hitler's invasion in 1941: 7.5 million Ukrainians were fatalities.

Lenin (lawyer-turned-revolutionary) converted the tsarist regime
to communism after the 1917 revolution and subsequent civil war.
During Stalin's reign, from the late 1920s to the early 1950s, terrorism was perfected. The dictator—"You can't make a revolution
with silk gloves," Stalin preached—created an artificial famine in
which 5 million perished. An estimated 20 million to 50 million more
died in purges and the gulag and mass dislocation. Still another 27
million died in the Second World War, which Moscow calls the

Great Patriotic War; around 600,000 died in the 900-day Nazi siege of Leningrad (St. Petersburg), Peter the Great's "window on the West" (1703) on the swampy delta of the Neva River. Women now outnumber men to a greater extent than in any other state.

Weather ranges from tropical to Arctic. Russia is richly endowed with minerals, but gold reserves to help meet foreign payments seem to be below long-held Western estimates. It is the world's leading oil producer and has the most known gas reserves and the largest forests. Oil- and gas-drilling rigs explore more than a mile below the tundra. Europe's longest river is the 2,300-mile-long Volga, the watery highway through the heartland, linking the Baltic and the Caspian Seas; via the Don River, it connects with the Black and the Mediterranean seas. A network of canals connects five seas. Plans were drawn to redirect three major down-north rivers to the parched Asian south, but the enterprise was rejected by the new, democratic organs in the late 1980s as too costly, monetarily and ecologically. The weather-bruised, 1,640-mile-long Ural Mountains, east of Moscow, with no peak now higher than 6,000 feet, break the Great Northern European Plain, which stretches eastward from western France. The northern coastline runs about 3,000 miles along the Arctic Ocean. The ice-free port of Murmansk, northwest of Moscow, is the world's largest city north of the Arctic Circle. About 70 percent of the population of 285 million lives in the European region. There are 50 million Muslims. About 600,000 people are in the space program, which launched Sputnik in October 1957, and then the first man in space. The military has the largest stockpile of nuclear weapons (27,000) on the planet. A nuclear-weapons production complex nearly 1,000 miles east of Moscow is the most polluted spot on Earth. The Caspian Sea, which is the world's largest inland body of water, is about 92 feet below sea level; it has no outlet.

JAPAN consists principally of four volcanic outcroppings separating the northwest Pacific Ocean and the Sea of Japan. This xenophobic floating kingdom is smaller in area than California. South of the main islands is the planet's oldest (170 million years) ocean floor, beneath 18,652 feet of water. Japan has few natural resources, only a couple of drops of oil, and it has almost mined itself out of coal. It cannot feed itself. Yet it is a world superpower, interacting spatially over the whole globe with energy, skill, and ideas. This fascinating geo-

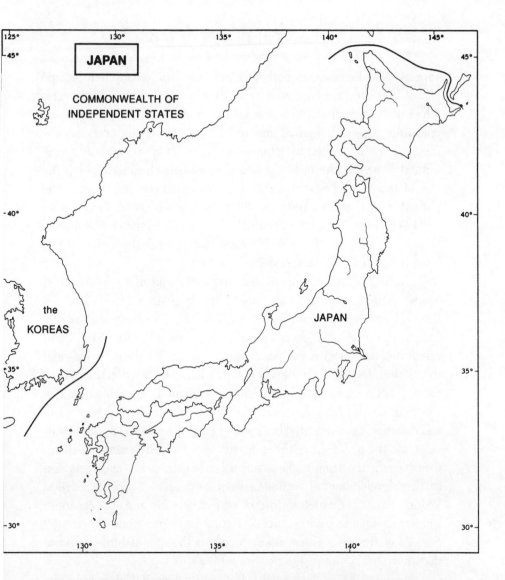

COMMONWEALTH OF
INDEPENDENT STATES

JAPAN

the
KOREAS

JAPAN

graphic entity could be mapped by showing raw materials pouring into the islands and high-tech products flying out. A map of Japanese ownership abroad of real estate, factories, and the like would be an awesome representation of the country's economic power. Long-range planning and fiscal prudence are the baselines of the economy, not quick profits and placated stockholders. Japan, Inc., has relatively little outlay for military hardware.

JAPAN: catapulted out of isolation in 1854 by menacing American warships, and institutionally democratized by American occupation forces after the Second World War, it has become an economic superpower. Japanese of both genders now live longer than anyone else. Life expectancy for women is 81.8 years; for men 75.8 years. The infant mortality rate is the lowest in the world. Japan is at the epicenter of world finance and technology, achieving economically much of what it failed to achieve militarily. It is second only to the United States as the largest garbage producer. (Japan is 1/24th the size of the United States and has a population one half that of the United States. But by the year 2000 the capital city of Tokyo alone will have a population larger than all of Australia.) Japan is the world's largest creditor country and largest foreign-aid dispenser. (The United States is the largest debtor country.)

There are myriad Japanese millionaires and many billionaires. Many of the corporations are among the largest on earth. Tokyo is the most expensive city in the world, but Tokyo workers are the best paid in the world. An accelerating investment in United States real estate has unnerved many Americans. (Great Britain, Canada, and the Netherlands are actually even more heavily invested in the United States.) The combination of a highly educated, disciplined, hard-working, docile, homogeneous, highly functional, distinctive society and dynamic, goal-oriented business organization has been the touch-stone for the astonishing three-decade success. The economic structure has shifted from exports toward a home-market emphasis and further development of technologies and products. Since the Second World War, after which a third of all industry lay in ruins, Japanese women have gone from a state of virtual unemployment to being 50 percent of the work force, a key factor in the diminishing birthrate that may shrink the population total by a third.

Japan's 3,000 rugged, relatively recently formed islands and islets in the northwest Pacific sit on the Ring of Fire along the east coast of mainland Asia. Much of the land is covered with volcanic ash and lava. There are 19 active volcanoes that are considered dangerous and any number of geologic faults that threaten calamitous earth-quakes. The archipelago trembles an average of four times a day. Japan has become the world leader in mud-flow control. Geologists believe that the basin that forms the Inland Sea was created by fault-ing, which split a single land mass into three of the four major islands.

The state has a density of about 844 people per square mile, because the population of 123.5 million lives on 2 percent of the land. Every arable acre is farmed intensively. Japan's rice farmers are the most government-protected farmers in the world. Building heights are limited by the risk of calamitous quakes. Land speculation is a social ill, because the country prizes harmony. In the 1990s, about $3 trillion will be invested in public works to spruce up the state. The sea is the central artery of culture and commerce. It is harvested for 10 million tons of fish annually. Natural resources are negligible, so there is heavy dependence on outside sources. The islands north of Japan—the Northern Territories—were seized by the Soviet Union in the last week of the Second World War and have been a source of severe tension between Tokyo and Moscow. Japan is the number-one importer of oil. Tankers 300 miles apart form a perpetual train bringing energy to the country, which also imports 77 percent of its coal requirements. Shintoism, "the way of the gods," is the oldest surviving religion.

Less than 150 years ago, Japan was estranged by choice. United States Commodore Matthew Perry changed all that in the 1850s. His black ships and his cannon and Western economic canons introduced the shoguns to the nineteenth century. The doors opened, fresh air roared in. In March 1945, a United States firebombing raid killed nearly 100,000 Japanese. In August 1945, the "first nuclear war" killed another 100,000. As part of the United States–Asian defense shield, about 60,000 American troops are stationed in Japan, which has relatively little military power, in accordance with the agreement ending the Second World War. Sending minesweepers to the Persian Gulf after the war there in 1991 was the first overseas mission for Japanese armed forces since 1945. The Japanese–United States relationship is the single most important bilateral relationship in the world. The American commitment to defend East Asia was the *sine qua non* of Japanese independence and economic assertion.

CHINA: around 20 percent of the world's population lives in the People's Republic of China, on the eastern flank of Eurasia. In a land of enduring traditions and huge cities, the country remains primarily rural; urbanization and industrialization are still far down the road. Hundreds of millions of peasants tend their crops quite as peasants have done for centuries. China may be the oldest civilization in the world. It maintains an isolation that makes it relatively impervious to outside influences. Beijing has a nuclear arsenal. **KOREA:** The peninsula jutting out of the strategically vital East Asian mainland has had a chaotic history. The two Koreas, divided since

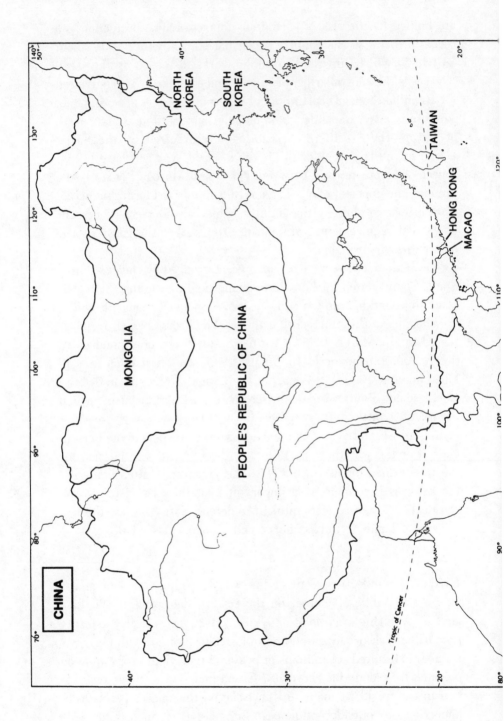

CHINA

MONGOLIA

PEOPLE'S REPUBLIC OF CHINA

NORTH KOREA

SOUTH KOREA

TAIWAN

HONG KONG

MACAO

Tropic of Cancer

the Second World War, are feeling each other out about reuniting, pooling their resources and strengths, and becoming a joint super-power. In December 1991, they forswore any armed force against each other and agreed to end formally their 1950s war against each other. Both had become members of the United Nations in September 1991.

MONGOLIA, a big-sky country, an enormous pasture looking to industrialize and shift to a market economy, is a central Asian wedge between the Commonwealth of Independent States on the north and the People's Republic of China on the south. Relatively isolated, Mongolia is almost as large as Arizona, Texas, Louisiana, and New Mexico combined. It is the least densely populated country on the planet: only four people per square mile. An alliance with the Soviet Union went back to the 1920s. In 1966, Mongolia signed a mutual assistance pact with Moscow. But socialism has been abandoned.

Less than 1 percent of the expansive terrain—prairies, ranges, plains, peaks—is fit for cultivation. Here are the half million square miles of the Gobi Desert, a broad depression, part sand, part steppe. (Gobi is a Mongolian word meaning gravel and rock debris, and it denotes all the deserts and semideserts of the vast, windswept Inner Mongolian plateau, which stretches across North China.) The country's lowest geological point is 1,814 feet above sea level. The continental climate is rigorous, the precipitation low. About half of the 2.1 million Mongolians are nomadic herders, more at home in traditional movable felt tents, or gers, on the grasslands and steppes than in conventional housing. The horse is the mode of transportation. There are a dozen times as many livestock as people. Chief exports are wool, cloth, and hides. There are substantial mineral deposits.

About 3.5 million ethnic Mongolians live across the border in China, and another half million live in what was Soviet Asia. At one time, 40 percent of the male population entered the celibate life of Lamaist monasteries. In the 13th century, Genghis Khan led the Mongols in the creation of an empire extending from the Pacific Ocean to Hungary. After decades of being ignored by Communist historians, Khan has resurfaced as a venerated Mongolian hero. In

1994, the country resumes using its old script as the official written language, replacing the Cyrillic alphabet in use since the 1940s.

NORTH KOREA, the self-isolated Democratic People's Republic of Korea, mines natural uranium and may have nuclear devices by the mid-1990s, which has been heightening concern in Seoul, Tokyo, Beijing, and Washington. The population of 22.5 million is homogeneous and totally obedient. Kim Il-sung, the "Great Leader," has been president since 1948. Twelve camps hold more than 100,000 political prisoners. North Korea suffered 2.5 million casualites in the three-year civil-war stalemate with South Korea ("the coldest war") in the early 1950s.

Slightly smaller than Mississippi, the northern half of East Asia's 400-mile-long Korean peninsula, which is the length of Florida, is mountainous and forested. The country is heavily industrialized, but the economy is moribund. There are hydroelectric resources and abundant minerals, such as lead, tungsten, magnesite, graphite, iron, and copper.

Self-reliance is the Communist dictatorship's guiding principle. North Korea and South Korea hint they are committed to reunification, and halting progress is being made. Rapprochement could generate an immensely powerful rival for Japan.

PEOPLE'S REPUBLIC OF CHINA, which is slightly larger in area than the continental United States, is the world's longest-surviving civilization and has the largest population of any country in the world: 1.13 billion, 93 percent of whom are Han Chinese. Fifty-five national minorities make up the rest of the population. The itinerant population of upwards of 50 million poses serious problems for the government, and there are over 100 million farmers in the rural work force who are underemployed or unemployed. For 17 centuries, the Chinese culture was the most efficient in applying knowledge of nature to useful purposes. "Let China sleep," Napoleon warned. "When she awakes, the world will be sorry." A national birth-control program limits most couples to one child; social pressure for compliance is effective. China, which has an economy growing relatively rapidly, has the world's largest army: 3 million men are under arms;

another 5 million are in the reserves. It tested a hydrogen bomb in 1967 and is active in missile weaponry.

Nearly a quarter of the land is limestone rock. Loamy yellowish-brown loess covers a vast area. China's densely populated eastern half is walled off on the west by mountains, plateaus, and deserts and is one of the best-watered regions in the world. Rain-forest belts in the south bar easy passage into southeast Asia. Coal is the most abundant fuel. The Yangtse, which is 3,960 miles long, the world's third-longest river, has been designated the boundary between severe and moderate climes, so housing south of the river is constructed without regard for heating. The Yellow River, the cradle of Chinese civilization, has supported an agricultural society for more than 7,000 years; 11 percent of the country is being cultivated. Western China is rich in resources but has not caught up economically with the prosperous east. Famines and floods cause millions of deaths. In 1961, about 25 million Chinese starved to death. In 1991, floods forced relocation of 10 million. Earthquakes also are a periodic disaster: 2 million died in a 1927 quake, upwards of 800,000 died in a 1976 quake.

China has been ruled by Communists since 1949 and is officially atheistic. Fifty types of crime are punishable by death. After the Red Guards' cultural revolution, which began in 1966, there was a decade of eased rule and enlivened economic reform, which initiated free enterprise. China has long held prices artificially low to prevent social unrest. After "an unchecked flood of bourgeois liberalization," students and democratization were gunned down in Beijing's Tiananmen Square in June 1989, to uphold "the people's democratic leadership." Dissidents continue to urge the iron-fisted, bureaucratically inert government to generate a degree of federalism or face a further decline in central control and even economic disintegration. More than 40 million subsist below the poverty line of $42; per capita income is $350. The tourist industry was expanding before the student protesters were crushed in 1989. Like Cuba, China has made remarkable strides in public health, longevity, and infant mortality.

SOUTH KOREA: lurching toward democracy and away from government repression, it is billed as "the second Japan." Slightly larger than Indiana, this industrially supercharged republic wants to be as economically self-reliant as its offshore superpower neighbor, with which it has had testy relations since being brutally colonized and annexed by Nippon for the years 1910–1945. At the end of the Second World War, and after a millennium of shared history, Korea (once "the land of good manners") was divided by the victorious Allies into two occupation zones: the Soviet Union's, north of the 38th parallel; the Americans', south of the 38th parallel.

On June 25, 1950, North Korea invaded South Korea. The 1953 peace treaty has not been signed, and the two Koreas are officially still at war: between them on the strategically located northeast Asian peninsula is one of the world's most militarized borders. More than a million United States soldiers have served in South Korea since 1945; 40,000 have been on duty with a nuclear arsenal. In 1990, high-level diplomatic contacts between the Koreas were initiated at the very time that South Korea and the Soviet Union were courting each other in first-ever discussions. South Korea's concerns are cultural and economic cooperation; North Korea's are political and military situations.

South Korea's homogeneous population of 45 million is nearly twice North Korea's. Twenty percent of the people are Christian. Seoul, the capital city, has the largest Presbyterian and Methodist congregations in the world. Violent protest has a long history. There are frequent student demands for wider democracy and reunification. The economic boom was built on low wages. Winters are cold, summers hot and humid.

HONG KONG (a non-state) is an affluent, trade-driven, 413-square-mile British crown colony in the South China Sea consisting of land adjacent to China and an island off the southeast coast of China: the island itself, the Kowloon Peninsula, and the New Territories on the adjoining mainland. This hub of world commerce is living with anxiety. In 1997, China takes over Britain's last Asian outpost as a Special Autonomous Zone. Many of the 5.8 million Hong Kongers don't trust the Chinese and their promise of "one country, two systems"; the well-educated and the successful have plans to emigrate. Already, scores of pregnant women in their resourcefulness have traveled to North America to give birth; each baby automatically becomes a citizen of the United States or Canada, and the family gets an inside track on immigration.

Will Hong Kong be able to maintain the status quo—the social and economic system—and revel in its laissez-faire capitalism, and enjoy freedom and unrestricted pleasure? Such is the concentrated wealth of the colony—it's been described as "an absolute money machine"—that it has more Rolls-Royces per acre than any other place in the world. It also enjoys the highest per capita consumption of cognac. Prosperity boomed in the wake of the 1949 Communist takeover of the mainland. One million Chinese found a haven here. Hong Kong is a free port (the best deep-water port on the China coast) and the world's second-busiest port. It is Asia's principal tourist destination. Suffrage has not been available to the masses. There was no voting on the proposed takeover and on the raising of the Chinese flag, a white baukinua flower against a revolutionary red background.

MACAO (a non-state) is the oldest European settlement in the Far East: it was a Portuguese trading post more than four centuries ago. About 40 miles southwest of Hong Kong, it consists of a maritime enclave, a peninsula, and two small islands at the mouth of the Canton River on the southwest coast of the People's Republic of China. Macao is only six low-lying square miles, yet 432,000 people—almost everyone an ethnic Chinese—live here, all but 2 percent of them in an urban environment. The highest natural site is 571 feet.

Macao is Chinese territory under Portuguese administration, and Portuguese is still the official language, though most people speak Cantonese. By negotiated agreement, the way of life and the capitalist system will be maintained for the half century following turnover of Macao to Chinese sovereignty in 1999, two years after the United Kingdom returns Hong Kong to Chinese sovereignty. The region already relies on China for drinking water and for much of its food supply. Gambling casinos, massage parlors, and smuggling have given the free port—"Hong Kong's playroom"—its notoriety, attracting more than 4 million tourists annually.

TAIWAN (a non-state) has all but ended its 40-year state of war with the People's Republic of China, finally regarding the Chinese

Communists as a political entity that controls the mainland, and choosing to move toward closer ties with Beijing. Normalization of informal relations began in the late 1980s.

Taiwan, formerly called Formosa, is the world's 13th-largest trading nation. Its foreign-exchange reserves ($75 billion) are the world's largest, exceeding even those of Germany and Japan. Separated from China by the 100-mile-wide Strait of Formosa, the island is about the size of Switzerland or of Connecticut and New Hampshire combined. There are steep, craggy mountains in the east, intensely cultivated fertile farming lands in the west. Every year, there are four or five crops from the same land, and yields are among the highest in the world. Rice is the biggest crop, but tea, sugar cane, and pineapples are abundant.

After the Second World War, a defeated Japan returned Formosa to China, which had in 1895 ceded it to Japan. Communist China regarded Taiwan as a renegade province, and the Nationalists on Taiwan saw the People's Republic of China as being one country with two governments. In 1987, martial law was lifted after 38 years, as was the ban on contact with Beijing. Reunification is a probability.

The $3,000 per capita income of the 20.3 million Taiwanese is 10 times higher than that of the mainland Chinese. High economic growth rates are expected to continue. The huge middle class demands political participation. Taiwan was ousted from the United Nations in 1971 when the People's Republic of China was admitted as a member; this was the deal China had demanded.

THE SOUTH PACIFIC: the Pacific Ocean, which is 12 million square miles larger than all the land area in the world put together, is dotted with tens of thousands of islands that add little to the planet's total land mass. Most of the islands are volcanic or coral, and uninhabited. The three lengthy island chains are Melanesia, Micronesia, and Polynesia, which are in politico-geographical transition.

FIJI, after a military coup had toppled the 322-island state's democratically elected government in 1987, declared itself a republic and was expelled from the British Commonwealth. Forming of new electoral rolls and boundaries delayed the first postcoup election until

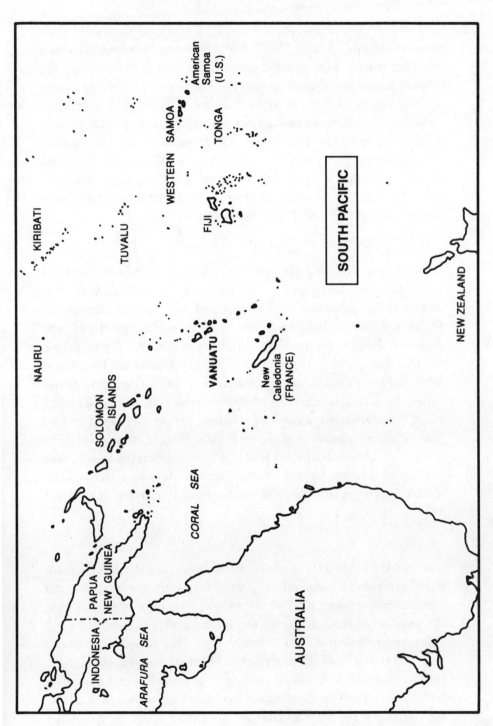

mid-1992. Most of the population of 758,000 are descended from laborers shipped in from British India between 1879 and 1916, and most are Hindu. Tensions and conflicts between the Hindus and the Fijian Kamaks are of a serious nature. The topography, whose square mileage equals the area of Massachusetts, varies considerably: from mountains to valleys to rain forests to fertile plains. Twelve percent of the land is arable. One island, Viti Levu, has more than half of the islands' total land area. Fijians by law own 83 percent of the land in the communal villages. The principal crops are sugar cane, bananas, and ginger. It was on Fiji, in 1979, that the crested iguana, one of the rarest reptiles, was discovered.

KIRIBATI is a tropical Micronesian republic of 33 islands and coral reefs whose combined area is less than that of New York City. It is located at the bisection of the International Date Line and the equator in the mid-Pacific. Indigenous are copra, breadfruit, pandanus, and papaw. Rainfall is erratic and the country is poor; per capita income is $460. The population of 69,000 relies on economic aid from Great Britain, New Zealand, and Australia. In 1985, Kiribati gave Soviet tuna-fishing vessels access to its 200-nautical-mile zone, tuna being the major exploitable resource. The island's previously exploited natural resource (phosphate rock, on Banaba) was exhausted in 1979. Tarawa, a chain of Kiribati islets in a reef around a lagoon, was the site of a major United States Marine victory in a bloody four-day battle with occupying Japanese troops during the Second World War.

MARSHALL ISLANDS, which spread over 4,500 watery square miles, are a proving ground for United States nuclear weapons, and so populations (which total 42,000 people) have been evacuated from the mid-Pacific republic during atomic and hydrogen bomb tests and also during missile trials originating in California. Kwajalein, 78 miles long, and the target of the practice ICBM firings, is the world's largest atoll. A group of 32 lagooned atolls (69 square miles) and about 870 reefs east of the Caroline Islands and north-northwest of Kiribati, the Marshalls are an internally self-governing sovereignty, but the United States is responsible for their defense and provides economic assistance; the two countries signed a Compact of Free Association in 1986. (Honolulu is 2,000 miles northeast of the islands' capital of Majuro, and Guam is 1,300 miles northwest.) The Marshalls were held with absolute sovereignty for nine years by the Japanese. American forces conquered heavily fortified Kwajalein and Einiwetok in early 1944, during the Second World War.

MICRONESIA, which was first settled by seafarers from southeast Asia, consists of 2,000 outcroppings of earth extending across 1,800 miles of the western Pacific east of the southern Philippines, but it has less land area than Rhode Island. The region has been dominated in turn by Spain, Germany, and Japan, and—following the Second World War—by the United States, which administered Micronesia as a trust territory under the United Nations. A Compact of Free Association with the United States was signed in 1986. The Federated States of Micronesia are internally self-governing, but the United States runs the military-defense operations. Yap, Ponape, and Truk are island groups; the fourth principal area, Kusaie, is a single island. A role for traditional leaders and customs is recognized by each spot's constitution. A new capital has been built southwest of Kolonia in the Palikir Valley on Ponape, which has an annual rainfall averaging up to 250 inches a year. The population hovers around 100,000.

NAURU has one of the world's highest per capita incomes, $19,512, all derived from the mining and exporting of the essence of decayed marine organisms and bird droppings—phosphates—the lone resource of one of the world's smallest states (only eight square miles). Nauru, once called Pleasant Island, lies in the western Pacific, just south of the equator and halfway between the Hawaiian Islands and Australia. When the phosphate reserves are depleted at the end of the century, the 9,000 Nauruians (67 percent Protestant, 99 percent literate) will be supported by government-established trust funds. (In 1968, the island won independence from a three-nation trust.) All food, fuel, and water must be imported. There is no capital city, and there are no taxes.

SOLOMON ISLANDS, sighted by a Peruvian expedition in 1568, are a series of warm, rain-drenched, volcanic, rugged islands and coral atolls totaling 10,640 square miles, an area slightly larger than Maryland, northeast of Australia in the South Pacific Ocean. They include the Santa Cruz Islands, Guadalcanal, New Georgia, and part of Papua New Guinea. The Solomons achieved worldwide headlines during the Second World War as the scene of violent battles between American and Japanese military. A British protectorate until 1978, the country subsists on yams, rice, and bananas, but most food must be imported. There are 90 local languages in the population of 324,000. More than 90 percent of the people are Melanesian; 24 percent practice South Sea Evangelical.

TONGA, for 70 years a British protectorate, is an aggregate of 172 subtropical South Pacific islands, mostly coral reefs, occupying an area smaller than New York City. It is set at the beginning of time, so to speak, on the international date line. The island groups are Ha'apai, Tongatapu, and Vava'u. On the 77 percent of cultivatable land, coconuts, cassava, citrus fruits, and sweet potatoes are grown. Tonga receives foreign aid from Australia and New Zealand. A kingdom since 1970, it has a population of 100,000, almost exclusively Polynesian. Most Tongans live on the main island of Tongatapu. Suffrage is limited to literate adults over 21 years of age, and the males must be taxpayers. Many would agree that Tonga was put on the map during the coronation of Queen Elizabeth when Queen Salote, every inch of her six feet and four inches the queen, rode in an open carriage through London's pouring rain, smiling graciously as if she were enjoying every soaking minute of it. (She probably was.) There is virtually no crime in Tonga.

TUVALU is a 360-mile-long chain of nine coral-reef islands in the South Pacific, the former Ellice Islands, or Lagoon Islands, north of Fiji. The land area is nine square miles, which is less than half the size of Manhattan. The highest point is 15 feet. The population of 9,000 Polynesian Protestants earns a meager living by weaving mats and baskets for export. The principal crop is coconuts. Britain and New Zealand provide extensive economic aid. Until it achieved independence in 1978, Tuvalu was a territory of the British Commonwealth.

VANUATU, "the land," consists of 80 hot, rainy, densely forested southwest Pacific islands 1,200 miles northeast of Brisbane, Australia. Slightly smaller than Connecticut, it was once the Anglo-French condominium of the New Hebrides, which was named after the Hebrides Islands in the Atlantic Ocean west of Scotland and jointly administered by France and the United Kingdom. Vanuatu has been an independent republic since 1980. The population of 160,000 is mainly Melanesian, who speak a variety of dialects, and is spread through the 5,700 square miles. Bislama is the national tongue. Only 20 percent of the people are literate. Vanuatu is one of the few countries where the life expectancy of men (56.2 years) exceeds that of women (53.7 years). Meat canneries and fish-freezing support the economy.

WESTERN SAMOA, which is about the size of Rhode Island, comprises the Polynesian South Pacific islands Savai'i and Upolu. Western Samoa was once a German colony, then a New Zealand mandate, then a New Zealand–United Nations trusteeship. The state became fully independent in 1962. Western Samoa is basically mountainous and volcanic, but on the arable land bananas, coconuts, and tropical fruits grow profusely. Ninety-eight percent of the population of 182,000 is literate. There is limited suffrage.

NEW CALEDONIA (a non-state), tropical southwest Pacific islands (8,548 square miles) about 1,000 miles east of Australia and about 1,000 miles northwest of New Zealand, has few resources, but its nickel deposits make it one of the world's largest producers of the metal. Once a penal colony, New Caledonia is a French overseas territory. It imports many products from France. The population of 152,000 is 43 percent Melanesian, 37 percent French, 60 percent Roman Catholic, and 91 percent literate. The Melanesians, or Kanaks, press for independence.

SOUTHEAST ASIA is a physiographically complex, hot, wet, rainforested land of endless cultural diversity, having been invaded by many far-flung outsiders who left their imprint. Colonialism prevailed

until the end of the Second World War. The societies are highly
structured, with national consciousness often very old. The terrain
is variegated, ranging from barren to productive, from mountain barriers to rich valleys and raw coastlines. Agricultural production is
high, especially in rice. The region produces some oil, but it depends
on the Middle East for nearly three quarters of its crude. Indonesia
consists of 13,667 islands and is the world's fifth-most-populous sovereign state—around 200 million people. The Philippines consists of
7,100 islands. Population is dense everywhere except in Laos.

BRUNEI—or, formally, Brunei Darussalam, a name meaning Brunei, Abode of Peace—has the feel of a colonial outpost, but it does
have the second-highest per capita gross national product in the world:

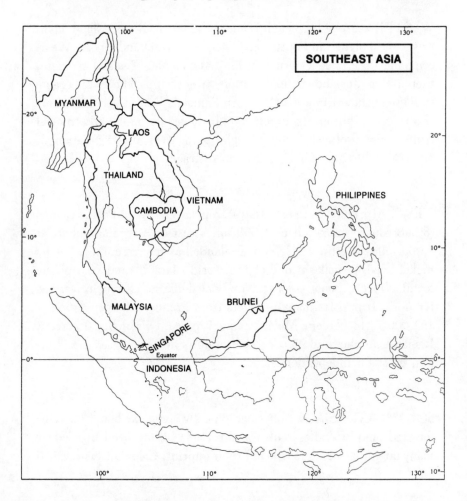

$25,500. The extraordinarily shy, powerful, Anglophile Sultan of Brunei is the wealthiest man in the world: $31 billion! Sources of the wealth are oil and natural gas reserves and substantial foreign investments. In 1987, the Sultan donated, at the request of the United States, $10 million to the Nicaraguan contras. He also has established one of the most enviable welfare states for the population of 250,000.

Brunei, once a powerful empire, now amazingly isolated and remote, is a tiny country, hardly bigger than Delaware, a two-parcel, petroleum-rich jungle perched on the northwest coast of the world's third-largest island, Borneo. (Brunei held sway over Borneo for three centuries and long attracted adventurers, pirates, and headhunters.) Brunei shares the island with the Malaysian states of Sabah and Sarawak, to the east and west, and the Indonesian state of Kaliman-tan, to the south. The northern border is on the China Sea. Brunei has no concert halls, no discos, no galleries, no nightclubs, and no liquor. Magellan visited here in 1521. Independence from Britain came in 1984. Half the population is under the age of twenty-one.

INDONESIA, the fifth-most-populous country in the world, is the largest nonfederated state in the world, and it has the largest Muslim population. About 90 percent of the population of 192 million is Muslim, yet Indonesia is not a Muslim country. Indonesia has had a patchwork history of Malayan, Hindu, Muslim, and Dutch influences, in addition to a pervasive native mysticism. Indonesia is an archipelago of 13,667 tropical islands spread 3,200 miles over the planet's equatorial belly, from continental southeast Asia to the tip of Australia. It spans the principal sea routes between the Far East and Europe and forms a natural barrier between the Pacific and Indian oceans. The straits between the islands, which are part of a mountain chain mostly hidden under the sea, are strategic military chokepoints.

"Unity in Diversity" is the national motto. Indonesia was for-merly the Dutch East Indies. The colonials never united the islands under a central administration. Only since the Second World War have the islands constituted a republic. During the 1960s, ethnic, political, and religious violence erupted and hundreds of thousands of Indonesians were slaughtered. The Communist Party was wiped out. When Indonesia invaded and annexed as its 27th province

East Timor (formerly Portuguese Timor), in the Lesser Sunda Islands, about 100,000 Timorese were murdered or died of starvation. Chinese-language schools and all Chinese printed matter are banned, yet Indonesian Chinese own about 75 percent of the private capital.

Four out of every five Indonesians work the fertile soil, which is enriched with volcanic ash. Three crops a year are possible. The country aspires to be a second-generation Asian Tiger, or Dragon, taking its place alongside the developing economic powerhouses of South Korea, Singapore, Taiwan, and Hong Kong. It is already the region's largest producer of oil. Economic policy is formalized in a series of five-year plans. About 2.5 million acres of the world's second-largest rain forest are being destroyed annually by loggers and farmers. Great teak forests have been decimated. Export earnings help to keep the economy stable. The island of Java, which has 300 thunderstorms a year and is the world's most densely populated island (more than 110 million), is a bit larger than Louisiana and pocked with 61 volcanoes, 17 or them active. The island of Sumatra is atop one of the world's largest fields of natural gas. There are more plant species on the island of Borneo than in all of Africa.

LAOS—the Laos People's Democratic Republic—had its physiography reshaped during the Vietnam War. United States aerial raiders interdicted Vietcong supply routes through Laos with more than 2 million tons of bombs. (There were 580,944 sorties.) The Communist-run, rice-farming, mountainous, jungled, tropical, Buddhist-dominated country, which is slightly larger than Utah, is landlocked in the Indochina Peninsula in southeast Asia. It is strategically located, a buffer zone bordered by Vietnam on the east, the People's Republic of China to the north, Myanmar (Burma) on the west, Thailand to the south and southwest (border clashes erupt periodically), and Cambodia (Kampuchea) to the south and the southeast. In 1991, Laos adopted its first constitution in 16 years of Communist rule; it calls for a leading role for the party.

Only 4 percent of Laos is arable. Agriculture is concentrated in rice, corn, tobacco, coffee, and cotton. Because the substantial mineral deposits of tin and the large timber reserves have yet to be exploited to their potential, Laos is one of the world's poorest states. Until 1954, when it won its full independence, the country had been a French protectorate for a half century. Since 1975, when the short-lived coalition government and the monarchy were abolished, more than a tenth of the 3.9 million population have fled Laos, crossing the Mekong River to refugee camps in Thailand.

MALAYSIA, which is two-thirds jungle and beleaguered by heavy rainfalls, annual monsoons, and a humid climate, has long been known for its rubber plantations and tin mines—and for being divided into two widely separated regions. The country produces 35 percent of the world's tin—the largest open-cut tin mine in the world is on the outskirts of the capital, Kuala Lumpur—and more natural rubber than any other state. Malaysia became a major Far East petroleum supplier, and today has a prosperous, fast-growing economy.

Lying principally on the lush Malay Peninsula at the southeast tip of Asia and just north of the equator, Malaysia is slightly larger than New Mexico. Eleven states—West Malaysia—are on the southern portion of the peninsula; 400 miles east, across the South China Sea, and on the northern coast of Borneo is East Malaysia, consisting of two former British colonies: Sarawak, once the private domain of a family dynasty, and Sabah. Islam has been entrenched for five centuries. There is tension between the rural Malaysians (59 percent of the diverse population of around 17 million) and the urban Chinese. (In 1965, Chinese-dominated Singapore split off from Malaysia and became a sovereign state.) The country produces steel, electronics, tropical timber, pepper plants, and tea, a crop with tradition. Malaysia is the world's number-one maker of room air-conditioners and condoms. Japan has become an important player in the economy. The highest mountain is Kinabalu, on Sabah, at 13,455 feet.

MYANMAR, one of the most isolated states in the world, was Burma until its name was changed by the repressive military government in 1989. The largest country on the southeast Asian mainland, the Texas-size state is under the totalitarian rule of a military mafia: torture, persecution, and other human-rights violations make Myanmar the darkest state in the region. When it seized power, the junta (called SLORC, for State Law and Order Restoration Council) mowed down thousands of unarmed students. It closes down colleges and universities whenever there is the whisper of a pro-democracy uprising. "Capitalist mercenaries" have been expelled and borders closed. The military controls much of the wealth. Thousands also have been slain in the rebellious northern territories. The population is destitute (as it was during Japan's occupation in the Second World War). Hundreds of thousands have been forced to move from cities to satellite towns in the outskirts. A dissident under house arrest was honored with the Nobel Peace Prize in 1991.

Monasteries and pagodas dot the landscape in this Hinayan Buddhist country. In Mandalay, there are 100 monasteries and about 70,000 orange-robed monks and novices: the city symbolizes devotion to religion and ritual. Myanmar has been the source of 80 percent of the world's teak, but the last of the great timber is being ravaged to help finance further military obscenities. The state, which has a population of around 41 million, in 70 ethnic groups, is surrounded by mountains and the Bay of Bengal. It borders on the People's Republic of China, Laos, and Thailand to the east and Bangladesh and India to the west. (Burma won independence from Britain in 1948.)

The climate is tropical, hot, wet, and hospitable to monsoons. The north-south river valleys and the delta of the Irrawaddy River are fertile. The river, Myanmar's principal transportation route, drains 65 percent of the country and is navigable to shipping for 930 miles. Farmers have been encouraged to try alternatives to coca leaves. Self-sufficient in food, the state has been the continent's most productive rice granary. There may be as many as 30,000 varieties of rice in the northern areas. Beans, sugar cane, and peanuts are grown. Minerals include oil (on a small scale), lead, silver, tin, and precious stones. Smuggling accounts for about a third of the gross national product.

PAPUA NEW GUINEA, north of Australia, with which it was once geologically joined, includes the eastern, or tail, half of the tropical, dragon-shaped South Pacific island of New Guinea (the world's second-largest island—its western half is Indonesia's province of Irian Barat), plus 600 other islands, including the Bismarck Archipelago and Bougainville, which has one of the world's largest open-pit copper mines. The aggregate is somewhat larger than California. Mostly mountainous and forested, and with coasts that are swamps and jungles, Papua was a stepping stone between Asia and Australia for the great migrations millennia ago. Because so much of the island is topographically formidable and isolated, and the road system is so fragmented, it probably has more airstrips per capita than any other country.

The region was settled more than 10,000 years ago, and many of the 3.7 million (mainly Melanesian) ethnically diverse, fiercely independent Papuans are still in a developing stage. There are more than 700 linguistic groups, and most of the tongues are mutually incomprehensible. The common language is pidgin. There is little ethnic tension. Faiths are based on spirit and ancestor worship, and sorcery is practiced. Nearly a century of colonial rule—by imperial Germany, then Great Britain, finally Australia—ended with statehood in 1975. The Second World War put New Guinea in the headlines as United States General Douglas MacArthur advanced with Allied armies from Brisbane in Australia to Port Moresby and Hollandia in New Guinea en route to his liberating return to the Philippines.

PHILIPPINES are an archipelago of 7,100 islands and rocks off the coast of southeast Asia, the only Catholic state in the region, and one of the world's most strategically important land masses. The Philippines are volcanic, mountainous, forested, and swept by periodic typhoons. They are astride sea lanes linking the western Pacific and Indian oceans. Slightly larger in land area than Nevada, the islands stretch 1,100 miles, the funnel of trade for the western Pacific. Manila Bay is the finest harbor in the Far East. After the Spanish-American War, which ended 333 years of Spanish control, the Philippines were an American colony until independence was granted on July 4, 1946. The Japanese military had struck a massive blow in 1941, occupying the islands until General Douglas MacArthur's return in late 1944.

Ambitious social and economic problems are being tackled in the wake of the 20-year reign of the late Ferdinand Marcos, who instituted martial law and plundered the treasury shamelessly for personal self-aggrandizement. There are canyonwide divisions between rich and poor. A few landowners have acquired massive wealth, while almost three quarters of the population of 65 million live in direst poverty, unable to satisfy basic needs. Mestizos make up 2 percent of the population but garner 55 percent of the personal income. Mineral deposits include gold, iron, petroleum, copper, and cobalt.

People without land tend to migrate to urban-squatter settlements. The government must generate 750,000 new jobs each year and increase food production by 40 percent in the next decade. Land reform was a priority in Corazon Aquino's "people power" presidency.

About 43,000 Filipinos were employed at United States military bases, which housed about 16,000 troops and added $1.1 billion a year to the economy. (Nuclear devices were stored there.) During the Korean and Vietnam wars, as many as 60,000 American troops were stationed at Clark Air Base, a logistical hub. The bases were a part of the United States Pacific defense system.

Communist guerrillas—the New People's Army—are spirited. Muslims have fought a secessionist war in Mundanao. Most of the people are Malay in origin, but there are more than 75 ethnolinguistic groups. All but 5 percent of the population live on the 11 largest islands. Some people like the Tasadays live so remotely, they have only recently been discovered, and disturbed, by the outside world. About 75 indigenous tongues, including eight major ones, are spoken. The official one is Tagalog.

SINGAPORE—sharp, successful—is one of the Asian powers whose economic prowess has earned them the nickname of the Four Tigers, or the Four Dragons. Strategically located at the convergence of major sea lanes in southeast Asia, the island-city-state at the southern tip of the Malay Peninsula experienced an average annual 9 percent economic growth between the years 1965 and 1990. "Singapore" once conjured up images of a somnolent and indolent place in the mysterious Far East, vaguely astir with romance and intrigue and pursuing pampered pleasures at the hospitable beckoning of the Raffles Hotel.

Singapore was a British colony until 1959, when it became a self-governing state within the Commonwealth. It joined the federation of Malay in 1963 and seceded in 1965 to become a commercially aggressive independent republic. The Singapore of Somerset Maugham is indeed but a faded page from an irrecoverable past. Today, the port is a bursting metropolis, a world financial center, a leading center for tin, rubber, and refined petroleum. The state occupies less space than New York City and has a density of 11,190 people per square mile. Minicities are being planned to reduce congestion and improve the quality of life.

Almost all of the 3.3 million population is Chinese, but there are four official languages: Tamil, Malay, English, and Chinese. The literacy rate is 85 percent. Life expectancy for women is 76.2 years; for men 69.9 years.

THAILAND is a mountainous, heavily forested, tropical, rice-exporting Buddhist kingdom about the size of Texas on the Indochinese and Malay peninsulas. Its location at a crossroads between India and China gives it a long history of dealing with foreigners, and it is the only southeast Asian country never colonized by a European power. Vietnamese troops were repulsed in the 1980s. The government was overthrown by the military in 1991, martial law was imposed, parliament was dissolved, political parties were suspended, and the constitution was revoked in the name of democracy. Corruption is rampant, and a throwaway consumer culture is burgeoning. The estimated population is 55.6 million.

AIDS has increased at an alarming rate, and drug trafficking is heavy. Water shortages and pollution problems abound. Air quality in the traffic-clogged, factory-dotted capital has deteriorated significantly in recent years; dangerous high levels of lead in blood samples prompted introduction of unleaded gasoline. Bangkok is congested, with an estimated 3 million people living in 1,500 slums.

The country is a world player in rice, tin, rubber, and tapioca, and is becoming a competitor in jewlery, gemstones, and processed food products like frozen prawns. Per capita income has reached $1,500. Tourism continues as the largest source of foreign income. The fast-growing economy was expected to expand by 8 percent in 1991. A flood of refugees from Myanmar, Vietnam, Laos, and Cambodia has descended on the country, and there has been an influx of rural workers into urban areas.

Some Thai communities go back 6,000 years. The country was called Siam until 1939 and, as Siam, was immortalized in the Broadway musical *The King and I*. Thailand was occupied by the Japanese during the Second World War. The bridging by European POWs of the river Kwai for the Burma-Thailand "death railway" was the focus of an Oscar-winning motion picture, *The Bridge on the River Kwai*, in 1957. (Three hundred and thirty-one thousand POWs were needed to build the 250-mile road.)

VIETNAM, which beat back two mighty Western armies in the 1950s and the 1960s–70s, is a war-weary, resilient, reunified south-east Asian state, one of the poorest on the Pacific rim. Its troubled history dates back centuries: Vietnam was battling the Chinese as Columbus was plotting a westward voyage to the Far East. In the last half century, North Vietnam's Communist forces have ousted both the French colonials and the United States military, conquered South Vietnam, routed another Chinese invasion, united the country, and established a socialist republic.

Vietnam is a long and narrow country, about the size of New Mexico. Its mountainous landscape constricts a southward advance by China, and Vietnam has no need to move into China. The 1,400-mile-long coastline lies on the Gulf of Tonkin and the South China Sea. Vietnam shares borders with China (which is nearly 100 miles north of the capital city of Hanoi), Laos, and Cambodia (Kampuchea). The terrain includes plateaus, the weather includes monsoons. Hanoi, which is on the same latitude as Havana and Honolulu, and is the oldest capital in southeast Asia (it was founded in A.D. 1010), was the seat of the colonial government of French Indochina from 1887 until 1954, then the capital of North Vietnam when the country was partitioned.

Seventy percent of the 66.9 million Vietnamese are in agriculture. Rice is exported in nonfamine years. Coffee and tea are grown in the south. The country is being touted as a promising oil producer; there may be reserves of 2 billion barrels. Vietnam is in dispute with Indonesia over the continental shelf in the South China Sea, and there persists a major territorial dispute with China over the Spratly Islands, east of Vietnam. The standard of living is low, with the per capita income $200. Unemployment, corruption, and smuggling are rampant. The state has flirted with free enterprise. The society had traditionally valued education, but illiteracy and school delinquency have grown alarmingly.

AUSTRALIA was once part of Antarctica, once connected by land bridge with New Guinea, and eventually sailed free to become an immense breakwater between the Pacific and Indian oceans. The country, which is a bit smaller than the United States in size, is hardly lived in. The population of 17 million is less than the number of babies born each year in India. The smallest of the seven continents

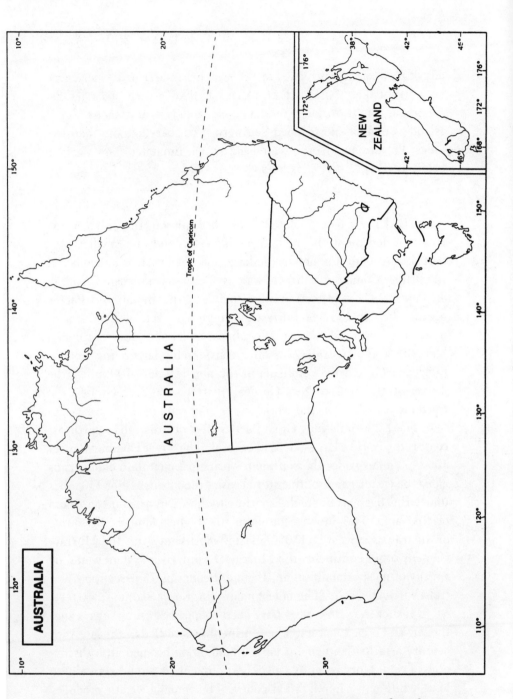

AUSTRALIA

AUSTRALIA

NEW
ZEALAND

Tropic of Capricorn

became England's penal colony in the late 18th century after the 13 American colonies were no longer available. The black aborigines had migrated from southeast Asia about 20,000 years earlier and spread throughout the severely arid continent. They remained isolated from outside influences until the English arrived and took over, and they are today strangers in a strange land. Until the southeast industrial city of Newcastle was rocked in 1989, no urban area had experienced an earthquake. **NEW ZEALAND:** Unlike mostly pancake-flat Australia, this two-island country in Oceania is mountainous. South Island's range of snowcapped Southern Alps soars to more than two miles. Polynesian Maoris, who love the sea, inhabited New Zealand before the first Europeans landed.

AUSTRALIA, a federal state, a British dominion, the only country that is a continent, is the world's largest island and smallest and flattest continent and one of the largest countries. It is a member of the British Commonwealth of Nations. Located southeast of Asia, in an important geostrategic position between the Indian and Pacific oceans, the world's second-driest continent (Antarctica, 2,000 stormy miles due south, is the driest) is almost as large as the continental United States. Australia was once considerably larger: the tropical jungles of Cape York Peninsula in the northeast are 100 miles (the Torres Strait) south of New Guinea; Australia and New Guinea, the third-largest island in the world, were once one.

The 1,250-mile-long Great Barrier Reef off Australia's northeast coast is the world's largest deposit of coral. The Great Dividing Range, 100–200 miles wide, the continent's principal mountain chain, runs along the eastern and southeastern coasts, then vanishes for 150 miles under the Bass Strait in the southeast, then emerges on the island of Tasmania. The Snowy Mountains, the highest land, is the source of Australia's longest (1,170 miles) and most vital river, the Murray. (On any other continent, the Murray-Darling river system would be worthy of no special attention, though it contains 75 percent of Australia's irrigated soil.) The tallest mountain, Kosciusko, is 7,310 feet.

Most of the rain forests have been chopped down by sugar cane, timber, and ore exploiters. The continent was settled in the late 18th century as a British penal colony. Immigration boomed after discoveries of gold. Until recently, successive governments tended to adhere to a "white-only" immigration policy. The commonwealth produces 30 percent of the world's wool and is a major exporter of wheat. Some of the 80,000 sheep stations are larger than Rhode Island. A close trading partner is Japan. Coal reserves match Saudi Arabia's oil in potential energy. There are 11 trillion feet of recoverable natural gas under the North West Shelf.

Most of the population of 17 million (only 5 Aussies per square mile) live on the east coast. Almost 50 percent live in the three major metropolises, Sydney, Melbourne (the financial hub), and Brisbane. (There are more Greek-speaking people in Melbourne than in any city outside of Greece. There are many Yugoslav and Irish Australians, too.) The remarkable aboriginal population of 200,000 in the arid outback is economically disadvantaged. There is only one city of size, Perth, in the whole western half of the country. The literacy rate is nearly 100 percent. Life expectancy for women is 78.8 years; for men 72.3. Twice as much money is spent on gambling as on national defense. Australians are the greatest consumers of alcohol in the English-drinking world.

Only 6 percent of the land is arable. The interior is often savagely hot. Western Australia, a third of the colossal continent, is one of the harshest of habitats. Everywhere, the lack of water imposes strains. Distinctive animal species are found in this topsy-turvy menagerie, flabbergasting zoologists: the koala bear, the platypus, the dingo, the euro, the kangaroo, the Tasmanian devil. The many unusual and unique species were cut off in the continent-building processes that separated the "island" from Pangaea. Nearly all the half million seed-bearing plants in the world are descended from a single pepperlike plant that grew in Australia 120 million years ago.

NEW ZEALAND—Rudyard Kipling's "Happy Isles" and one of the world's first welfare states—is principally two big, mountainous, remote South Pacific islands: North Island, the larger, and South Island, spectacularly lofty; together, they are 1,000 miles long, no more than 280 miles wide—about the size of Colorado. Geologically, the isolated sealocked pair are microcontinents, riding separate, colliding tectonic plates 1,200 miles east of Australia, 4,500 miles west of South America, and about halfway between the South Pole and the equator. They were first sighted by a European navigator in 1624. The rugged Southern Alps extend almost the entire length of South Island: 17 peaks are over 10,000 feet tall and Mount Cook, at 12,349 feet, is one of the tallest mountains in the Pacific.

The population of 3.4 million (more than two thirds live on the more temperate North Island) is outnumbered 23 to 1 by 70 million sheep, which may explain New Zealand's preeminence as the world's leading exporter of wool products. Only 2 percent of the land is arable, but the nation is an agricultural cornucopia. Most of the world's kiwi fruit is grown on North Island. Eighty-one percent of New Zealanders are descendants of Europeans. Maoris, of Polynesian descent, are 12 percent of the population and famous for their wood carvings. (Seagoing Polynesians settled here in the sixth century.) Auckland, on volcanic North Island, is the principal port and it was the first capital. Wellington, on South Island, is the capital today.

Strongly antinuclear, New Zealand, which has been independent of Great Britain since 1947, does not allow nuclear-armed or even nuclear-powered vessels to use its port facilities, generating a rift with the United States. New Zealand has agreed to help defend Singapore and Malaysia from invasion. The social welfare programs in the almost classless society are noteworthy. Kiwis (the nickname for New Zealanders) were the first to adopt noncontributory old-age pensions (in 1898) and to begin a program of socialized medicine (in 1941). The literacy rate is just shy of 100 percent. Serious crime is a rarity. There are many indigenous animal species, and nearly a tenth of the territory has been set aside as national parkland.

SOUTH ASIA: the overriding fact of this underdeveloped, poverty-ridden region is its teeming population. Gargantuan India is a subcontinent with upwards of 900 million people; it will presently overtake the People's Republic of China as the world's most populous country. The Ganges River Valley is one of the most densely populated places in the world. (The 1,560-mile-long Ganges River is the most sacred Hindu river.) The topography of Bangladesh leaves the country vulnerable to cyclones and devastating floods. The island groups in the restless, beautiful, violent (waves may build 70 feet high in the southwestern reaches), and sometimes mystifying Indian Ocean between India and Africa are prime natural sites for strategic military bases, another factor in any assessment of prospects for stability and peace—the why of where. The annual number of new HIV infections in all of Asia is expected to exceed Africa's huge total sometime in the 1990s.

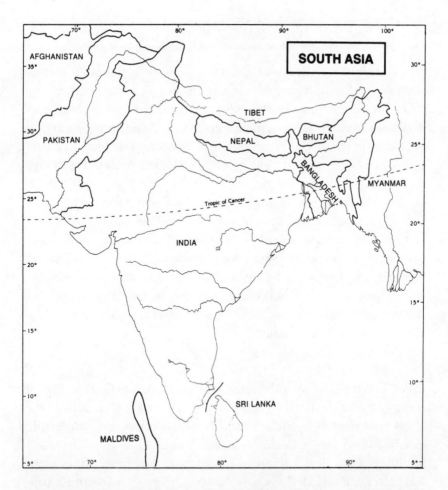

SOUTH ASIA

AFGHANISTAN, whose most profitable cash crop is opium, has been called a narrow sword cut in the hills. It is an underdeveloped Asian crossroads about the size of Texas and hemmed in by the Soviet Union, the People's Republic of China, Pakistan, and Iran. No invading force has mastered the mountainous state—a collection of many often-hostile tribal nations—and no other state offers so many prime opportunities for guerrilla warfare. The legendary 35-mile-long, twisting, dual-road Khyber Pass on the eastern frontier soars upwards of 3,500 feet, but the floor is only 40 feet wide in places; steep boulder-raked slopes have provided cover for raining death and destruction on invading Moguls, Persians, Greeks, and Turks. Desert regions influence the south.

In late 1979, Soviet troops stormed in from the north, bolstering the pro-Soviet Marxist coup in Kabul and quelling a Muslim uprising. They damaged or destroyed more than 75 percent of the villages, and 1.5 million people have been killed. Millions of Afghans (25 percent of the population) fled over the mountains to sanctuaries abroad. (Refugees still strain neighboring Pakistan's economy, threatening its fragile democracy.) In the eight-year war, the Western-financed *mujahidin* freedom fighters, holy warriors for Islam, firing rocks and Western-supplied, hand-held missile launchers, stalemated and repulsed the Soviet military machine. The Geneva Accord in 1988 led to withdrawal of the Red Army. To subdue the Afghans, it is said, famine would have to be provoked in the countryside.

Afghanistan has deep-rooted ethnic and religious complexities and a tumultuous patchwork of fiefdoms. Three quarters of the population of 12 million are Sunni Muslim, and Islamic laws determine beliefs and life-styles, religious and secular. Over 80 percent of the adults have no formal schooling, and the literacy rate is only 12 percent. Per capita income is $240. Life expectancy for men is 42.5 years; for women a little less.

BANGLADESH, the vulnerable southeast Asian Golden Bengal state, is one of the basketcases of the world: misery is endemic. One of the most densely populated agricultural countries, the flat, fertile, marshy, world's largest delta land begs in order to survive. The world has little geopolitical interest in Bangladesh. The birth rate is high. The average Bangladeshi is only 16 years of age. The destitute population of 118 million (it expands by 100,000 babies every week, despite government-sponsored family planning) may reach 219 million in the next 30 years. The people, who are mainly Muslim, are chronic victims of diseases and floods, food and medical shortages, June-to-September monsoons. Many Bangladeshis live in straw and mud huts.

Bangladesh, which has little land more than 30 feet above sea level, is about the size of Wisconsin, which has one twentieth the population. It is semitropical and river traversed. It has been hit by 55 cyclones since 1900 and 8 of the world's 10 worst floods since 1980. In 1970, a cyclone and a tidal wave claimed some 300,000 lives. In 1987, floods put much of the country under water and left 2 million homeless. In 1991, a cyclone and "waves as high as mountains" left 139,000 dead. It is said that the only way that Bangladesh can escape the violent storms is to move to another part of the world. New research challenges the claim that flooding, the most dangerous and frequent natural hazard, is related, in part, to rapacious deforestation in the Himalayas to the north, massive cuttings for firewood that otherwise would help to soak up the rains. The average temperature is 84° F., over 100° F. in the wet summers. The average annual rainfall is 100 inches.

The first governments took power through coups or assassinations. The turbulent republic had been founded in 1971 with military support from India, gaining its independence from culturally and geographically different Pakistan (1,000 miles across India to the west) and changing its name from East Pakistan. In the genocidal civil war related to the separation from Pakistan, 1 million Bangladeshis died and 10 million fled into India. Peasant farmers harvest the tough, tall, jute plant (at least half of the world's output), tea, and rice; 22 million of the 26 million acres under cultivation are given over to rice. A reliable irrigator in the dry season is the Brahmaputra River. Per capita income is $150.

BHUTAN, the last of the Himalayan Buddhist monarchies, is nicknamed "Land of the Thunder Dragon" because of its violent storms. The kingdom, which is the size of Vermont and New Hampshire combined, is isolated amid some of the world's highest mountains and between two states that keep an eye on it, India and the People's Republic of China, which took over nearby Tibet in 1950. The exotic south Asian country had sealed itself off from the rest of the world for most of its existence. Limited numbers of tourists have been admitted since 1974. In 1949, after the departure of the Raj, India assumed Britain's role as subsidizer and foreign-affairs director. The king, who was enthroned in 1953 and was the heir to a 1,300-year Tantric Buddhist tradition, abolished slavery and the caste system, emancipated women, and initiated land reform.

While largely mountainous, Bhutan has thick subtropical forests in the south and alpine meadows and fertile valleys richly nourished by monsoon rains in the north. Only 2 percent of the land is arable, yet the kingdom is self-sufficient in food production. The economy is underdeveloped; handicraft production is the chief industry. Mineral resources include silver, iron, and mica. Transportation and communications are primitive.

The present monarch, "the precious ruler of the Dragon people," has had four wives, all sisters. Seventy-five percent of the 1.5 million Bhutans are Mahayana Buddhists, who live enveloped in incense. An ethnic Nepalese minority in the south has been rebelling against

Bhutanese culture. Per capita income is $102. Suffrage is one vote per family.

CAMBODIA (or KAMPUCHEA), which is about the size of Tennessee and may be the most isolated country in the world, suffered mass famine, violent human-rights abuses, forced emigration, and intermittent carnage during the mid-1970s' despotic reign of Prime Minister Pol Pot's Khmer Rouge, which was supported by the People's Republic of China. The killing fields were the site of the systematic murder of an estimated 3 million Cambodians; most of the victims were educated or employed in the professions or had links to the former government. The brutal reorganization of society included the devastating evacuation, dislocation, and destruction of urban populations. Only 11 percent of the population of 6.9 million still lives in cities. Twenty percent of the babies die before their fifth birthday.

Located in southeast Asia, Cambodia was a French protectorate until 1953. It is surrounded on the Indo-Chinese Peninsula (as it has been known historically) by Thailand on the west, the Gulf of Thailand on the south, Laos on the north, and Vietnam on the east. Invading Vietnamese armies expelled the Khmer Rouge. A reconciliation among the warring groups in 1991 may lead to a permanent political settlement.

Because so much of the agriculture was destroyed, almost all of the country's food must be imported. Rice is the major crop. The climate is tropical, with high humidity and rainfalls and temperatures. Seventy-five percent of the country is forested. In the northwest, a long escarpment separates Cambodia and Thailand. The national religion is Hinayana Buddhism.

INDIA, the world's largest democracy, is the planet's second-most populous country—around 900 million people today and increasing by at least a million babies a month—and one of the poorest. By the year 2000, more people will be living in the city of Calcutta than in all of Canada. The government adopted the world's first nationwide population-control policy in 1952, classifying birth control as a top priority, but it has not proved successful. About a third the size of the United States, the Asian giant is woefully divided by religions, castes, ethnic groups, and languages, and by the expanding rift between the affluent and the poor. By the end of the decade, a million Indians may die of AIDS. (Nearly 50 million Indians are homosexual.) It is a state of many nations. About a half billion Indians cannot read or write. Some 120 million children work more than a 12-hour day to earn their daily bread.

After independence from Great Britain in 1947, this "jewel in the crown" was bloodily partitioned, its eastern and western extremities amputated to form the two halves of Pakistan. Resettlement of roughly 10 million Hindus and Muslims led to the slaughter of a half million. Buddhism was founded here, but Hindus dominate India. Sikh separatists demand an independent state. (Extremist Sikhs murdered India's prime minister in 1984, and Tamil separatists murdered her son, a former prime minister running for re-election, in 1991.) India is a nonaligned member of the nuclear club, and it has a large, disciplined army.

There are three principal land areas: rich agricultural soils in the north (there are more than 400 inches of rain in the northeast annually), the fertile, densely populated Gangetic Plain, and the plateau-peninsula extending south into the Indian Ocean. Their climates range from Arctic cold in the frozen Himalayas to tropical heat on the sultry shores of Great Nicobar Island. India is the oldest source of diamonds. Natural resources remain undeveloped. The largest water project in history calls for construction of 30 major dams along the 800-mile-long Narmada River.

MALDIVES, a palm-garnished, flat, hot, poor Asian country, is a union of 1,190 islands (only 115 square miles of land) clustered in 19 archipelagoes in the Indian Ocean 400 miles southwest of Sri Lanka and south of India. The country has been described as a coral necklace on a velvet blue sea; it is also a series of maritime chokepoints. There is no natural elevation higher than about 15 feet. A slight rise in the level of the shallow ocean waters would wipe the republic off the map. Two thirds of the capital city of Male, where most Maldivesians live, was covered with seawater in April 1987; high waves were generated by a storm off Australia, 3,500 miles to the southeast. Monsoons bring heavy rains to the region annually. Breakwaters and seawalls have been built with foreign assistance.

The population of 211,000, for which malaria is a constant threat, is of mixed Singhalese, Indian, and Arab stock (1,756 people per square mile) and is 95 percent literate. Per capita income is $475. Islam is the only religion, and the Maldives, which were British until 1965, has close ties with other Islamic countries. The 815-year sultanate was abolished in 1968.

NEPAL, tucked between Tibet and the colossal powers of India and the People's Republic of China at the top of the world, is the sequestered last Himalayan kingdom outside of India's control since Sikkim was absorbed in 1975 and Bhutan virtually became a ward. In 1990, the world's only Hindu monarch, who rules by divine right and is revered by many Nepalese as the reincarnation of the Hindu god Vishnu, lifted a 30-year ban on political parties, which led to a pro-democratic movement and an independent government. The Himalayas are a sacred presence. Remote, landlocked, earthquake prone, and reputedly the birthplace of Buddha, Nepal has a population of 15.5 million and one of the least-developed economies in the world.

The terrain is beautiful. There are three distinct topographical regions: a narrow, flat, and fertile plain in the south; the lower Himalayas and swiftly flowing mountain rivers forming the hill country in the center; and the high Himalayas framing the border with Tibet in the north. The climate ranges from subtropical in the south to severe winters in the northern mountains. The landscape is largely deforested, denuded, which seems to contribute to catastrophic flooding during the monsoon season.

The state has become dependent for fuel and other necessities on trade routes through India, which at one time closed 19 of 21 border crossings in a crippling economic blockade. For centuries this land of the Abominable Snowman had been virtually sealed off. Even today there are few access roads. Much of the local economy is based on barter, while tourism and handicrafts are the major sources of foreign exchange. Literacy is 29 percent. Per capita income is less than $200. Nepal is slightly larger than Arkansas.

PAKISTAN—"land of the pure"—was carved away as a separate Muslim state during the bloody partition of India in 1947. The geographic anomaly was divided into two provinces nearly 1,000 miles apart on opposite sides of India. In 1956, Pakistan in the west, and predominantly Punjabi, became the world's first Islamic republic.

(East Pakistan proclaimed its independence as Bangladesh in 1971.)

Legislation may make the Koran the supreme law, subjecting all aspects of life, from social behavior to civil liberties, to Islamic tenets; there are close ties with the mullahs of Iran. Martial Pakistan, which has a history of riots, coups, and wars, is a major player in the international narcotics trade, though the government has striven to reduce poppy growing. It actively seeks nuclear weapons—the army already has a vast array of arms; the United States justifies military aid as a preventive to development by Pakistan of nuclear devices. The state takes care of more foreign refugees than any other; millions poured across the border when the Soviet Union invaded neighboring Afghanistan.

Pakistan is one of Asia's poorest states. Its poverty is grinding. There has been an economic slide since industries and banks were nationalized in the 1970s. Population growth has been unrelenting, though the government endorses family planning. Labor is unskilled, per capita income is $340, the literacy rate is 26 percent. There are border tensions with India, and the two have warred over Muslim-dominated Kashmir. Punjabis compose the largest ethnic group in the population of 112 million. The caste system is defined by birth. After years of military dictatorship, there is only now a fragile democracy.

The republic, which is about the size of Texas and Ohio combined, is on the Arabian Sea in southern Asia and sandwiched between Iran, Afghanistan, India, and the People's Republic of China. Some northern mountains are more than 22,000 feet high. Fertile plains and a desert also are physiographic features. Vast areas in the west are inhospitable. Cotton production is developing a textile industry. The economy also includes leather goods, rice, and carpets. The largest city, the sweltering, swollen former capital of Karachi, its population 8 million, lies on a flat, dusty, dun-colored plain on the sea.

SRI LANKA—"Resplendent Land" in Hindu legend—has been ripped by decades of ethnic civil war between the ruling Singhalese (Buddhist) majority and the rebellious Tamil (Hindu) separatist minority. The Tamils have been demanding a separate state in the dense jungles in the northeast corner of the teardrop-shaped south Asian tropical island-state in the Indian Ocean, visited by Marco Polo 700 years ago. The Tamils, who have high education levels and linguistic skills, are 20 percent of the population of 16.9 million. The Singhalese, who have racial preference systems, admit they have mistreated the Tamils since the country's independence from Britain in 1948.

Before 1972, Sri Lanka was called Ceylon. The island, which is just north of the equator and is the size of West Virginia, is separated from mainland India by the shallow Palk Strait and a chain of shoals. (There are some 60 million Tamils in southern India.) The country is mostly mountainous, but possesses a broad coastal plain with enchanting vistas. Sri Lanka is a mecca for gemologists, but it is basically agricultural despite unreliable rainfall. At times, the island has had to import as much as half its food. Rice (one crop after another), tea, rubber, coconuts, textiles, and limestone help to support the economy. Sri Lanka has universal education.

TIBET (a non-state)—the Autonomous Region of the People's Republic of China—is regarded by China as an inalienable part of its territory. China's invasion more than four decades ago reasserted its age-old claim to the loftiest country in the world. About 120,000 Tibetans fled to India and Nepal, but another million have been murdered, and massive numbers of children have been shipped to China for education and indoctrination.

Tibet is the site of the illusory Shangri-la of imperishable youth and beauty, of life and love everlasting. Foreigners were barred until 1920, though Sherlock Holmes was said by Conan Doyle to have spent 10 years there. Chinese-built roads ended Tibet's isolation. Most of the 6 million people are Mongol in origin and live on a large plateau between The Himalayas in the south and the Kunlun range in the north. Tibet is the gateway to Everest's northern slopes. (Tibetans call the world's highest mountain Qomolangma—"goddess mother of the world.") The rarest resource is forestland. One of the world's great rivers, the Brahmaputra, begins as a glacial trickle in the west.

Life in Tibet is pastoral. Agriculture is extensive in the valley. There are large deposits of salt, gold, and radioactive material. The dominant religion is Lamaism, a variant of Buddhism, and the Dalai Lama was the spiritual and governmental leader for centuries. The present Dalai Lama, or god-king, Tenzin Gyatsu, the 14th incar-

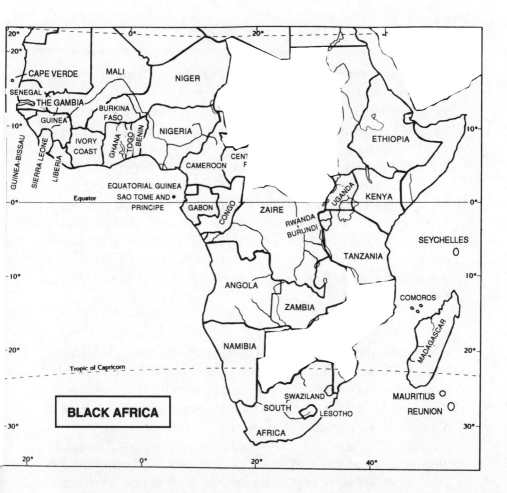

BLACK AFRICA

nation of the spirit of godliness, has been in exile since an unsuccessful uprising against Chinese occupation in 1959; he is the spiritual leader of 6 million Tibetan Buddhists and heads the expatriated government.

It is said that if the "liberation" of Tibet continues much longer, China will "liberate" the Tibetans right out of existence. Human-rights abuses are well documented.

BLACK AFRICA: once described as "the dark continent," the sub-Saharan region is bisected by the equator and characterized by huge resources (copper, diamonds, iron ore), dense rain forests, extensive grasslands, impenetrable jungles, much hunger, and ubiquitous disease, particularly malaria, cholera, sleeping sickness, and now AIDS—more than seven million cases. East Africa's Great Rift System has been shaped by uplift volcanism and faulting; it is one of the planet's most complex and conspicuous features. South Africa has great natural wealth and inviting weather and is controlled by a minority white population. At the Berlin Conference in 1884–1885, a dozen European countries and Turkey agreed that each of them could claim as its own any of equatorial Africa that it could conquer and develop. Only four of the continent's 52 countries today were independent in 1950. By the year 2025, the continent is expected to have a population greater than the combined populations of Europe, North America, and South America. The power of Islam in African politics—the march of Islamic fundamentalism—could establish governments antithetical to secular politics and Western cultures, posing a new challenge to Western policy makers after the collapse of Soviet Communism.

ANGOLA, which was formerly Portuguese West Africa, borders the Atlantic Ocean in southwest Africa. It ranks near the bottom of the world's states in providing its citizens with even the most basic needs. The formerly hard-line Marxist government has moved toward a Western-style, multiparty democracy.

The relentless civil war after independence from Portugal in 1975 killed more than 200,000 of the 8.6 million Angolans and left another 50,000 with amputations caused by land-mine explosions. The infrastructure was demolished and a million people displaced; the Portuguese who had run the colony left for their homeland; insurgency spread and the economy collapsed. Both Cuba and the Soviet Union had a strong presence in the independent "people's republic," and South Africa and the United States supplied millions in covert assistance to guerrilla forces. In April 1991, Marxism-Leninism was set aside in favor of social democracy. It was time to seek power at the ballot box. Angolans can now form political parties, hold political meetings, and demonstrate.

The country is somewhat larger than California and Texas combined. A fraction of the country, Cabinda, a northern tropical-forest enclave, is separated from the rest of the country by the short Atlantic coastline of Zaire. Angola is mostly desert or savanna, rising abruptly from the narrow coastal strip and becoming a vast plateau ranging between 3,000 and 5,000 feet high. It is a leading producer of coffee, and fishing and livestock raising are of some importance. Angola is potentially rich, with deposits of diamonds, iron ore, manganese, petroleum, and copper. Only 3 percent of the land is arable. Portugal entered Angola in the 15th century and arranged for most of the 3 million slaves shipped to Brazil. The official language is Portuguese, but most Angolans speak Bantu.

BENIN, which is slightly smaller than Pennsylvania, is a nonaligned, Marxist-Leninist west African people's republic on the hot, humid, rainy, former "slave coast" of the Gulf of Guinea. It is one of the world's poorest countries; per capita income is $290. The land is flat and covered with dense vegetation. The economy is agriculturally based—peanuts, cotton, tobacco—but largely underdeveloped. Benin, which was called Dahomey before 1975, has been independent of France since 1960, but French is still the official language. The first presidential election in 21 years took place in March 1991. Brigadier General Mathieu Kérékou became the first chief executive on the African mainland to lose power through an election; there had been pro-democracy unrest. The population of 4.7 million is mostly black and 70 percent animist. The literacy rate is 28 percent. Life expectancy for women is 46 years; for men 42 years.

BOTSWANA, a nonaligned South African multiparty democracy larger than France and only slightly smaller than Texas, is the world's largest producer of quality diamonds. Incredibly rich deposits were discovered after independence in 1966. (Jwaneng is the world's richest diamond mine.) This subtropical, landlocked republic, formerly the British protectorate of Bechuanaland, has had one of the continent's fastest-growing economies. In addition to mineral wealth (copper, coal, and nickel are also mined), there is a thriving cattle industry; cattle, in fact, outnumber the 1.3 million Botswanians two to one.

The country has offered sanctuary, tolerance, and freedom to refugees fleeing racial strife in neighboring states. A fifth of the land has been damaged by drought and overgrazing, and another 20 percent is set aside as national parks or game preserves. Botswana is the range for Africa's largest herd of free-roaming elephants. Bushmen, the original inhabitants, live mainly in the Kalahari Desert, the "great thirstland." Half of the population is under 15 years of age. There is an astonishing archaeological legacy, thousands of paintings created by bushmen and black tribes millennia ago.

BURKINA FASO, which was once the French colony of Upper Volta, is a military-run, landlocked tropical west African country south of the Sahara. It won both its independence and admission to the United Nations in 1960. The size of Colorado, it is desert-savanna-forest from north to south, and very poor: per capita income is $153. Only a tenth of the land is arable, and there are frequent droughts. The main exports are livestock and livestock products. Agriculture, such as it is, yields cotton, rice, peanuts, and sesame. France and other countries contribute aid. Several hundred thousand people migrate annually to neighboring countries in search of farm work. Mineral industrialization includes manganese, gold, and limestone. The population of 8.7 million is primarily animist and descended from warrior migrants. There have been at least four coups and numerous coup attempts in the last 10 years. The military-led government has promised a multiparty democracy, and voters approved a draft constitution providing for executive, legislative, and judicial bodies that would make the president less powerful; a presidential election would be held every seven years.

BURUNDI, underdeveloped and sometimes famine stricken, was part of the Belgian trust territory of Ruanda-Urundi before gaining independence in 1962. Landlocked and densely populated, the east-central African republic is near the equator, and it is the southernmost source of the White Nile. Lake Tanganyika in the northeast is 420 miles long and the second-deepest lake (4,710 feet) in the world. Because of its high altitude, Burundi enjoys a pleasant climate. About half the land (the state is roughly the size of Maryland) is arable, and there is a profusion of grassy highlands. Forests, the main source of fuel, are facing depletion.

Nearly everyone is employed in agriculture; coffee and cotton are exported and peat is harvested. The Hutus and the tall Watusi wield the most power in the population of 5.5 million, who include the Twa—pygmy hunters and descendants of the first inhabitants. Ethnic clashes have killed thousands. Tribal violence forced almost 50,000 Hutus into neighboring Rwanda in August 1988; however, all but 900 of the refugees had returned home within a year in one of the most rapid repatriations in modern African history. Per capita income is $273.

CAMEROON is an agricultural republic, somewhat larger than California, on the Bight of Bonny between west and central Africa. The terrain is varied: coastal plains, rain forests, plateaus, forested mountains, savannas, marshes—and 14 percent of the land is arable. The population of 10.9 million is also diverse; there are some 200 tribes. Most of the people live in the French-speaking eastern region, away from the hot, humid, coastal area, and more than half of all Cameroonians are animist. Oil is the major export. Other industries include aluminum processing and palm products. Clouds of toxic gas of volcanic origin emanating from Lake Nyos killed 3,000 people in 1986. Eighty percent of Lake Chad has dried up in the last two decades; former harbors are now miles inland.

CAPE VERDE is an Atlantic archipelago of 10 eroded, stark, volcanic west African islands and five islets—together, they are a bit larger than Rhode Island—off the northwestern bulge of Africa. Mainland neighbors are Mauritania and Senegal. Once a shipping center for slaves bound for the New World, the republic exports coffee, nuts, and sugar. Rainfall is low, drought and famine persist. Salt is among the few natural resources. Ten percent of the land is arable but largely infertile. The population of 364,000 is 71 percent Creole (mulatto), 80 percent Roman Catholic, 37 percent literate. Independence from Portugal was achieved in 1975.

CENTRAL AFRICAN REPUBLIC, which is located near Africa's geographical center, is the former French overseas territory of Ubangi-Shari. It was incorporated into French Equatorial Africa in 1910 and achieved full independence in 1960. A tropical forest is in the south, and a semidesert is set between Chad and Zaire on the north and the south, respectively; Sudan is to the east. The chief agricultural products are cotton, peanuts, coffee, and sisal, and the major export is diamonds. But this rural, underdeveloped republic of nearly 3 million, which is almost as large as Texas, is only 3 percent arable and steeped in abject poverty. A bloodless military coup overthrew a ruthless emperor in 1979; the state has been in the sway of the People's Republic of China.

CHAD, a landlocked republic in central northern Africa without a solid economic base, has been torn by chronic disorder—decades of civil wars and invasions—since gaining its independence from France in 1960. It was once a territory of French Equatorial Africa. Today, France and Libya back different ethnic and regional factions within the country. French troops have helped the government keep at bay invading Libyans, who covet a strip of northern Chad.

Famine and desertification overwhelm the economically poor country, and 20 percent of infants do not live to the age of one. Longevity is 40 years. The population of 5 million is Islam Arab or black African, and there are 200 distinct ethnic groups and about 100 languages. Per capita income is $88, and the literacy rate is 17 percent.

The northern region of this ethnic crossroads, which is about the size of Alaska, is desert; the southern region is grassland; only 2 percent of the country is arable. There is a serious deficit of water. Nomadic farming exists in the north and small permanent cultivations in the south. The chief crop is cotton, and the chief mineral is uranium; tungsten and sodium are also mined. The primary revenue source, however, is customs fees on imports.

COMOROS, a volcanic 863-square-mile, four-island archipelago in the Mozambique Channel between northeast Mozambique and northwest Madagascar, is a federal Islamic republic half the size of Delaware. It has been described as a "fourth-world" state; the soil is so rocky or depleted that nearly half of Comoros's foodstuffs must be imported. Per capita income for the population of 444,000, which is only 20 percent literate, is $248. The country, independent of France since 1975, is a leading producer of the essence of ylang-ylang, used in manufacturing perfume. Crops include vanilla, copra, and fruits. The coelacanth, the "living-fossil" fish, is known to exist only in the Indian Ocean off Comoros. (It was thought the coelacanth had been extinct for more than 70 million years until a South African fishing boat landed one in 1938.)

CONGO, which adopted a Marxist-Leninist stance in 1963, three years after gaining independence from French rule, is a tropical west-central African people's republic slightly smaller than Montana. The slave trade flourished here for four centuries. Subsistence farming occupies most of the available arable land, which is only 2 percent of the total. Crops include cocoa and tobacco; potash and zinc are mined. A hot climate, heavy rainfall, and dense forests dominate the northern countryside. The population of 2.3 million (45 percent Bakongo) is ethnically and linguistically diverse.

EQUATORIAL GUINEA is an abysmally poor group of five tropical volcanic islands in the Bight of Bonny and the Gulf of Guinea, off the west coast of central Africa and the province of Rio Muni, between Cameroon and Gabon on the mainland. Only 8 percent of the land is arable. The chief products are sweet potatoes, timber, bananas, and coffee. Per capita income for the population of around 353,000 is $250. A harsh dictatorship supplanted independence from Spain in 1968 and bankrupted the country, which now is heavily dependent upon foreign aid. Equatorial Guinea is the only black African state with Spanish as its official language. Life expectancy for women is 47.6 years; for men 44.4.

ETHIOPIA (formerly Abyssinia) is the oldest independent state in Africa and one of the oldest peopled places in the world. Once a Cold War flashpoint, it has the largest armed force (nearly 400,000 soldiers) in black Africa. It is also among the world's poorest countries: per capita income is about $140 annually.

In 1990, the once Soviet-allied government abandoned its Marxist socialism in favor of a freer economy and possible political change. Positioned with both Somalia and Sudan at the Red Sea-bordering Horn of Africa (the coastal regions were colonized by European powers at the end of the last century), Ethiopia has been bludgeoned and brutalized through the years by chronic drought and economic instability, by soil erosion and deforestation, by political tension and violence. "Red terror" in 1977 slaughtered thousands. Millions have starved to death, yet paradoxically parts of Ethiopia produce agricultural surpluses; the United States is the biggest single donor to international food-relief operations. Politics have often kept emergency rations from reaching famine victims. (It is said to be easier to buy a rifle here than a loaf of bread.)

Nearly the size of Alaska, Ethiopia has over 40 ethnic groups, but it is predominantly a Christian state, and it has been one since the fourth century. The country was invaded by Mussolini's Italy in 1935, and from 1936 to 1941 it was formally annexed to Italy and organized with Eritrea and Italian Somaliland as Italian East Africa. It was liberated by British military in 1941. The last emperor was Hailie Selassie, "the lion of Judea," who was overthrown in 1974. Intense Ethiopian-Somalian enmity was based on religious rivalry; it is now based on rival nationalisms. Through a decision of the United Nations, Eritrea was attached to Ethiopia; Eritrea offers access to the Red Sea, a potential chokepoint for oil tankers and naval vessels between the Indian Ocean and the Mediterranean Sea. Eritrean rebels have fought for independence, as have the Tigreans of the north. The civil wars are the continent's most protracted armed conflict.

Ethiopia's central plateau is split diagonally by the Great Rift System. It is also crossed by the Blue Nile. Many mountain peaks are more than 7,000 feet high. Only 12 percent of the population of 50 million is urban. Major finds of human ancestors here have included "Lucy," the earliest known hominid to walk upright.

GABON is a prosperous, underpopulated, hot, humid republic the size of Colorado on the equator at the Gulf of Guinea south of the western bulge of Africa. The landscape is timeless, the jungle aboriginal. The region was first inhabited by pygmies. Slave and ivory trading flourished. Gabon was part of French Equatorial Africa until 1960. The country is oil-rich; oil, in fact, represents about 80 percent of export earnings. There are thick rain forests, and tropical hardwoods are exported. Mineral wealth includes manganese, uranium, iron ore, and gold. The land, which is only 2 percent arable, lies mostly within the Ogooúe River basin; the mighty Ogooúe is the main artery for traffic. Key economic ties have been with the People's Republic of China and with the United States. More than 40 ethnic groups in the population of slightly over a million include Fangs, Eshiras, Bapounons, and Tekes. Gabon has moved toward a multiparty political system.

THE GAMBIA is virtually a 200-mile-long, snakelike enclave penetrating Senegal near the western bulge of Africa. Its average width is only 18 miles and its highest elevation 174 feet. The first British outpost in black Africa, The Gambia gained its independence in 1965. Rural farmers and seasonal laborers from Senegal harvest peanuts (the country's chief export), rice, and palm products. Per capita income is $255. There have been severe famines. Ninety percent of the population of 799,000 is Muslim. The literacy rate is only 12 percent. (English is the official language.) Life expectancy for women is 44.1 years; for men 40.9. A world-famous son is Kunta Kinte, the ancestor-hero of Alex Haley's *Roots*.

GHANA is a compound of real estate; it comprises the former crown colony of Gold Coast, the former trusteeship territory of British Togoland, and the former British protectorates of Northern Territories and Ashanti. It is on the Gulf of Guinea in southwestern Africa and is not quite as large as Oregon. Ghana won its independence within the British Commonwealth in 1957 and became a republic in 1960. It has a swampy coast, the interior is forest and savanna, and the climate is hot and humid. Low fertile plains are crossed by rivers and artificial Lake Volta. Cacao, hardwoods, manganese, industrial diamonds, gold, and bauxite are produced, and cocoa is the main (70 percent) export. Almost all 15 million Ghanians are black. The majority are animist or Christian. Ethnic diversity is wide, and more than 50 languages and dialects are spoken. The official language is English. The literacy rate is 30 percent, and per capita income is $420.

GUINEA has rich soil and vast mineral reserves (including one of the world's largest bauxite deposits), yet this 94,925-square-mile, mostly rural farming country on the Atlantic coast of west Africa has a per capita income of only $305. The only urban area is the capital city of Conakry. Once a trading center for gold and slaves, Guinea in 1958 became the first of France's colonies in west Africa to attain independence; it was the only French colony to reject membership in the French community. Eighty-five percent of the population of 7.1 million is Muslim, 48 percent is literate. There are nine official languages. Life expectancy for women in this military-run republic is 41.8 years; for men 38.7 years.

GUINEA-BISSAU, which gained its independence from Portugal in 1974, is a low-lying, underdeveloped, tropical west African republic on the Atlantic coast. Per capita income is $165. A swampy coastal plain covers most of the country (13,948 square miles), which is about the size of Connecticut and New Hampshire combined. Peanuts, corn, and cotton are the principal products. Only 15 percent of the population of 975,000 is literate. Ethnic groups include Balanta, Fula, and Mandinga. Guinea-Bissau is unique in its policy of suffrage for everyone over the age of 15.

IVORY COAST (Côte d'Ivoire), thus named for the elephant herds that once roamed the land, became with independence in 1960 a one-party dictatorial state with a Western-oriented free-enterprise economy. President Felix Houphouet-Boigny has served seven five-year terms. Ivory Coast has been the most prosperous of tropical African countries and a leader of the pro-Western bloc in Africa. The standard of living is the highest in black Africa, and per capita income is one of the highest. The capital of Abidjan, the small country's chief port, has been a showcase of economic prosperity and stability. But Ivory Coast has also been cynically described as "just another failed African country subject to nepotism and corruption, with one of the highest per capita's foreign debts."

It is situated on the Gulf of Guinea and under the continent's southwestern bulge. Its immediate neighbors are Liberia, Guinea, Mali, Burkina Faso, and Ghana. Slightly larger than New Mexico, it is endowed with watered farmlands, hardwood forests, and a tropical climate. Coastal swamps merge into inland savannas. Forests cover the western half of the country, and there are low mountains in the northwest. Less than a tenth of the land is arable, but the economy is primarily agrarian. The former French colony is the world's largest producer of cocoa and Africa's largest coffee producer. Pineapples, bananas, rubber, cotton, and timber are also produced, and diamonds and manganese are mined.

Most of the 11.8 million people, nearly all black, live in huts in small villages. There are millions of immigrant workers, and there has been a flow of refugees from Liberia. There are more than 60 ethnic, potentially divisive groups, including the Baule, Bete, Senoufou, and Malinke. Sixty-three percent of the people are animist.

A new capital city, Yamoussoukro, is being planned for an island location.

KENYA has one of the highest population-growth rates of any country in the world. Increasing by about 4 percent annually, the population may reach 46 million by the year 2020. (It is now about 25 million.) Kenyans belong to more than 40 different groups, each with its own language and culture, though Swahili links them all. "Pull together" is the national slogan for this former British protectorate on the Indian Ocean coast of east Africa.

It is not a totally harmonious society, though it has been described as an island of peace, political stability, tolerance, and breathtaking economic development in an ocean of political strife, civil war, and famine. A breakdown of traditional family structures has occurred. By some estimates, 50,000 children roam the streets of the capital city of Nairobi, which has a population of around 2 million. Asian families have dominated the economy. Remains of the earliest humans, dating back more than 2 million years, have been found here.

The country, which is a bit smaller than Texas, and a de facto one-party state since 1969, has close ties with the West. For more than a decade, the United States has had the right to use the port of Mombasa for the support of military operations and for troop rest and recreation. Mombasa, which is predominantly Islamic, is Kenya's second-largest city (600,000 residents). Nairobi is cosmopolitan, the takeoff point for safaris.

The northern three fifths of Kenya is semidesert. There are savannas and highlands. Equatorial heat is air-conditioned seasonally by monsoon winds. Only 4 percent of the land is arable. Sisal, tea, and corn are the main crops. Among the minerals are diatomite, barite, feldspar, and garnets. Mount Kenya, 17,058 feet high, is a major tourist attraction. The country is 38 percent Protestant, 47 percent literate.

LESOTHO, once called Basutoland, is completely surrounded (or trapped) in the east-central part of the Republic of South Africa. An independent kingdom since 1966, it is a tableland high plateau, an enclave rising toward mountains in the east. The "lowest" site is nearly a mile high. It is largely an agricultural and stock-raising country. Diamonds are mined, but Lesotho lives off the sale of blankets. The country, which is slightly larger than Maryland, has provided a sanctuary for rebel groups fighting to overthrow Pretoria's apartheid policies, but it has been forced to depend almost entirely on South Africa for economic survival. About 200,000 adult males among the 1.8 million Lesothoans, all blacks, spend up to nine months a year in South African mining, farming, or industry. Per capita income is $281.

LIBERIA is the only black African state to have escaped colonialism, having held both France and Great Britain at bay. The "founding fathers" were freed American slaves backed by private interests in the United States. In 1847, it became Africa's first independent country. Liberia is today the closest United States ally in west Africa. During the Reagan Administration (1981–1989), the country was the largest per capita recipient of American aid to the sub-Sahara, the only state in the region where United States military planes could land with simply a day's notice.

Slightly smaller than Pennsylvania, Liberia is at the southwestern extremity of the warm, humid, coastal bulge of the continent. The country is blanketed with tropical rain forests. A coastal plain rises to a rolling plateau and low mountains near the borders with Sierra Leone, Guinea, and Ivory Coast. Six major rivers run west to the Atlantic Ocean.

The rubber industry has been the major employer; it built the world's largest contiguous plantations. Western companies also developed iron-ore mining. A vast international merchant fleet is registered under the Liberian flag.

Twenty percent of the population of 2.6 million is Muslim; ninety-five percent are members of indigenous tribes, most of which hold traditional beliefs. English is the official language, but at least 20 local languages of the Niger-Congo group are spoken. Rice is the leading cash crop. Per capita income is only $362, though the country is rich in minerals (diamonds and gold are mined). There is a skilled work force. The literacy rate, however, hovers around 20 percent.

Until recently, the state had been relatively free of ethnic strife.

More than 150,000 Liberians fled to the forested hills of Guinea and Ivory Coast to escape the 1990 civil war. The brutal, corrupt reign of Samuel K. Doe siphoned off the country's wealth, an egregious example of economic mismanagement and prodigious self-aggrandizement, and it did not observe human-rights standards. Liberia was the first African country ever suspended from the International Monetary Fund and from World Bank borrowing.

MADAGASCAR, or Malagasy Republic, is the fourth-largest island in the world, excluding Australia. About the size of Texas, or half again as large as California, it is at once one of the world's richest states in its natural wonders and one of the poorest in terms of economic development. Madagascar, in fact, is one of the world's 10 most impoverished states. The environment is threatened. Four fifths of the land is barren, burned over by subsistence farmers and cattle herders. Eighty percent of the population of 11.5 million lives off the land.

There is a diminishing supply of the basic diet of rice. Per capita income is $210. Vanilla, yielded by orchids, is a major export. Better economic conditions are stemmed by the annual population growth of nearly 4 percent; as it is, 122 of every 1,000 babies die before their first birthday.

An achingly beautiful site "moated" by the Indian Ocean off the southeast coast of the African mainland and the 250-to-625-mile-wide Mozambique Channel, this island of emerald foliage and hills has unique, bizarre, and marvelous biological diversity. The phenomenal variety of plants may harbor potential wonder drugs. Madagascar is the naturalist's promised land, a living laboratory for evolution. The world's largest known bird, the Aepyornis, disappeared in the waves of extinctions ending some five centuries ago. Scientists estimate that 150,000 of the 200,000 identified living things on the island, which was probably wrenched from the African mainland by tectonic forces 160 million years ago, are found nowhere else.

The 1,000-mile-long island has been independent of French rule since an armed revolt in 1960. Most of the people are of mixed African and Malay-Indonesian descent; there may be Polynesians. Arabs established colonies here from the 9th to the 14th centuries. The language is tongue-twisting polysyllabic. For the most part, the Africans live on the coasts, the Asians on the island plateau. The average life expectancy is 50 years. Hitler thought about shipping Germany's Jews here.

MALAWI, the fourth-poorest country in the world, is a pro-Western, matriarchal, small, narrow, southeast African republic sitting along the Great Rift Valley. Rural and traditional village customs prevail. Malawi was the British protectorate of Nyasaland until 1964. There was only one paved road and no university or any industry. The 457,747-square-mile "land of fire," the area of Pennsylvania, has limited resources but is self-sufficient in food. Fish provide the main protein for the population of 8.8 million. Most of the people, who by nature are friendly, happy, and optimistic (90 percent are Chewa), are subsistence farmers, often using slash-and-burn methods; some are miners in other countries part of the year. The work force is unskilled. Per capita income is $160 (it was $30 in 1964).

The population tripled between 1950 and 1990, inflated in part by the 700,000 who fled guerrilla wars in neighboring Mozambique; the refugees were received with orderliness and hospitality. The infant-mortality rate is astronomical (150 per 1,000 live births). Malawi was the first black African country to establish diplomatic relations with South Africa. The literacy rate is 25 percent. The incidence of AIDS is high. Life expectancy for women is 45.4 years; for men 42.7 years. Seventy-five percent are Christian. The official languages are Chichewa and English. A leader in wildlife conservation, Malawi in 1980 established the world's first freshwater, primarily underwater national park, to protect cichlids, a family of fish.

MALI is an impoverished, drought-ridden, sub-Saharan, one-party republic about the size of California and Texas combined. Landlocked in the interior of west Africa, it is half desert; only 2 percent of the land is arable. Saharan sands continue to advance on historic wetlands. President Moussa Traore seized power in a coup in 1968, eight years after Mali's independence from France, and was overthrown in 1991 by a multiparty democracy movement. Five centuries ago, Timbuktu, the capital, was a city of a million, a center of Muslim culture and extraordinary erudition and advanced medicine; its population today is a little more than 20,000. The official language is

French, but most of the 9 million people communicate in Bambara, a market language. The literacy rate is 9 percent and the per capita income rate is $190. A Mali woman can expect to live 43.6 years; a man 40.4 years.

MAURITIUS, thrown up from the ocean floor some 7 million years ago, is a serene, volcanic, tropical island the size of Rhode Island, set in the Indian Ocean 500 miles east of Madagascar. Mountain peaks surround a central plateau. Almost all of the arable land is covered with sugar cane. Indentured workers brought by the British from India replaced the African slave laborers imported earlier by the French. Ships of all countries are welcome. United States sailors pump a million dollars into the economy on a single liberty call. The population is 1.1 million: 51 percent Hindu, 7 percent Tamil, 27 percent Creole. English is the official language, though most Mauritians speak a French patois acquired during the 90-year settlement here by the French in the 18th and the 19th centuries. Knitwear is a major export. The extinct, flightless dodo bird waddled here.

MOZAMBIQUE into the 1990s suffered a vicious war between the avowedly Marxist government and right-wing guerrilla insurgents. The Soviet Union armed the government, which was militantly opposed to white rule in neighboring South Africa, and South Africa armed the guerrillas. Millions of Mozambicans were displaced, and another million, including 750,000 children, died, many starving to death. The economy was crippled. The United States and the Soviet Union finally agreed to bring the warring factions together.

Mozambique is a rural, underdeveloped, socialist republic, about the size of Texas, on the southeast coast of Africa and west of the island of Madagascar. Formerly the colony of Portuguese East Africa, it gained its independence in 1975 after a 10-year armed struggle by the Mozambique Liberation Front. The Portuguese carried on a lively slave trade here until slavery was abolished in 1878. About half the state is a coastal plain. The economy includes copra, cotton, cement, cashews, sugar, and tea. Coal, asbestos, bauxite, beryl, and gold are mined. The impoverished population of 14.3 million, mostly black African, belongs to about 10 groups. The literacy rate is 7 percent. Per capita income has been as low as $3 a week. Life expectancy is 46 years for women, 44 years for men.

NAMIBIA, sparsely populated, mineral-rich, the world's fifth-largest uranium producer, raised its blue, white, red, and green flag on March 21, 1990, and became one of the world's newest sovereign states. The black majority had won independence from illegal control by South Africa. Composed under the supervision of election monitors from the United Nations, the democratic, enlightened, nonracial constitution is the most liberal in Africa: an elected government, an independent judiciary, an extensive bill of rights, curbs on executive power. The constitution is considered the country's most precious possession. The United States spent $100 million to supervise elections, yet Namibia is the target of sanctions by American business.

The largely desert country, on the southwestern coast of Africa, is twice the size of California or almost as large as Germany and France combined. One percent is arable; drought is continuous. It is strategically placed between Angola and South Africa. Walvis Bay and its surrounding area are still legally controlled by South Africa, a certain source of conflict in the future. There are only 1.4 million Namibians, two thirds of whom are Ovambo (Bantus), who live in the north. The South West Africa People's Organization (SWAPO), with Communist support, fought a 23-year bush war with Pretoria for control of the territory. (In 1973, the United Nations General Assembly recognized SWAPO as "the sole and authentic representative" of the Namibian people.)

The fragile economy, heavily dependent on mining, is bolstered by the presence of diamond, coal, copper, zinc, lead, and tin mines. Uranium oxide is exported from one of the world's largest deposits. At the time of independence, with Namibia's goal a functioning and prosperous democracy, the average annual income of a white Namibian was $16,000; of a black $63. SWAPO controls the electronic media; the print media remain largely free to do what they want.

NIGER is a landlocked northern African republic about three times as large as California. It has been run by a military government since independence from France in 1960. With mountains and deserts in the north and a narrow savanna in the south, arable land is scarce, drought is commonplace. The majority of the 7.5 million population lives along the Niger River in the south, is black African and Sunni Muslim, and belongs to culturally diverse units. Peanuts, cotton, and animal products are exported, and there are extensive iron-ore deposits. The economic crisis is perpetual. Per capita income is about $200. Only 10 percent of Nigerans can read and write. Life expectancy for women is 44 years; for men 41 years. French is the official language.

NIGERIA is the continent's most populous country: at least 115.3 million. Oil resources underwrote an ambitious economic development program, but the $100 billion realized in oil sales in the 1970s was depleted through personal corruption, much of it being diverted to private bank accounts in Switzerland. Foreign debt is in the tens of billions of dollars. Nigeria expelled 2 million foreign immigrant workers to neighboring countries when the oil boom went bust. There were more than a million casualties in the provisional military government's civil war with the eastern region (Biafra), when it tried to secede in 1967–1970. More than 200 black African groups thwart national unity.

Four east-west regions divide the politically unstable country: a coastal mangrove swamp, a tropical rain forest, a plateau of savanna and open woodland, a semidesert. Thirty-four percent of the land is arable. Twice the size of California, Nigeria lies on the west coast of the continent's midsection on the Gulf of Guinea and borders on Benin, Niger, Chad, and Cameroon. Per capita income is $805. The de facto capital is the seaport Lagos; the future capital will be Abuja, in central Nigeria. The country is the world's largest exporter of kernels and palm oil, and counts among its other resources cocoa, rubber, cotton, hides, and tropical woods. A transition from military leadership to civilian rule is expected at the end of 1992.

RWANDA has the highest natural fertility rate among women in Africa—8.6 live babies per woman—and the highest population density in sub-Saharan Africa. The population of 7.3 million lives on almost every available inch of the tiny country's hilly land. Government appeals to limit family size have had little effect. Two of the cities are believed to have the highest rate of urban HIV infection on the continent.

Rwanda is the source of the Nile and one of the last refuges of the mountain gorilla. There is a chain of volcanoes in the northwest. The grass uplands and the hills are farmed, but the land is subject to erosion. Eighty-nine percent of the people are Hutu, 1 percent is pygmy. (About 175,000 pygmies live in the tropical rain forests of seven African countries.)

Rwanda was granted limited self-government by Belgium in 1961, and the Republic of Rwanda was recognized by the United Nations the next year. Once agriculturally self-sufficient, the country's population boom is outstripping the capacity to produce food. Coffee, tea, sisal, and cotton are grown. The economy depends on cattle raising and the mining of tin, gold, and wolframite, a mineral used as a source of tungsten.

SÃO TOMÉ AND PRÍNCIPE is a cocoa-growing west African island-state hardly larger than New York City. Set in the Gulf of Guinea 125 miles west of Gabon and virtually on the equator, it is part of an extinct volcanic mountain range. There are lush forests and croplands, yet the country is one of the least developed, with a per capita income, for a population of 121,000, of $345. Independent of Portugal since 1975, São Tomé and Príncipe completed the transition from a one-party state to a multiparty democracy in 1991. The land once held large plantations. Early settlers included Portuguese convicts and Jewish refugees.

SENEGAL, an oasis of hope for human rights, personal freedom, and the democratic process, was the first French outpost in black Africa. It won its independence and membership in the United Nations in 1960. Senegal was once the center of extensive slave trading—labor for the New World. It is mainly a flat, rural land of subsistence farmers which surrounds the drought-plagued enclave of The Gambia on the Atlantic coast. There are swamps and jungles in the southwest and low rolling plains. An irregular weather pattern brings severe drought to the Sahel, or desert borderland, in the northwest.

Agriculture, fishing, and stock raising are major pursuits; peanuts and seafood are chief exports. Phosphate is mined. Industry has been privatized and government bureaucracy streamlined. The transportation system is efficient, and there is a free press. Nearly all energy must be imported. Per capita income is $360.

The culture is French oriented; the capital city of Dakar is African chic. There are nearly 20 legal political parties, and there has been a turbulent period of strikes, rioting, and election violence. Polygamy is allowed, but women are encouraged to obtain an education. Ninety-two percent of the 7.5 million population is Muslim. The literacy rate is 10 percent. Senegal sent 500 troops to Saudi Arabia during the Persian Gulf War in 1991.

SEYCHELLES is a group of 92 coral atoll and granite islands strung out in the Indian Ocean 700 miles northeast of Madagascar. Many of the islands are unable to support human life, but the others are popular with tourists—jet-setters bask on the strips of beaches. Nowhere else are there midocean islands of granite. Seychelles has been called the world's smallest continent; it may be a remnant of an ancient land mass; it has a continental shelf. Seychelles, once a pirates' haven and a refuge for exiles, won its independence from a reluctant Great Britain in 1976. Most of the Asian-African-European population of 70,000 lives on Mahé (the largest island), is Roman Catholic, and speaks French-based Creole. Mahé boasts the 2,969-foot-tall Mount Morne Seychellois. Chief exports are shark fins, patchouli (an East Indian shrubby mint that yields an essential oil used in the making of a heavy perfume), vanilla, cinnamon, guano, and tortoise shells.

SIERRA LEONE is one of the world's largest producers of industrial and commercial diamonds, but per capita income is $258. The economy of this west African tropical republic, which is slightly smaller in size than South Carolina, is battered by corruption and inflation. The population of 4.1 million is about equally divided among Christians, Muslims, and animists. Life expectancy for women is 36 years; for men 33. Besides diamonds, exports include iron ore, palm-oil products, bauxite, rutile (a reddish-brown mineral consisting of titanium dioxide and a little iron), and coffee. Coastal mangrove swamplands have been cleared to increase agricultural production. In the 17th and the 18th centuries, the slave trade was brisk. Freetown, the capital, was established in 1787 as a refuge for freed slaves from Great Britain. Sierra Leone has been an independent state within the British Commonwealth since 1961, and English is the official language.

SOUTH AFRICA, where humankind learned to use fire a million years ago, has been arguably the most racist state in the world. Its policy of segregation, or apartheid, prevented the spread of national prosperity by reserving in the Group Areas Act nearly 87 percent of the land for the white minority, which is only 13 percent of the total population of 38.5 million. More than half of the black population— that is, 2 million more people than the entire white population— lives in "informal housing," or shacks. Elimination of major apartheid laws would allow all South Africans to live where they choose. (South African law even discrminated against the Indian spiritual leader and former lawyer Mohandas K. Gandhi because he was not white-skinned.)

Ten subsidized "homelands" were established for blacks, who were forcibly moved into them and stripped of citizenship. Johannesburg's sprawling township of Soweto (2.5 million blacks) is the largest urban concentration in the country; there is bitter, bloody factional violence among the rival black political groups. In June 1991, race classification, the legal foundation of apartheid, was ended; racially based legislation is to be no more. The process of ending apartheid totally has been prodded by the economic, jackhammer blow of worldwide sanctions and the increasing assertiveness of the black majority. The sanctions, however, hit blacks hard, eliminating old jobs and failing to create new ones. Despair and deprivation engendered by the lack of opportunity have contributed to rising crime and instability. Ten corporate conglomerates control nearly 90 percent of the shares listed on the Johannesburg Stock Exchange.

South Africa, which the British acquired after the Boer War with the Dutch (1899–1902), is the only African country with coastlines on both the Atlantic and the Indian oceans. It lies at the southern tip of the continent and is about twice the size of Texas. The richest economy in Africa was shaped by the discovery of gold and diamond deposits, which are hauled to the surface on the backs of black labor. Platinum, uranium, and chromium are also mined. Because the issues of poverty, joblessness, and attending violence afflicting the black majority have not been addressed, new investments are not being attracted. South Africa, in such an atmosphere of violence, could become ungovernable unless the economy improves. There are few major rivers or lakes, and rainfall is sparse in the west. The interior plateau is vast.

SUDAN, where African and Arab ways mingle and clash, seems to be an unmanageable Islamic country of 25 million people, 500 tribes, 115 languages, and 60 political powers. A Sudanese proverb goes, "When Allah created the Sudan, Allah laughed." The United States Department of State gives the maximum level of hardship pay to its employees in the capital of Khartoum, where the White and the Blue Niles converge to form *the* Nile, the watery thread that holds the continent's largest state together. A fourth the size of the land area of the United States, Sudan suffers the biblical plagues of famine, locusts, war, flood, and pestilence. Only 6 percent of the arable land is cultivated. When a flood brings on a bumper grain harvest, labor and transport crises waste half the yield. Roads are all but nonexistent. Hundreds of thousands of Sudanese are always on the verge of starvation. In 1988, about a quarter million southern Sudanese died from famine and illness. Sudan was once the "breadbasket" of Africa. (The United States is now the largest donor of food.)

A 17-year civil war produced a million casualties. The current ruthless, intolerant regime follows an ideology of militant Islamic fundamentalism. It deflects emergency shipments of food and bombs civilian relief sites. It is responsible for a third of all overdue payments to the International Monetary Fund. When Sudan sided with Iraq during the 1991 Persian Gulf War, the gulf states severed nearly all aid.

Known as the Anglo-Egyptian Sudan until 1956, the state lies at the eastern end of the harsh, barren Sahara. Christian blacks live in the swampy south, Arab Muslims live in the north. Eighty percent of the people cannot read or write. Per capita income is $375. Most of the major export, cotton, is grown between the two Niles in the world's largest irrigated farm, el Gezira, 2.1 million acres.

SWAZILAND is a landlocked monarchy in southeastern Africa near the Indian Ocean. It is bordered on the south, west, and north by the Republic of South Africa and on the east by Mozambique. It was a protectorate of the British until full independence was gained in 1969. The terrain is largely plateau and mountains. Europeans own nearly half the land, and European-owned mining operations (principally coal and asbestos) supply about half the kingdom's income ($790 per capita). Exports also include animal and timber products, sugar, cotton, and fruit. Swaziland has approximately the same population figure as Baltimore, Maryland; nearly half of the 756,000 residents are tribal religionists. Political parties are banned. The administrative capital is Mbabane; the legislative capital is Lobamba.

TANZANIA, where man's earliest ancestors roamed and made the first tools, and superb prehistoric artists painted on the sheltered surfaces of cliffs and on rock faces in the remote central plateau region, is a huge, seedy country. Twice the size of California, the equatorial east African land fronts on the Indian Ocean and is encircled by Kenya, Uganda, Rwanda, Burundi, Zaire, Zambia, Malawi, and Mozambique. Tanzania was formed when the Republic of Tanganyika and the island-republic of Zanzibar (the Isle of Cloves) overthrew Arab rulers in 1964 and united as a republic. (Zanzibar, with 700,000 citizens and a decaying economy, retains internal self-government.)

Tanzania possesses highlands, a 12-mile-wide caldera (which was formed when a volcano exploded 3 million years ago; it is now an animal preserve), windy coastal plains, and a central plateau, but two thirds of the country cannot be cultivated because of tsetse-fly infestation and the lack of water. Yet, agriculture provides 90 percent of employment. Sixty percent of the children are malnourished. Per capita income is less than $250, and less than 50 percent of the people are literate. There are many tribes, but not one tribe is large enough to dominate an election. Most of the 130 ethnic groups speak a distinct tongue. The single-party country of 25.3 million (with one of the globe's lowest population densities) is guided by a policy of self-reliance, but it is highly dependent on foreign aid. (The capital, Dar es Salaam, is said to have the largest potholes on the continent.) In a spectacular wildlife scene, millions of large animals roam the expansive ecosystem of savannas and woodlands. Twenty-million-year-old Lake Tanganyika, in the west, is larger than Belgium and is the continent's deepest lake (4,710 feet). The southern half of Lake Victoria, the second-largest freshwater body in the world, is in Tanzania. Kilimanjaro (19,340 feet), near the Kenya border in the northeast, with three steaming craters on its summit, is Africa's highest mountain.

TOGO, once a center of slave trafficking, is a military-run republic at the Gulf of Guinea on the southern coast of west Africa. President Gnassingbe Eyadéma suspended the constitution in 1967 and began one-party rule. Freedom of assembly is controlled tightly and an atmosphere of fear prevails. Togo is a mountainous land tucked among Ghana, Burkina Faso, and Benin, and it has a hot and humid climate. The French section of Togoland became the republic of Togo in 1960. Productive land is hard to come by, but there are two savanna regions. The main export crop is cocoa. Coffee, millet, manioc, and yams are also grown. Togo has one of the world's largest phosphate reserves. The leading manufactured product is shoes. For all that, it is still a poor country. More than half the population of 3.5 million blacks is animist.

UGANDA, a variegated country fragmented by warring guerrilla factions, is a landlocked, army-run east-central African republic slightly smaller than Oregon and straddling the equator. Temperatures are modified by the altitude: the lowest point is the Nile River Valley, 2,000 feet above sea level. The land is volcanic and fertile; rainfall is plentiful. Cotton and coffee are the major exports. The government becomes impoverished when there is a sharp drop in coffee prices. Enough copper is mined to export. Uganda was once known as the "pearl of Africa."

Corruption and ruthlessness set a record in the anarchic rule (1971–1979) of the dictator Idi Amin, who slew 300,000 of his people at whim and expelled 80,000 Asians because of their "economic imperialism," before he himself was expelled by Tanzanian troops and Ugandan exiles. Two thirds of the population of 17.1 million is Roman Catholic. The more than 40 distinct ethnic groups include Ganda,

Nikoke, Gisu, Soga, Purkana, Chiga, Lango, and Acholi; there has been much tribal violence. The pastoral Karamajong warriors in the harsh, rainless northeast believe they own the world's cattle; girls in their early teens are offered as brides for the price of 60 cows.

In recent times, the former British protectorate (Uganda has been independent since 1962) has been suffering the tragedy of a very high incidence of "the slim"—the word Ugandans use for the AIDS disease. The state has had only one doctor per 22,000 people.

ZAIRE, formerly the Belgian Congo, made familiar to the world by the *Heart of Darkness, The Niggers of the Narcissus,* and other novels by Joseph Conrad, is both Africa's most unstable country and the most pro-American country in the sub-Sahara. Government corruption and serious economic difficulties have dominated the continent's third-largest state in the past two decades. The West has had to intervene to protect its interests, helping to repulse invaders from Angola. For 26 years, President Mobutu Sese Seko held absolute power and amassed a personal fortune; in 1991, he responded to demands for democracy by agreeing to a coalition government.

Equatorial Zaire occupies the major part of the Congo River basin and a small part of the Upper Nile basin, with a sliver extending west to the Atlantic Ocean. It is the world's chief source of cobalt. It is also rich in other mineral resources, such as copper (large deposits), diamonds, and petroleum. But per capita income is $150. There are 250 ethnic groups. When the impoverished country, which is about the size of the entire United States east of the Mississippi River, gained its independence in 1960, there were fewer than 50 college graduates among the Zairians. Fifty percent of the population of 35.3 million is Roman Catholic. About 150 children of every 1,000 born die before the age of five. The capital city of Kinshasa was formerly called Leopoldville.

ZAMBIA, formerly Northern Rhodesia, is in one of the major copper-mining regions in the world, but its own reserves of copper—at present the chief export—will be exhausted by the end of the century. The region was ruled by the British until it joined with the Federation of Rhodesia and Nyasaland in 1953; independence was gained in 1964.

The landlocked, subtropical republic in south-central Africa is about the size of Texas and borders eight other countries. There is a vast plateau, with mountains to the northeast and the east. The Zambezi, which rises in the northwest, is a major African river. Only about 7 percent of the land is arable; tobacco, corn, and peanuts are grown.

During the 18th and 19th centuries, the population was ravaged by Arab and Portuguese slave-procuring. All but 1 percent of the 7.6 million Zambians are black African and most are animist; there are 73 ethnic groups and more than 80 languages. Per capita income is $410. Severe droughts occur, deflecting food sufficiency. The country's population growth of 3.7 percent is exceeded on the African continent only by Kenya's, and half the population is under 15 years of age. Zambia is in dire economic straits, which have induced a sharp increase in nutrition-related child mortality. The incidence of AIDS is increasing sharply (the son of the president was a victim) and a national emergency program on AIDS education has been urged. In late 1991, the president of 27 years was resoundingly defeated in a multiparty election, a rare defeat by popular vote of a senior African head of state. There had been pent-up frustration at a dramatic fall in living standards and a weariness with a paternalistic style of leadership.

ZIMBABWE, formerly Southern Rhodesia, is slightly larger than Montana. There were 7.2 million blacks and 220,000 whites when the majority population took control of the government in 1980, renaming the state. (Rhodesia, which was named for Cecil Rhodes, a statesman who made a fortune in South African diamonds, had been created to help spread British civilization throughout Africa.) Today's population is 10.2 million, of whom only 100,000 are white. There is widespread ancestor worship and almost everyone believes in "the spirits."

Landlocked in southern Africa, this economically strong subtropical country has high plateaus, an excellent system of railroads and paved roads, and large-scale cattle ranching and manicured fields of tea. Coal, high-grade chromite, asbestos, and copper are among the 50 minerals mined. A quarter of the people are urban. Per capita income is about $900. Breathtaking Victoria Falls ("smoke that thunders") is a mile-wide break in the 1,700-mile-long Zambezi River, and the capital, Harare, originally named Salisbury by British pioneers more than a century ago, is one of the world's loveliest cities. Commercial agriculture is dominated by the tiny white minority, which continues to hold much of the best farmland. Periodically, there is serious unrest. Zimbabwe, which means "venerated houses," is identified by some experts with the biblical Ophir, site of the fabled King Solomon mines. About 10 percent of the land is set aside for parks and game preserves.

REUNION (a non-state), which has belonged to France since 1665 and has the highest standard of living of all the islands in the Indian Ocean, is a tropical, cyclone-prone, volcanic overseas department (969 square miles) east of Madagascar and southeast of Mauritius. It was once called Ile Bourbon. The Reunionese are a mixture of French, African, Malagasy, Chinese, Pakistani, and Indian. Sugar is the chief export. Unemployment tends to be high.

NORTH AFRICA/SOUTHWEST ASIA: a regional system characterized by economic inequality, it is called the Arab world, but much of the population is not Arabic—it is overwhelmingly Islamic. This realm has been the wellspring of world history, a source of scientific discovery, a land mass of intense discord. The map of the Arab Middle East was drawn after the First World War by the victorious Allies when they seized power from the Ottoman Empire, a ramshackle Muslim domain held together by the glue of Islam. There are frequent territorial disputes and boundary frictions. Precipitation is unreliable, with a great deal of fluctuation.

Enormous oil reserves give southwest Asia substantial world influence. After the Second World War, Western oil companies negotiated concessions with the Persian Gulf rulers to tap crude for a royalty of about 20 cents a barrel. The Sahara, the exact size of which ebbs and flows, is so vast a desert (sometimes nearly 4 million square miles) that the 48 contiguous states of the United States could fit snugly within. Some of the dunes are 65 stories high. Twenty-two countries are at risk of being engulfed by the Sahara when it spreads. The Fertile Crescent, the seat of early civilization, has been blighted by deforestation and overgrazing, and is today part of a semidesert extending from Gibraltar to India. The Middle East in this century

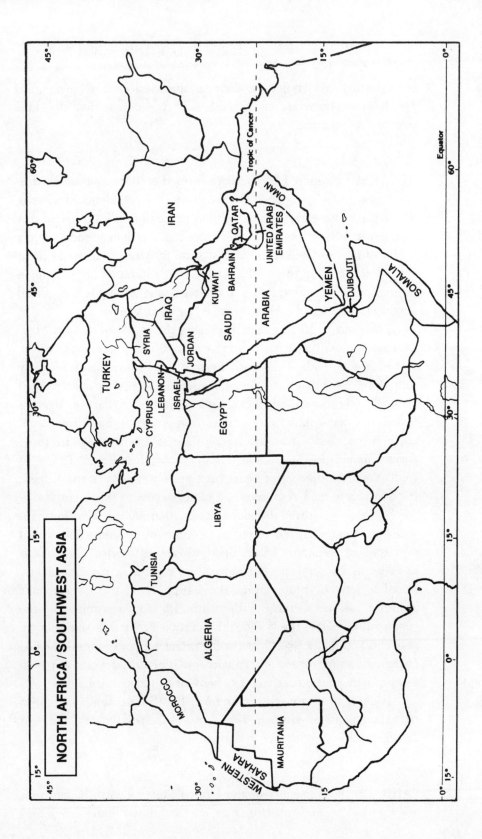

NORTH AFRICA / SOUTHWEST ASIA

has been divided by religions, alliances, and ideologies, not to mention greed. Most Arabs are comfortable with the notion that all Arabs form a single nation.

ALGERIA, conquered by the French in the 1830s and annexed in 1842, won its independence in 1962 during the presidency of Charles de Gaulle. The eight-year revolution preceding self-autonomy cost 600,000 Algerians their lives and propelled a million French settlers back to France. The second-largest state in Africa, Algeria is about a third the size of the continental United States. The 25 million population is of Arabic and Berber stock, and Arabic is the official language. The population growth is 3 percent yearly.

Nine out of 10 Algerians live along the fertile northwest Mediterranean coastline, which extends for 620 miles and embraces the 50-to-100-mile-wide Tell, with its moderate climate and adequate rain. Two major chains of the Atlas Mountains enclose a dry plateau area. The central and southern regions are desert. Part of Algeria's Sahara has major mineral resources (iron ore, phosphate, zinc, lead, oil, and gas) and a few oases with date and cork trees. In 1967, Algeria joined the Arab countries in war against Israel. It was in conflict with Morocco for years over neighboring Western Sahara; the aim was to gain resources and better access to the Atlantic.

After nearly three decades of stagnation and socialist misman-agement, Algeria has moved out of an economic quagmire and toward a progressive economy based upon private enterprise. Television-satellite dishes serve up an information explosion in this formerly closed society. In 1988, savage riots erupted over the government's austerity measures, rocking the authoritarian government of the state's ruling party, the National Liberation Front. Massive debts, a shriveling economy (world prices for petroleum and natural gas had collapsed), and rampant official corruption produced a crisis in 1990 and brought on the first free elections since independence.

Islam is the state religion and Muslim (Sunni) fundamentalism has gained political strength. Two thirds of the population is not even 25 years old.

BAHRAIN, headquarters of the United States Central Command, and the docking and refueling base for the United States Navy contingent that has been patrolling the Persian Gulf since 1949, became in 1932 the first country in the gulf-oil business. Occupied by the Portuguese for a century, and once a British protectorate, Bahrain is the smallest and the poorest of the gulf states and the third-most-densely populated country in the world. The state consists of a group of 33 flat, hot, humid, limestone, and sandy islands (only five are inhabited), one fifth the area of Long Island, midway between the tip of the Qatar Peninsula and mainland Saudi Arabia. The only "mountain" is 443 feet high. A $1 billion, 15-mile-long causeway has linked Bahrain and Saudi Arabia since 1987.

At the present rate of production, oil reserves will last another 10 to 15 years, gas reserves another half century. Offshore oil is being sought. The largest nonpetroleum economic activities are aluminum smelting and banking. Bahrain was the world center for pearl traders and is on the main round-the-world air route. The population is 500,000. (There are 165,000 aliens.) The Sunni Muslim sect, which includes the ruling family, predominates in the urban area, and the Shiite (70 percent of the population) in the countryside. Shiites have close ties with Iran, which has laid claim to Bahrain in the past. The state has the highest literacy rate in the Arab world: 74 percent.

There is a move afoot to make the island sheikdom the "Florida of the gulf." During the Persian Gulf War of 1991, Bahrain allowed the coalition access to its military facilities and provided rest and recreation for coalition troops. Summer temperatures reach 110° F.

CYPRUS is an island-state in the eastern Mediterranean Sea off the southern coast of Turkey. Its copper deposits were prized so highly in ancient times that the island was controlled successively by Egyptians, Assyrians, Phoenicians, Greeks, Persians, Romans, and Syrians. (The word "copper" comes from the Latin word *cuprum—aes Cyprium*, the Cyprus ore.) Copper is still mined. The island's forests have been repeatedly destroyed and resuscitated. In the 19th century, Turkey ceded Cyprus, which is smaller than Connecticut, to the British as security for a loan.

The population of 700,000 is more than four fifths Greek; the minority is Turkish. Strife between the two nationalities has a long history. After the Second World War, violence erupted when the Greeks pressed for self-rule and union with Greece. An invasion by Turkish troops in 1974 resulted in a formal segregation of settlement patterns, which was recognized only by Turkey. On November 15, 1983, the Turkish Cypriot minority unilaterally declared the independence of the Turkish Republic of Northern Cyprus. Thirty thousand Turkish troops occupy the breakaway northern third of the island in violation of international law. The sandy-beached southern section reflects its Greek heritage by its religion, language, and culture. Nicosia serves as the capital of both factions. The 118-mile-long cease-fire line consists of mine fields, barbed-wire fences, and sandbags. A wide plain lies between two mountain ranges, which rise to weather-impacting Mount Olympus, formerly Mount Troodos, at 6,403 feet above sea level. Cyprus is rich in agricultural production, mainly in wheat, olives, tobacco, and wine.

DJIBOUTI is a hot, parched, sandy, barren desert-republic of 329,000 at the juncture in the war-ravaged Horn of Africa between the Gulf of Aden and the Red Sea. A country of shepherds, it gained its independence from France in 1977, but about 4,000 French troops are still stationed here to keep the peace and buoy the economy. The (chiefly) Muslim republic, which is the size of Massachusetts, is hemmed in by Ethiopia, and there is persistent hostility between the Muslim sects, Afars and Issas. Lake Assal is 502 feet below sea level; Moussa Ali is 6,768 feet high. Salt in vast amounts is the principal natural resource. There is a project to harness geothermal energy for electricity and fresh water. Unemployment is high. Per capita income is $400, making Djibouti one of the world's poorest countries. The literacy rate is 17 percent. Because the earth is coming apart and oceanic crust is being extruded as dry land, Djibouti is expanding geologically several centimeters a decade.

EGYPT, one of the world's oldest and most distinguished civilizations, is the hub of the Arab world and the most populous state in the Middle East, the second most populous (after Nigeria) in Africa. History begins with the records of ancient Egypt. One third of all Arabs are Egyptian. The country may be the staunchest Arab ally of the United States, which directs it $2.3 billion in annual aid in recognition of its neutralizing 1979 peace treaty with Israel ending three decades of strife. The United States also forgave Egypt a debt of $7 billion for supporting the coalition forces in the 1991 Persian Gulf War. Egypt believes that the welfare of Arabs can be realized only in an atmosphere of regional peace and stability. The region's twin nemeses are economical backwardness and political radicalism.

The northeast African republic, which is about the size of Arkansas, Oklahoma, and Texas combined, is bounded by the Mediterranean and Red seas and by Israel, Sudan, and Libya. About 95 percent of the population of 55 million (90 percent of whom are Sunni Muslim) live on about 4 percent of the land: in the 550-mile-long Nile Valley and along the 103-mile-long Suez Canal, which opened in 1869. The population is expected to reach at least 100 million by the year 2020; there will be more people living in the capital city of Cairo than in all of Canada. Most of the country is desert or desolate and barren. Part of the Western Desert, which covers five times the area of Massachusetts, is hyperarid; generations pass without rain.

Cairo, the continent's largest metropolis, is one of the world's most polluted cities. Traffic there is so noisy, it is impossible to hold a conversation in the street. Thousands squat among tombs in Cairo's City of the Dead, a huge cemetery. The economy includes petrochemicals, beans, limestone, gypsum, cement, and the chief export, cotton. The Aswan High Dam on the Nile provides irrigation for more than a million acres. Egypt's brisk tourist trade is drawn by the ageless and seemingly indestructible pyramids.

IRAN, the "land of the Aryan," is one of the most arid regions on the planet. Summers are long, hot, and extra dry, and the state must import more than 65 percent of its food. The birthrate is among the world's highest; the population of 59.5 million might more than double in the next quarter century. Iran straddles a zone in the Middle East and southern Asia known for seismic catastrophes triggered by colliding tectonic plates. It is bordered by the erstwhile Soviet Union, Afghanistan and Pakistan, Iraq and Turkey, and the Persian Gulf and Indian Ocean.

In ancient times, Iran ("Persia" until 1935) was the heart of a great empire that extended east to India and as far northwest as the Danube. During the Second World War, the Allies used Iranian truck routes to supply the Soviet Union. Wealth ($100 billion between 1974 and 1978) is generated by oil, which was discovered in 1908.

The immensely popular Iranian prime minister Muhammad Mussadegh nationalized the oil industry in the early 1950s, but he was imprisoned by the Shah (who was encouraged by the United States) for his liberal policies. The country has been isolated by the world community since the much-revered Ayatollah Ruhollah Khomeini's "holy war" revolution in 1979 ousted the Shah from "the peacock throne" and transformed the cosmopolitan, Western-oriented society into a militant Islamic republic ruled, and economically mismanaged, by radical clerics. Iran fought invading Iraq to a standstill in an eight-year war in the 1980s and supports Iraqi Shiite Muslim rebels and Kurdish dissidents battling Iraqi president Saddam Hussein's Baath Socialist government. Iran has been striving for a more moderate tone since Khomeini's demise.

The country's highest geologic point is 18,386 feet (Mount Demavend) and lowest point is 92 feet below sea level (the Caspian Sea), 200 miles northwest of Tehran. One of the scandalous diplomatic imbroglios of modern times involved the sale by the United States of arms to Iran to support illegally the cause of the contras in Nicaragua.

IRAQ, the legendary locale of the Garden of Eden, is roughly coextensive in southwest Asia with ancient Paddan-Aram and with ancient Mesopotamia. President Saddam Hussein sees himself in the history of the area on a par with Hammurabi and Nebuchadnezzar. "The cradle of civilization," which embraced the Fertile Crescent of the Tigris and Euphrates rivers, was the site of the first city-states and generated the legal, agricultural, architectural, and phonetic-writing (cuneiform) skills of the inventive, biblical Akkadians, Sumerians, Babylonians, Parthians, and Assyrians. Here were the fabled hanging gardens of Babylon.

Much of the land is desert or wasteland. It slopes from mountains 10,000 feet above sea level to reedy marshes in the southeast. The temperature range is 120° F. to below freezing. Today's Iraq, which is about the area of California, was created after the First World War by the British from a remnant of the Ottoman Empire. Foreign control was overthrown in 1958. Having assembled one of the more Orwellian police states of the century, Saddam Hussein's Iraq aimed to be the dominant political and military force in the volatile Middle

East. The megalomanic dictator built the region's largest army, equipped with ballistic missiles and chemical and biological capabilities, and it may have had nuclear ingredients. Iraq went deep into debt during its eight-year war with Iran (a million dead) in the 1980s for control of the waterway Shatt-a-Arab, which is formed by the confluence of the Tigris and Euphrates.

The secularist state is water rich and has large crude-oil reserves (100 billion barrels). The oil industry was nationalized in the 1970s and took in $200 billion in the 1980s. Iraq invaded neighboring Kuwait in August 1990, to settle a long-standing boundary dispute (Kuwait denounced as being invalid the borders drawn on a map by a British civil servant in the 1920s), and was all but destroyed in turn by seven months of Western-Arab state coalition sanctions, six weeks of punishing air attacks, and a 100-hour land war. Some of the infrastructure was bombed into the preindustrial age, and oil-export operations were curtailed.

Three quarters of the population of 18.8 million are Muslim Arab. The capital of Baghdad is a booming city of more than 4 million. Iraqi women, who are 25 percent of the work force, are progressive and guaranteed equality under Baath Party doctrine.

ISRAEL, whose security is always at risk in spite of its military superiority—the democratic country is surrounded by 21 Arab-Muslim states and is one third of 1 percent the size of those states combined—was established as a Jewish state in a partitioned Palestine by the United Nations in May 1948. The industrialized, heavily armed, parliamentary land in the historic Jewish homeland at the eastern end of the Mediterranean Sea has fought for its survival every day since birth. On the very first day of Israel's existence, seven Arab states attacked their new neighbor. The wars in 1948–1949 and then again in 1967, 1973, and 1981 were victories for the nascent sovereignty. Israel integrates all Jewish refugees into its society, and they can claim immediate citizenship.

Arab terrorists are a constant threat to Israeli control of the annexed West Bank, East Jerusalem, the Golan plateau, and the water-starved Gaza Strip, all "occupied" when Israel routed the invading Egyptian, Syrian, and Jordanian armies in the 1967 Six-Day

War. The future of the United Nations–termed "Palestinian territories," in which the Israeli government has settled civilians and the military for greater security, and for strategic reasons, is the subject of ongoing, intense, worldwide political debate, though Israel by the legally binding United Nations Security Council Resolutions 242 of 1967 and 338 of 1973 is entitled to administer the lands until peace is made. Israel returned the Sinai to Egypt when the two governments signed a mutual peace treaty after Egypt had reopened the Suez Canal to Israeli shipping. An Arab-led economic boycott against Israel began decades ago; many Japanese concerns still won't sell products to Israeli buyers.

Most Israelis in the population of about 5 million, boosted by a million from the Soviet Union recently, are Jewish immigrants since World War Two or descendants of Jews who had settled in the region. The Jews are divided by socioeconomic differences based on ethnicity, income levels, and cultural heritage. Israel comprises much of the biblical region of Palestine, known in ancient days as Canaan. About the size of Maryland, it is the capital of four world faiths: Jerusalem at a crossing of major north-south and east-west routes is the Holy City of three major religions—Islamic, Christian, and Jewish—and Haifa and Acre are the holy shrines of Baha'i. After the Six-Day War, Jerusalem was reunited under Jewish rule for the first time in 18 centuries.

Israel is poor in natural resources, but it has become "the miracle" in the harsh, rocky desert, the envy of its Arab neighbors. The land averages fewer than five inches of rainfall a year, but Israel produces much of its own food requirements through improved irrigation techniques and soil conservation. The kibbutzim can be credited in large measure with wresting high productivity from the bleak soil. Diamond cutting is a major industry. Israel shares with Jordan the 51-mile-long Dead Sea, which is actually a salt lake, and the lowest point on the surface of Earth: 1,296 feet below sea level. Minerals include potash, sulphur, and manganese. Because of the military's defense needs, Israelis are among the world's most heavily taxed people. All but 10 percent of the work force is in the public sector. Voter turnout averages 80 percent.

JORDAN is another result of British imperialism. King Hussein's state, which was created from the land of Palestine when it was under the British Mandate Authority, is surrounded by Syria, the strategic oil fields of Saudi Arabia, Iraq, and Israel's "occupied" West Bank, where the border is one of the most heavily guarded in the world. It is only the lack of political power that prevents the majority Palestinian population (some 60 percent of Jordan's population of 4.1 million are Palestinian Arab) from creating at least a partial homeland in Jordan. The Palestinians' high birthrate lays additional stress on the kingdom.

Relations with Israel were strained when Israel diverted water from the Jordan River to the Negev. In the brief Israel-Arab war of 1967, Jordan lost the Old City of Jerusalem and the territory west of the Jordan River. (Only two countries, Pakistan and Great Britain, recognized Jordan's annexation of the West Bank and the northwest shores of the Dead Sea in 1950 in the wake of its invasion of Israel the day after the British mandate ended in 1948.) Jordan, which has been a base for the Palestine Liberation Organization, continues to try to balance its long-term friendship with the United States (Queen Noor, who is of Arabic descent, was American born and educated at Princeton) with sympathy for Iraq. During the Gulf conflict, Jordan processed about a million refugees.

Nearly 90 percent of the country is barren. Slightly larger than Indiana, this 10-millennia-old Hashemite kingdom has few natural resources. It has only enough water to grow crops on a sixth of the land, and there is no outlet to the Mediterranean Sea. The only Jordanian port is on the short Aqaba Gulf coast, at an erstwhile village that has become an industrial center of 40,000 people. Fertilizer manufacturing is the major industry. Jordan's largest source of income is the billions dispatched home by Jordanians laboring in the Gulf states. There is a high priority on education. One out of every three Jordanians is a student, and some 60,000 study abroad. Running north from the Gulf of Aqaba is Jordan's Rift Valley, one of the planet's major geological faults. English-educated King Hussein has been one of the true survivors of the Middle East maelstrom, ruling his country for nearly four decades.

KUWAIT, which must import almost all of its food, is largely a desert on the Arabian peninsula, wedged between Iraq and Saudi Arabia in the northwest end of the Persian Gulf in southwest Asia. It was part of the Ottoman Empire, whose provinces were carved up after the First World War by two victorious Allies, Great Britain and France, both of which imposed their own political values and interests. Kuwait became independent of Britain in 1961.

Kuwait was one of the world's largest oil producers, with reserves among the most extensive, before Iraq's invasion in 1990 and the six-week Persian Gulf War in 1991. Iraqi sabotage kept the wells from dispensing a significant number of barrels again until 1992; the fires consumed about 5 million barrels a day. Oil was discovered in 1938 and became the lifeline of the emirate, with its political system based on privilege. Paradoxically, the tax-free Muslim sheikdom supported Iraq with about $35 billion in dictator Saddam Hussein's eight-year war with Iran in the 1980s, and over the years it has aided other Arab states. Per capita income for the 2.1 million Kuwaitis (the majority are Sunnis) was $19,680.

A broad array of educational and social programs in the gilded welfare society elevated the standard of living, and there is a history of tolerance and relative political openness. Kuwaitis have been a minority in their own country because of the influx of foreign laborers to serve as the mainstay of government, technical, and other (usually gritty) services. The population declined to a half million when foreigners fled during the Gulf War; Jordanians and Palestinians may not be rehired.

Slightly smaller than New Jersey, the state has been ruled since its founding in 1756 by the Sabah family, which was chosen by the first Arab settlers. Iraq came to claim Kuwait as its 19th province, but British troops forestalled any Iraqi invasion. In 1990, Iraq renewed its claim and, infuriated by Kuwait's over-production of oil, invaded and turned the country into a penal colony and torture chamber. The cost of rebuilding since the war has been estimated to be upwards of $10 billion.

The subtropical, undulating, hereditary kingship is extremely hot and arid. Summer temperatures reach 120° F.

LEBANON, beautiful and once prosperous, was fractionalized and battered by anarchy from 1975 to early 1991, by religious and political civil strife, principally between Muslims and Christians. Christian insistence on maintaining the ancient regime was a root cause of the

16-year war. The presence of Muslim-supported Palestinians prompted the Christians to form their own militias. The Maronites are the largest of 16 Christian sects. The Druse, adherents of an esoteric branch of Ismaili Shiite Islam, are about 10 percent of the estimated population of 3.4 million. (There has not been a nationwide census in more than a half century.)

A wall of barricades and earth mounds cut the capital of Beirut into antagonistic halves. The law of the jungle prevailed, and at least 150,000 Lebanonese died. Prewar tourism had provided a fifth of the country's annual income. Throughout Lebanon, which is slightly smaller than Connecticut, fiefdoms are controlled by warlords and private armies. Israel maintains a buffer zone in the south, and Syrian troops occupy two thirds of the country, including Beirut, serving as the referee of conflicts. Over time, 14 armies have fought over Lebanon, which is strategically situated in southwest Asia at the eastern end of the Mediterranean Sea and between Israel and Syria.

Lebanon was part of the Ottoman Empire between 1516 and the First World War. It became an independent state in 1920 and was administered under French mandate until 1941; French troops withdrew in 1946. Since the creation of Israel in 1948, nearly 400,000 Palestinians have lived in Lebanese refugee camps and terrorist bases. Baalbek, in the fertile Bekka Valley, has been a bastion of radical Islam, and the Palestine Liberation Organization has been active within Lebanon, which is often described as a "headless society." The country must import most of its food and industrial goods. There are no oil deposits.

The Litani in the Bekka is the only river in the Arab Near East that does not cross a national boundary. It runs between two north-south mountain ranges, the Lebanon and the Anti-Lebanon, which historically have been places of refuge. Beirut was once known as "the Paris of the Middle East," attracting sophisticated travelers. The Lebanese are nothing if not resilient; there are signs of an energetic effort to rebuild the shattered country.

LIBYA has been an independent North African state since 1951 and oil rich since 1959. Under junta-led Colonel Muammar al-Qaddafi's erratic and baffling rule since 1969, it promotes Arab unity and Palestinian causes. It threatens its neighbors Jordan, Egypt, and

Tunisia, but its sparse population (4.2 million) cannot fulfill the ambitions of Colonel Qaddafi. The revolution was based on Islamic law as Qaddafi interprets it. In its ancient history, the region was ruled by Carthage, Rome, and the Vandals. In 1986, United States bombers raided several terrorist-related targets in retaliation for a Libyan-sponsored terrorist bombing of a West Berlin discotheque that had killed a United States serviceman. (In 1805, the United States routed the Barbary pirates at Tripoli.)

Ninety-two percent of the republic, which is larger than Alaska, is Sahara or semidesert; it took the discovery of oil to propel the country out of the depths of poverty. The planet's highest recorded temperature was 136.4° F. in Azizia. Libya is bounded on the north by the Mediterranean Sea, on the east by Egypt, on the southeast by Sudan, on the south by Chad and Niger, on the west by Algeria, and on the northwest by Tunisia. Almost all Libyans, a mix of Arab and Berber, live in a narrow coastal zone along the sea. The economy includes wheat, textiles, dates, grapes, and olives. There is a large chemical-factory complex in the desert near Rabta, about 40 miles south of Tripoli, the capital.

MAURITANIA is an Islamic republic the size of California and Texas combined. It lies in the hot, dry western Saharan shield of crystalline rocks in northwest Africa. The northern four fifths of the country is barren desert. The earth is so sere, livestock starve to death. The 2 million Sunni Muslims (mostly Moors of Arab-Berber origin) suffer long periods of drought, which cripples the economy. Almost everyone has given up the nomadic life and settled in cities. (The Maghreb, Africa's second-richest region—oil, gas, phosphates, iron—extends geographically from Mauritania to Libya.) Formerly a protectorate and a colony of France, Mauritania won its independence in 1960. It built the continent's first desalination plant. The chief crops are dates and grain, and the chief industry is iron mining. There are few paved roads. Literacy is 17 percent.

MOROCCO has been aggravated by the pressures of population growth and by hardships worsened by austerity programs. The people have suffered riots over economic hardships, bleak job prospects, and unfulfilled government pledges. The 26 million citizens of this northwest African state, which is larger than California, live mostly west of the Atlas Mountains, which are a protective shield against the Sahara. Morocco, which rims the continent's northern edge for 1,500 miles, is also bounded by the Mediterranean Sea, Algeria, and Western Sahara. It is separated from Spain by the Strait of Gibraltar, and its western flank is on the Atlantic Ocean. Morocco and Spain have discussed "bridging" the Strait. A conflict with Mauritania over the northern, phosphates-rich section of Western Sahara continues.

Morocco is already a world leader in the export of phosphates; the mining of coal and cobalt is a key industry. Eighteen percent of the land is arable, but there are periods of drought. Olives, citrus fruits, and grapes are under irrigation. The city of Fez is the country's religious and intellectual center, and Rabat is the capital. But coastal Casablanca is the most famed of Moroccan cities, being the locale of meetings of heads of Allied states during the Second World War and the focus of one of the best-loved movies of all time. Almost all Moroccans are Arab-Berber, and 30 percent are literate. King Hassan enjoys particularly warm relations with the United States.

OMAN, whose goal is to be the Singapore of the Middle East, is a strategically located, oil-rich, pro-Western sultanate about the size of New Mexico. It wraps around the Arabian Sea and the Gulf of Oman and is between Yemen and the United Arab Emirates on the southeast Arabian Peninsula in southwest Asia. Oman controls the Musandam Peninsula along the vital Strait of Hormuz. During the 1991 Persian Gulf War, the United States military prepositioned equipment and used airstrips here. The politically volatile region was known as Muscat and Oman before 1970.

Most of the land is desert, flat, and arid, but Mount Sham rises to a peak of 9,957 feet. Per capita income is $8,915. (Oil superseded fishing and farming as the main source of revenue.) Revolutionary internal strife has led Oman to devote a fourth of its budget to defense. There are 1.5 million Omans, mostly Arab Muslims of the Ibadi sect. A significant foreign population includes Indians, Pakistanis, and East African blacks. Until the late 1980s, tourism was not allowed. Money earned from oil exports helps to derrick Oman out of medieval isolation. Ugliness of any kind distresses the sultan. A person can be fined for driving a dirty automobile or leaving a building unpainted.

QATAR, whose income is derived almost exclusively from petroleum products, is a small, stony desert-peninsula jutting from Saudi Arabia on the west coast of the Persian Gulf in southwest Asia. Vegetation is scarce. After Iraq's invasion of Kuwait in 1990, Qatar granted multinational forces military access. The country is no larger than Rhode Island and Connecticut combined, and it has the smallest population (a half million) of any of the Gulf states (foreign workers increase the population considerably). Some oil revenues have been used to improve housing, transportation, and public health. Qatar declined to join the United Arab Emirates in 1971, but it cooperates with the UAE in development schemes. Defense expenditures constitute 13 percent of the gross national product. About 96 percent of the people are Muslim. The literacy rate is 60 percent, but there is no suffrage.

SAUDI ARABIA, a welfare state, Muslim, conservative and profoundly anti-Communist, is rich beyond the dreams of Croesus, all because of oil. Its bonanza of known oil reserves (257.6 billion barrels), a quarter of the Earth's, should last another half century; deposits include the world's largest offshore oil fields. Oil, which was discovered in 1938, a mere six years after the devout, puritanical, sharply stratified Islamic kingdom was founded as a modern nation-state, has transformed a primitive land the size of the United States east of the Mississippi River, a land with a fierce warrior society, into a technologically advanced population wherein a relatively few are rewarded with a sybaritic life-style exceeding anything to be found in the Arabian Nights.

In his absolute rule, the king has had to maintain a delicate balance between conservative religious forces, who tend to be anti-Western extremists, and liberals eager for a more open society. But all the trappings of wealth must still be displayed against one of the grimmest backdrops on the planet. The entire state, which is situated on most of the Arabian Peninsula between the Persian Gulf and the Red Sea in southwest Asia, is limited by the harshest of landscapes, mostly a series of vast, hot, unremitting deserts; there are some mountains and green plains. The Rub al-Khali is the world's largest sand desert, about as big as Texas. There is no natural body of sweet water. Some of the deserts, where summer temperatures hover at 120° F., go years without rain; Saudi Arabia has an average rainfall of only four inches. Only 1–2 percent of the country is under cultivation.

Islam is practiced at its strictest, and the closed society is built on traditional religious and tribal values. Social restrictions and political restraints are mandated rigidly. Women wear veils and are forbidden to drive automobiles. The Saudi flag proclaims, "There is no God unless only one God. Mohamed is the Prophet of the God." The state is "custodian of the Two Holy Mosques," Mecca and Medina. Militant fundamentalists have grown in number, and may be using religion to change the established order. The commercial center, Jiddah, is the most expensive city in the world. Rapid modernization has seen the building of a half million homes, the world's largest desalination plant, the addition of 2 million telephone lines, the opening of a new primary school every day for the last 20 years, the planting of more than a million palm trees, and the erection of scores of gleaming skyscrapers in the capital city of Riyadh.

Saudi Arabia is the dominant member of OPEC, and through its financial clout it has made its moderate position on crude-oil prices prevail. In 1973, it reduced its oil production by 30 percent and presently quadrupled the price so that in one year its GNP rose 250 percent. The House of Saud has 5,000 princes and princesses. The oligarchy funds poorer states, invests in foreign securities, and has spent more than $100 billion on defense since 1970. It supported Iraq's eight-year war with Iran in the 1980s and was the base for 700,000 coalition troops in the Persian Gulf War against Iraq. Extremists angry about Riyadh's decision to call in the coalition would have preferred that Allah defend the country.

Saudi Arabia is underpopulated. Although there never has been a census, some estimate that the country is home to only 16 million Saudis. The state has no suffrage.

SOMALIA is a small, socialist, non-Arab republic, a strategic prize during the Cold War on the eastern edge of the ethnically fractious Horn of Africa, its northern coast on the Gulf of Aden at the entrance to the Red Sea. State building in the Horn was aborted after the Second World War by European powers creating nation-states with arbitrary borders. Somalia wraps like a boomerang around the eastern border of Ethiopia. The northern port of Berbera offers direct access to the Red and Arabian seas. Somalia had one of Africa's most corrupt and brutal regimes under a president who came to power in a military coup in 1969 and was removed by clan-based guerrilla groups in 1991. Somalians regard a gun as a basic necessity.

Only 2 percent of this sliver of a state is arable. Poverty is extreme, with per capita income an abysmally low $110. The only non-Arab-state member of the League of Arab Nations, it was formed in 1960 through the union of the former territories of Italian and British Somaliland when they became independent. In the strategic calculus of superpower assets and debits, it became one of the United States' handsomely financed Cold War allies after being deserted as a military and economic client of the Soviet Union, which switched its favor to Ethiopia, a former American ally. Somalia has the longest airstrip in sub-Saharan Africa; it was built by the Soviets. The country has its eye on annexing Ogaden, the huge eastern region of Ethiopia.

Fierce clan loyalties divide the Somalis. The capital of Mogadishu has been reduced to rubble. There have been flagrant human-rights violations by the government, including the killing of Isaaks, the largest clan in the north. Years of brutal civil war have killed at least 60,000 people. About 75 percent of the population of 8.3 million are camel-herding semi-nomads. Everyone is a Sunni Muslim, a rare instance of religious homogeneity, and everyone speaks the same language.

SYRIA is the dominant power of the northern Arab world. State power is tightly controlled by the Alawite minority. Once a client of the Soviet Union, Hafez al-Assad's dictatorship maintains a military presence in Lebanon. Assad (his name means "protector of lions") has ruled for two decades and seeks a regional hegemony that would vindicate the expansionist Assyrian Empire of ancient times. He is a former fighter pilot. In opposing Israel and Muslim sects, he has been charged with gross human-rights violations and the sacrifice of countless lives.

Ancient Syria linked the civilizations of Mesopotamia and the Nile Valley. The capital city of Damascus is the oldest continuously inhabited city (4,000 years) on the globe. The country is a land of desert, mountains, fertile plains, and irregular rainfall. Fifty percent of the work force in the population of 12.6 million is in primitive agriculture. Islam is the most powerful cultural force in this mountainous southwest Asian state (the size of North Dakota), which fronts on the east coast of the Mediterranean Sea and is surrounded by Turkey, Iraq, Israel, Lebanon, and Jordan. Syrian-Turkish enmity is fueled by the claim to the *sanjak* of Alexandretta. (Jews have been living here for centuries in notable peace and freedom.)

Oil pipelines crossing from Iraq and Saudi Arabia are a substantial source of revenue. (Syria, itself, has oil reserves of 1.7 billion barrels.) Glassware, brassware, olives, and sugar also spark the economy. Two losing wars with Israel left tensions with Jerusalem unresolved. During the Persian Gulf War in 1991, Syria performed a delicate balancing act in supplying 20,000 troops to the Allied coalition. Nearly half the population is under the age of 20.

TUNISIA is an oil-exporting, agricultural republic with a centrally planned government and a markedly middle-class society with equitable land distribution. Crops include dates, cereals, olives, and grapes. The leading export is phosphates. Iron ore, lead, zinc, and silver are also mined. About the size of Missouri, Tunisia is sandwiched between Algeria and Libya on the northern coast of Africa and bounded on the east and the north by the Mediterranean Sea. Wooded fertile lands lie to the north, arid lands to the south.

Tunisia maintains a balance between pro-Western and Arab persuasions, despite the mingling of Islamic upsurges and economic pain familiar in many parts of the Arab world. Most of the 7.9 million Tunisians are Arab and Muslim, and there is a definite move toward an Islamic government. More than half the European population emigrated when Tunisia gained independence from France in 1956. (After the surrender of France to the Nazis in 1940, Tunisia remained loyal to the Vichy regime.) The country has promoted women's rights. Unemployment hovers around 33 percent. Prudent management of the debt is improving the balance of payments. Tunisia was once the headquarters of the Carthaginian Empire and part of the Turkish Ottoman Empire.

TURKEY has thrown off its traditional "backward" ways and become the first truly independent country in the Afro-Asian world after Japan. The only Middle Eastern state other than Egypt to recognize Israel, it is also the region's only state linked to the West by a military pact (NATO). Turkey is eager to join the European Community.

The United States and Western Europe see the country as a secular, prosperous role model for adjacent Muslim states and for nearby republics in the late Soviet empire. Turkey was founded in 1927 after collapse of the Ottoman Empire and has been under both military and civilian rule. The presidency is strong, the legislature is weak. Torture and attacks on freedom of expression continue. In the southeast, warfare continues with a separatist Kurdish minority. The country took a forceful stand against Iraq in the Persian Gulf War of 1991.

Twice the size of California, Turkey is a two-continent republic (half urban, half rural) in southeastern Europe (3 percent) and southwestern Asia (97 percent), divided by the Dardanelles strait, sandwiched between the Black Sea, the Aegean Sea, and the Mediterranean Sea, and bordered by Bulgaria, Greece, the ex-Soviet Union, Iran, Iraq, and Syria. Turkey and Greece are at odds over control of the nearby island of Cyprus and the continental shelf of the Aegean Sea. Turkey has over 20 mountains that are 10,000 feet high, and there are rolling plains, fertile coastal plains, wide plateaus, and pastures. In the west were the now-vanished cities of Sardis, Smyrna, and Pergamum.

Turkey is a large wheat producer. The chief industry is mining: coal, copper, chrome, and lignite. Sardines, steel, automobiles, barley, borate, and antimony also boost the economy. A vital Iraqi oil pipeline runs across southern Turkey to the Mediterranean. Extensive irrigation of the southeast through diversion of the historic Euphrates River would transform subsistence agriculture into a vast commercial enterprise. Almost everyone in the population of 56.7 million is Sunni Muslim. Ethnic and religious tensions are ongoing.

UNITED ARAB EMIRATES, the former Trucial States under British protection, has been since 1971 a federation of seven independent desert sheikdoms formed to provide unity and strength after the British had ended their treaties. It is rich in oil, which was first exported in 1962. Reserves are 98.1 billion barrels. Per capita income for the population of 1.6 million is $23,242. Most of the people are Sunni Muslim. Hundreds of thousands of Pakistanis and Indians work in the Emirates, which is the size of Maine. The UAE is located on the Arabian Peninsula and at the Persian Gulf. The barren, flat coastal plain gives way to uninhabited sand dunes. In common with some of its Arab neighbors, the Emirates has been increasing its defense spending. Each of the sheikdoms has its own emir, or Arab chieftain.

YEMEN until the 1990 merger, 10 weeks before Iraq's invasion of Kuwait, was two Yemens: North Yemen, which was relatively well-off and pro-Western, and South Yemen, which was very poor and pro-Soviet. Their boundaries were never clearly defined by the partition, and there was constant friction. Yemen abstained from the Arab League resolution condemning Iraq's invasion.

Dominating the entrance to the Red Sea to the west in southwest Asia, and bounded by Oman to the east and by Saudi Arabia to the north and by the Arabian Sea and the Indian Ocean to the south, the elbow-shaped territory in the southwest corner of the Arabian Peninsula is a bit smaller in area than France and has a population of about 9.8 million. Along the Red Sea coast, groves of palm trees line the long stretches of pure white sand and water. But Yemen is mostly mountainous, much of it well watered. The climate is very hot and very dry. Only 1 percent of what was South Yemen (which was about the size of Nevada) is arable; it produces coffee, grains, and qat. Its culture is secular, and the most valuable resource is the port of Aden. Crude oil and food must be imported.

South Yemen was the Arab world's only avowedly Marxist state, a revolutionary center. Labor emigration—1.7 million South Yemenis alone work in Saudi Arabia—was a factor in the decline of agriculture. The smaller North Yemen, which was about the size of South Dakota, isn't oil rich, either. Dominated by tribes and religion, the population is Muslim Arab (Shiites and Sunnis). Historically, the republic was a region of farmers, ruled for four centuries by the Ottoman Empire. The horn of the black rhinoceros has been coveted here for ornate curved-dagger handles. Resource exploitation is financially prohibitive. Most foreign earnings come from remittances from Yemenis working abroad. A major export is mocha coffee, named for the ancient trading port.

Sanaa, the capital, boasted the world's first skyscrapers, 10- to 12-story buildings dating back 2,000 years. Yemenese scholars added logarithm to mathematics. Nearly 3,000 years ago, the kingdom of Sheba was located here.

WESTERN SAHARA (a non-state), once a Spanish colony, is the only decolonized African territory (other than Eritrea) not to gain automatic independence. A mineral-rich, phosphate-exporting desert nation in northwest Africa, it has been a battleground for decades between Morocco (which wants the phosphates) and the Polisario Front guerrillas, who have been until recently heavily supported by Algeria and seek independent state status. The Moroccan Army built 1,550 miles of sand and stone walls to keep the Polisario's forces at bay. (Mauritania gave up its claim to a third of the country in 1979.) A United Nations plan has called for a referendum by the inhabitants on independence or union with Morocco. The population of 165,000 (only two people per square mile!) is mostly nomadic Arab and Berber. Rainfall is negligible.

SOUTH AMERICA: "South America is a place I love, and I think, if you take it right through from Darien to Fuego, it's the grandest, richest, most wonderful bit of earth upon this planet. . . . Why shouldn't somethin' new and wonderful lie in such a country?"—Sir Arthur Conan Doyle, Sherlock Holmes's creator, in *The Lost World*, 1912.

In its relative position with North America, the world's fourth-largest continent (a dozen sovereign states) is really Southeast America. The Peruvian capital of Lima on the Pacific coast of South America is farther east than Savannah, Georgia, which is on the east coast of North America. There is no *South* America due south of Chicago or New Orleans or Kansas City or Dallas.

The geography reflects the mixed cultures of Spain and Portugal and the indigenous Indian population.

Early peoples included Incas, Aztecs, and Mayans, highly developed societies. The countries have little to do with one another. Interaction is inhibited by vast distances (South America is a thousand miles longer than the United States is wide), physical barriers (the Andes and the Amazon jungle), and complex, heterogeneous, and pluralistic cultural constraints (Native Americans are still a large part of the population, and the Catholic Church is the prevailing power). Almost all of the world's cocaine is produced and refined in Bolivia, Colombia, and Peru. Around 60 percent of the population live in towns or cities as the process of urbanization speeds up. In 1945, the population of Latin America (South America and Central

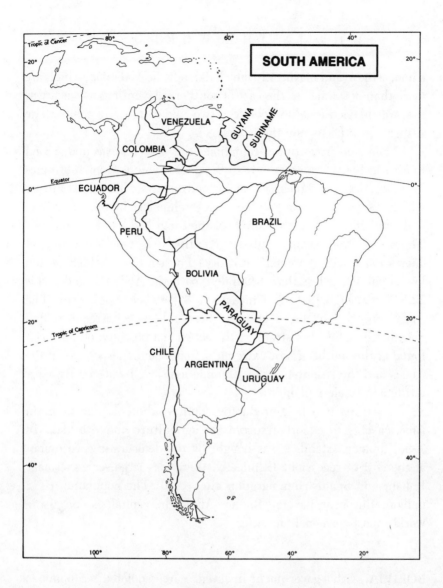

America) was around 132 million; it will be around 600 million at the end of this decade.

ARGENTINA has a history of militarism and rule by strong dictatorship. Latin America's most modern country was traumatized by army rule in the 1970s and the 1980s; relentless, hyperinflationary spirals; a Gargantuan deficit; and about 10,000 murders. Argentinians are notoriously tax hating; not long ago, only 1 percent filed returns. In 1982, the country clashed with Great Britain over the two-island, damp, windy (and maybe oil-rich) Falklands, or Malvinas, which are more than 300 miles to the east of southern Argentina; it was a war that was likened to a fight between two bald men over a comb. (The British continue to rule the islands.)

Land-rich Argentina is four times the size of Texas and second in area in South America only to Brazil. It extends from subtropics in the north to long, bleak Patagonia and subarctic Tierra del Fuego, which is separated from the Antarctic Archipelago by the "most mad seas" and winds "as if the bowels of the earth had set all at liberty." The world's southernmost city is Argentina's Ushuaia, the capital of Tierra del Fuego National Territory. The continent's highest and lowest points are less than 100 miles apart: the Andes' Aconcagua is 22,834 feet high, Salinas Chicas is 138 feet below sea level. The Andes provide an almost absolute barrier between Argentina and neighboring Chile, constraining the occasional hostility; oil has been found at the southernmost extremity of both countries. Between the Andes and the Atlantic is the 100-mile-long, well-watered Pampas, fertile and treeless plains.

Argentina is a beef-producing country, the only one in Latin America able to export a surplus in agriculture and livestock. In 1990, financial stability was bought by the democratic government at the cost of recession. Ethnically, there was large-scale German, Italian, and Spanish immigration after 1880. The population is 32 million. European-flavored Buenos Aires, the capital, is one of the world's most cosmopolitan cities.

BOLIVIA, with a government-involved "narco-addicted" economy, is South America's poorest state, a once Inca-run landlocked region in the central Andes. Per capita income is $536. (In 1985, consumer prices inflated 8,170 percent.) Bolivia is very large, about the size of California and Texas combined. Sixty thousand coca producers grind out 100 tons of cocaine—"white gold"—a year, and one in five working Bolivians makes his living in the production and distribution of coca and cocaine. (The United States has tried to stifle the trafficking.) There is constant political instability and guerrilla activity.

Tin is the country's major mineral; silver, copper, zinc, and gold also are mined. The republic has two capitals: Sucre, the legal capital, and the world's highest capital, La Paz, the seat of government. La Paz is perched at over 12,000 feet in the magnificent snowcapped mountains, where the air is about 35 percent thinner than at sea level and it is difficult to start a fire. Bolivia was named for Simón Bolivar (1783–1830), the independence fighter who ousted the Spanish. In a series of wars, Bolivia lost its Pacific seacoast to Chile, the oil-bearing Chaco region to Paraguay, and rubber-growing areas to Brazil.

Indians are the majority of the 6.6 million population, who are concentrated on the 600-mile-long, wind-whipped altiplano, or plateau, more than two miles high in the mountains. There are semitropical forests in the east-central region. Almost everyone is Roman Catholic. Suffrage begins at age 18 for a married person, at age 21 for a single person.

BRAZIL, which was named for a triangular nut-bearing tree, is the world's fifth-largest sovereign state, and the largest in Latin America by far. It is larger even than the 48 contiguous states of the United States. Much of Brazil is dense, wild, and cut by enormous rivers. The Amazon contains nearly a fifth of all the fresh water that runs over the surface of the planet. It is the world's most voluminous river, each of 20 tributaries longer than the Mississippi River. Brazil is the only country in the world that stretches across both the equator and a tropic (Capricorn). North to south, the distance is the same as from London to Tehran. Occupying the eastern half of South America, Brazil abuts all but two of the continent's 11 other countries (Chile and Ecuador). The Atlantic coastline is 4,603 miles long, and the rain forest is the world's largest. A section of the forest the size of Connecticut is destroyed annually. The biologically diverse Amazon basin (which is the size of five Texases) covers half the country and has more species of plants and animals (humankind's genetic treasury) than any other place on the planet. The Xingu National Park, isolated at the state's geographic center, has long held a central position in Brazil's psychology as a preserve of primeval Brazil, with its forbidding climate. Hundreds of Amazon tribes have been decimated through diseases spread by contact with whites and by cultural extinction through assimilation. (The town of Ariquemes is the malaria capital of the world.)

Much of the country's considerable wealth derives from an infinity of grazing lands. Eighty percent of Brazil is owned by only 5 percent of the population of 151 million. Once a sleeping giant, today

Brazil is a burgeoning industrial powerhouse, Latin America's largest economy, and a world leader in hydroelectric production. It is the only Latin American state with large mineral sources, which include 30 percent of the world's iron-ore reserves. The Amazon Basin may overlie a prodigious pool of oil and house hundreds of billions of dollars' worth of gold. Brazil is the world's leading producer of coffee. It also has the largest foreign debt in the developing world—more than $100 billion—and the out-of-control inflation rate has hit nearly 2,000 percent. Finances are in dire condition. The state plunged into the United States Export-Import Bank's riskiest-loan category. Economically, the 1980s were a "lost decade." A 21-year military dictatorship, with extensive press censorship, ended in 1985; it had all but ignored the poor. Fernando Collor de Mello became the first directly elected president in 29 years.

The population includes 50 million needy children and adolescents, 7 million of whom are abandoned and live in the streets, many of whom are vulnerable to murder—a tragic social dilemma. Roman Catholicism is the principal religion. Abortion is illegal, and birth-control devices are not widely available. The country can be divided into three major regions: the poor, agricultural northeast; the urban-industrial southeast; and the resource frontier of Amazonia. Half the population lives in the southeast region favored by climate and resources; the region produces 80 percent of Brazil's industrial output and three quarters of its farm products. The United States is Brazil's number-one trading partner. São Paulo is the most populous city (15 million). São Paulo is the country's industrial and commercial center, but it is Rio de Janeiro with its magnificent Copacabana beach and its laissez-faire joy-making that draws tourist dollars. About a million Japanese, who first came as farmers, live in São Paulo, more than in any other city outside of Japan.

So that the bloated bureaucracy would be redistributed, Brazil's capital was moved from the overpopulated east coast to the hinterland to promote opening up the frontier. Brasilia was planned for 600,000 people; the capital now has at least 1.6 million. Gasahol and ethanol propel the automobiles. Brazil was for three centuries under Portuguese colonial rule; it is the only Portuguese-speaking state in the western hemisphere. Not the least of Brazil's exports are the samba and the bossa nova.

CHILE is an elongated (2,650 miles) and slim (100 to 221 miles) state on the southwest coast between the Pacific Ocean and the altitudinous Andes and includes Easter Island, the Juan Fernández Islands, and thousands of other islands. In total area, Chile is larger than Texas. It has been wracked by earthquakes and tidal waves. The country has the most successful economy on the continent. Tourism, manufacturing, industrialization, wineries, and intensive agriculture are contributing to prosperity. Seventy percent of the land is arable. The population of 15 million is concentrated in a rich central valley, "the garden of Chile," but a third of the people live in poverty.

The northern desert is endowed with deposits of copper and iodine and all of the world's natural sodium nitrate. Forests cover 26 percent of the land, and timbering is a major industry. Grazing lands lie to the south, but there are no rivers of size. Lascar in the Andes is the highest active volcano in the world, at 18,077 feet. An 8,741-foot-high mountain peak will be the site of the world's largest telescope, designed for scanning the edge of the universe. The Atacama Desert is second only to the polar plateau of Antarctica in aridity; it has never rained on the town of Calama.

Right-wing dictator Augusto Pinochet's 17-year military coup, which succeeded leftist Salvador Allende Gossens' early '70s rule, was marked by civil war, repressive government, civil disturbances, grave human-rights abuses, secret-police brutality, social inequality, and private enrichment—and by the creation of a free market. Pinochet's was the continent's last military dictatorship. A relatively democratic regime succeeded Pinochet's and is a notable success "in a bleak neighborhood." The regime has eased tensions, respected human rights, and merged democracy and the marketplace. (Pinochet is commander in chief of the army until 1998.) The telephone system works.

COLOMBIA has a climate of "backwardness" and violence heightened by drug trafficking and narco-terrorism. The violent Medellín cartel controls much of the world's cocaine, but the underworld in the modern city of Cali has become the leading exporter of the "white gold." The Medellín monopoly confronts Colombian authority head-on; the Calis, who control 80 percent of the European market, work within the existing political system and invest heavily in legal business ventures.

In recent years, Medellín terrorists have mowed down opposition politicians and murdered more than 50 judges. In 1990, there were more than a thousand kidnappings countrywide. The large guerrilla armies killed 20 police officers in two separate attacks over the 1991 New Year's holiday. The much-publicized Colombia–United States effort to vanquish the cocaine trafficking has yielded disappointing results; in fact, the cultivation of poppies is now compounding the drug problem with a flourishing trade in heroin. Forty years ago, about 200,000 Colombians died during rural banditry and urban disorders. Land and social reform have been promised but not implemented.

More than half the population of 32 million is mestizo, and nearly everyone is Roman Catholic. Colombia, which is about the size of Texas and New Mexico combined, is the only South American state with coastlines on both the Pacific Ocean and the Caribbean Sea. Situated at the northwestern bulge of the continent, it is bounded on the north by the sea, on the east by Venezuela and Brazil, on the south by Peru and Ecuador, and on the west by Panama and the ocean. Three ranges of the Andes run north-south. (A mist-shrouded 10,000-foot-high range in the east has been a stronghold of the largest guerrilla army.) One of the continent's greatest natural disasters was the 1985 eruption of Nevado del Ruiz. A mudflow 130 feet high spreading over 16 square miles drowned 23,000 Colombians. Fifty percent of the land is arable. The rain forest in the Choco region is one of the planet's wettest sites: an average of 400 inches (or 33 feet) inundate it every year. In addition to cocaine, the country is a leading world producer of coffee and oil, and it mines nearly all of the world's emeralds, the most expensive gem.

ECUADOR, once the northern bastion of the Inca Empire and part of Simón Bolivar's Grand Colombia (an embryonic United States of South America), pumps enough oil from its share of Amazonia—the Oriente region covering half the country—to have made it a member of OPEC since petroleum was discovered there in 1966. But declining petroleum exports generate severe economic problems.

Ecuador, which is about the size of Colorado, is split into three zones by two Andean chains: hot, humid, fertile plains, jungles, and a desert on the coast; temperate highlands between the polar-cold mountains; and rainy, steaming lowlands to the east. In only 20 years, the coastal lowland forest was all but chopped down, and the land went from primeval forest to farmland. The highlands support the bulk of the population of 10.3 million. Since 1952, Ecuador has exported more bananas than any other country. Other crops include rice, sugar, coffee, palm oil, cocoa, and balsa. Ecuador boasts 2,670 species of orchids and it has 50 volcanoes. Periodically, this Roman Catholic republic is shaken by earthquakes.

More than half the population is mestizo and mulatto, but economic and political power is wielded by Caucasians. Because of tempestuous rivalries, no Ecuadorian leader between 1925 and 1948 was able to complete a full term. In 1822, Ecuador was liberated from Spain's three centuries of cloistered rule; 10 years later, it acquired the eerie Galápagos Islands (five large ones, nine small ones, made famous by Darwin's visit) in the Pacific Ocean, about 600 miles west of the coast. Traditional, conservative Quito, the capital, is only 14 miles south of the equator (for which the country was named), and it is two miles above sea level; Quito is one of the oldest continuously inhabited cities in the western hemisphere.

GUYANA (formerly British Guiana, a crown colony on the northeast coast of South America) became an independent dominion in 1966, a republic in 1970. Four centuries ago, the region was explored by Sir Walter Raleigh. The population is 770,000: 51 percent East Indian, 43 percent black, a few whites and American Indians. The two main political parties reflect Guyana's ethnic diversity: there were racial riots in the 1960s. Nine out of 10 inhabitants of this socialist, tropical land the size of Idaho live along the fertile coastal plain, which has rich alluvial soil. Sugar, rum, rice, hardwoods (from dense forests), gold, diamonds, and bauxite are exported. Drugs, shrimp, furniture, and cigarettes bolster the ailing economy. Per capita income is under $500 a year.

PARAGUAY, one of two landlocked states in South America, is surrounded by Brazil, Argentina, and Bolivia. It has neither beaches nor Andes. Long a poor country, the republic has had the fastest-growing gross national product in the hemisphere. Smuggling has been an integral part of the economy; about half of Paraguay's trade goes unreported.

The size of California, Paraguay has a homogeneous population of 4.5 million, 95 percent mestizo. An influential minority are the 100,000 residents of German descent. In the 1700s, Jesuits converted the population to Roman Catholicism. The capital city of Asunción was founded by conquistadores and served briefly as the hub of Spain's empire in southern South America. In the 19th century, half the Paraguayans were slain in war with Uruguay, Argentina, and Brazil. In 1989, a military coup ended the 35-year dictatorship of General Alfredo Stroessner, son of a Bavarian brewer, a "state of siege" reign that perpetrated human-rights abuses and gave sanctuary to Nazi officers, including the notorious Josef Mengele, hiding from world justice.

The broad, north-to-south Paraguay River bisects the country, separating the flat, semiarid Gran Chaco (marshes and scrub trees) in the west, where only 3 percent of the population live, from the temperate, fertile, green highlands, forests, and lowland farms in the east, where most of the farming occurs. The bustling town of Ciudad del Este, bordering Brazil and Argentina, is often called the "Hong Kong of South America." Agricultural products include cotton, to- bacco, citrus fruits, and mate. Industries include meatpacking and the manufacture of molasses, rum, and alcohol. There is only one large lake. The new Itaipu Dam is a prodigious hydroelectric project, transforming the country into the world's largest exporter of elec- tricity.

PERU, home to diverse Native Americans for 20,000 years, is in economic straits, though it is the world's foremost fishing country and grows 60 percent of the world's coca leaf. Superimposed on a map of the United States, Peru would stretch from Chicago and New York City in the north to Miami in the south. Deep-rooted problems of economic instability, rampant poverty, widespread corruption, and the highest rate of political violence in South America create the prevailing conditions. Exploitation, discrimination, and brutality have become part of the Peruvian experience. Inflation has hit 7,650 per- cent per annum. In 1991, a cholera epidemic threatened the entire country.

More Native Americans (8 million) live in Peru than in any other country. Caucasians, who are only 15 percent of the population of 21.5 million, control much of the wealth of the continent's third-largest state. Nearly half the population is concentrated along the arid Pacific coast—a 1,400-mile-long strip that is one of the planet's driest stretches. Fifty percent of the capital city of Lima's population of 7 million live in shantytowns; there is infrastructure in Lima for only 1.5 million.

More than a decade of brutal insurgent violence by the Maoist Shining Path guerrillas, supported by rural Indians and mixed-race warriors, has failed to improve living conditions, while slaying tens of thousands of lives and claiming up to 40 percent of the national territory. (Lima's sabotaged electrical system was out of order for more than two months in 1990.) The impoverished hillside soil of the Upper Huallaga Valley is the world's largest source of the illicit coca leaf, the raw, wealth-generating material for cocaine. The valley, known as "Cocaland," is a hotbed of indiscriminate violence by guerrilla groups, drug traffickers, paramilitary forces, and government security troops. About a million Peruvians, mostly peasants, depend on coca-growing for their livelihood.

Peru is surrounded by Ecuador, Colombia, Brazil, Bolivia, and Chile. It lost its valuable nitrate region in the south in the 1883 War of the Pacific with Chile and Bolivia. Jungles and forests in the humid sweep of the green, steaming Amazon Basin cover more than half the state. The Andes occupy another 27 percent (Nevado Huascaran is 22,133 feet high). There are deposits of oil, coal, phosphates, iron, gold, zinc, and copper. Peru is the archaeological capital of South America. The Incan capital of Cuzco, "the city of the sun," is on the altiplano 350 miles southeast of Lima, which was founded in 1535 by Francisco Pizarro, the Spanish adventurer. The country lies atop an area where the Pacific crust slides beneath the continent; it has been the epicenter of earthquakes with magnitudes of seven and higher. An avalanche in 1970 killed 18,000; an earthquake killed 66,794 and flattened about 186,000 buildings—the worst natural disaster ever in the western hemisphere. Towns were wiped off the face of the earth.

SURINAME, a Dutch colony before becoming an independent state in 1975, is a multiracial jungled country wedged between Guyana, French Guiana, and Brazil on the northeast coast of South America. The population of 401,000 (6 people per square mile) is remarkably cosmopolitan; it includes Javanese, Chinese, Creoles, East Indians

(37 percent), Amerindians, and blacks descended from escaped slaves. Largely unexplored hills cover most of the tropical state, which is somewhat smaller than Georgia. Contract laborers are imported from the Orient to help mine bauxite, the principal ore of aluminum; gold is prospected. Rum, timber, and coffee are also exported. Shrimp, rice, and fruits are harvested. The government is run by the military; there was a bloodless coup in December 1990. Insurgents called the Jungle Commondo are mostly descendants of African slaves who fled the plantations in the 18th century.

URUGUAY is the only South American state completely outside the tropical latitudes. The northern portion is about 30 degrees south of the equator. About the size of the state of Washington, the republic lies on the Atlantic Ocean and is separated from Argentina by the Uruguay River. Brazil, of which it was once a part, lies to the north.

Uruguay is blessed with rich alluvial plains, which produce wheat, tobacco, rice, olives, and grapes for wine making. Sheep and cattle graze on the grasslands and play a vital role in the economy, which benefits from a temperate climate, a significant hydroelectric potential, the largest planted forest area on the continent, beautiful beaches, a high standard of living, and progressive social-welfare programs.

Recent repressive military governments jailed and tortured thousands of political prisoners, earning one regime the dubious distinction of having more political prisoners than any other state on the planet. Democratic traditions returned in the mid-1980s. More than 90 percent of the population of 3 million (half live in the capital, Montevideo) is of Italian or Spanish heritage. Eight percent are mestizo.

VENEZUELA is South America's second-largest oil exporter, and an OPEC member. The world's largest oil reserves may be in the Orinoco tar belt. The international 200-nautical-mile law allows the tropical federal republic to exercise authority over a significant part of the western Caribbean Sea it borders; its geostrategic position is linked to the Caribbean as a whole. Among Venezuela's dependencies are the Margarita and Tortuga islands and several smaller island groups. The plains support a thriving cattle industry, and coffee and cocoa grow on the slopes of the highlands, a continuation of the Andes. Many of the vast, mysterious, and remote cloud-covered mountains are the remains of mighty sandstone plateaus.

Venezuela's inflation rate has reached 80 percent, and most of the country's food supply must be imported. There are a high birthrate and a sharply uneven distribution of wealth in the continent's richest country. The traffic-death rate is high: 31.1 per 100,000 people. Sixty-nine percent of the population of 19.3 million is mestizo, and 96 percent is Roman Catholic. Angel Falls is the world's highest waterfall (3,212 feet); it was discovered in 1910 and then rediscovered by the American bush pilot Jimmy Angel in 1935.

MIDDLE AMERICA: underdeveloped north-south microstates and an intercontinental land bridge dominated by Mexico and Costa Rica. "Turbulent" describes the region's politics, history, and probable future. Middle America has been over-seered by its northern "uncle," the United States, which has intervened on numerous occasions. There are numerous vacation islands in the Caribbean Sea, creations of European colonialists of the 16th and 17th centuries. Three fourths of the Caribbean people are descended, at least partially, from African slaves. In 1961, only three of the islands were independent: Cuba, Haiti, and the Dominican Republic. Earthquakes repeatedly destroy cities on the mainland.

ANTIGUA AND BARBUDA—a two-island constitutional monarchy independent of Great Britain since 1981—are dry, tropical, flat, partly volcanic (Antigua), partly coral (Barbuda) tracts comprising 171 square miles. They lie 26 miles apart in the eastern Caribbean Sea, and some 250 miles east of Puerto Rico. They have a combined population of 64,000; 90 percent are literate. Most citizens are descendants of African slaves who were imported by the British to work sugar cane plantations. Many of the slaves developed crafts or became fishermen and enjoyed more independence and freedom than the slaves on other Caribbean islands.

Eighteen percent of the land is arable, sugar is still produced and refined, droughts are commonplace. White-sand beaches on Antigua and pink-sand beaches on Barbuda invite tourism, which rivals cotton production in economic importance. Issues of significance on

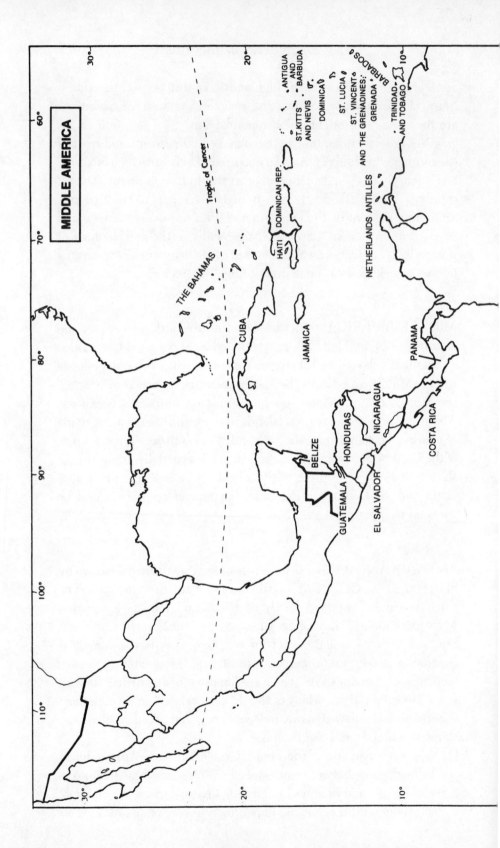

MIDDLE AMERICA

Tropic of Cancer

THE BAHAMAS

CUBA

HAITI

DOMINICAN REP.

JAMAICA

ANTIGUA AND BARBUDA

ST. KITTS AND NEVIS

DOMINICA

ST. LUCIA

ST. VINCENT

ST. VINCENT AND THE GRENADINES

GRENADA

BARBADOS

TRINIDAD AND TOBAGO

NETHERLANDS ANTILLES

BELIZE

GUATEMALA

HONDURAS

EL SALVADOR

NICARAGUA

COSTA RICA

PANAMA

"poor, inconsequential" Barbuda are said to be of little interest to Antigua and of no interest whatsoever to the rest of the Caribbean. Barbuda may be the only Caribbean island with a deer population; donkeys and feral cattle, guinea fowl and wild hogs also roam the bush. Columbus stopped by in 1493.

THE BAHAMAS are a tax haven about the size of Connecticut in the western Atlantic east of Florida—nearly 700 long, flat, coral, semitropical islands extending 760 miles northwest to southeast. Columbus discovered the Bahamas, which he thought were coastal islands near the Asian mainland. Only about 30 of the islands are inhabited, and only 2 percent of the land is arable. The highest point (206 feet) is Mount Alvernia, on Cat Island. About eighty-five percent of the population of 247,000 are descendants of slaves. The Bahamas were a British colony from 1783 to 1973, when they became an independent commonwealth. The principal resources are salt, timber, and aragonite. Tourism represents 70 percent of the GNP. A majority of the visitors come to gamble. The per capita income is $7,598. Nassau, an 18th-century rendezvous for pirates (notably Blackbeard), is the principal city and tourist resort.

BARBADOS is the generally level, coral, tiny easternmost island in the Lesser Antilles, West Indies. It has been independent from Great Britain since 1966 and is densely populated. Nearly 260,000 people live on only 166 square miles: 80 percent are of African descent, 70 percent are Anglican, 99 percent are literate. Tourism, sugar, rum, lime, and light industries support the economy.

BELIZE, once the base of Mayan civilization, and the last British possession on the American mainland (it was British Honduras until 1981), is a hot, humid, tropical, sandy-beached parliamentarian state slightly smaller than Vermont. British troops squelched neighboring Guatemalan territorial claims. A bridge between the countries of Central America and those of the "Carib" basin, Belize has the least population in Central America. Forty percent of the 176,000 Belizians are Creole, 33 percent are mestizo or mixed European/Indian. Six urban areas are tucked into a remote shore along the Caribbean Sea. It is dry and dusty in the north, ruggedly forested in the south. The economy has been dependent on forestry. Only a small portion of the arable land has been cultivated. Sugar is the chief export. The world's second-longest barrier reef lies in the sea from 8 to 25 miles offshore. Per capita income is $1,000.

COSTA RICA, which hasn't had an army since the 1950s, is largely an agricultural Central American republic, smaller in area than West Virginia and free from Spanish rule since 1838. Land ownership is widespread, and relative harmony reigns among all social classes. Unique for the region as well is the 90 percent literacy rate. Because the military is not the ultimate arbiter of politics, arguments are settled constitutionally instead of at the end of a rifle.

Because of its mountainous terrain and aversion to war, Costa Rica is often compared with Switzerland. The lofty central area is bordered by coastal plains on both the Atlantic and Pacific coasts. Cerro Chirripó is 12,530 feet high. The Caribbean lowlands are tropical. Coffee, the chief export, and bananas drive a generally prosperous economy, which also includes the manufacture of fiberglass, fertilizer, roofing, cement, hemp, cotton, and sugar.

In comparison with other countries in the region, there is a relatively large population of European descent. Ninety-six percent of the 3 million Costa Ricans are white or mestizo, 93 percent are Roman Catholic. Life expectancy is high for a Central American state: 71.9 years for women, 67.5 years for men. President Oscar Arias Sánchez provided enlightened and progressive leadership. He was awarded the Nobel Peace Prize for finding a non-violent path to peace in Nicaragua.

CUBA, which is 90 air miles southeast of Key West, Florida, has been ruled by authoritarian governments for more than five decades. After the Spanish lost the Spanish-American War in 1898, the 746-mile-long island became virtually a protectorate of the United States until 1934. Fidel Castro overthrew the corrupt, oppressive Batista regime in 1959 and shifted the Caribbean state from inequitable capitalism to dogmatic Marxism, with support from the Soviet Union. The United States deflected Castro's early entreaties, cut off diplomatic relations, and imposed a tight economic blockade. To this day, however, it maintains a naval and marine base at Guantánamo, on the southeast coast of the island.

Cuba is the westernmost of the West Indies and about as large as Pennsylvania. In 1961, the United States intervened disastrously when Cuban rebels, with CIA sponsorship, tried to overthrow Castro. The United States has tried at least seven times to assassinate the Cuban leader. In 1962, the Soviet Union's plot to install long-range ballistic missiles was thwarted by Washington, which sees the island as "our Antilles heel." Until late 1991, the Soviet Union's forward bases—"training brigades"—here projected power into the hemisphere, notably in the civil wars in Nicaragua and El Salvador. (President Thomas Jefferson had the opportunity to buy Cuba but didn't have the money for both it and the Lewis and Clark expedition into northwest America. In 1854, there were discussions to purchase Cuba from Spain.)

The Cuban economy has deteriorated. The ex-Soviet Union, with the demise of Communism at home, informed Castro that it would be removing its troops and all but eliminating economic assistance. Food production began to stagnate in 1986 when Castro abolished free peasant markets. In late 1990, desk workers were ordered from the cities, which have become crumbling skeletons, to state plantations to help increase food production. The ever-tighter austerity has been likened to wartime hardships. Fuel shortages are severe, 100,000 Chinese-made bicycles have replaced automobiles, power shortages are frequent, and rationing has been instituted. The eccentric autocratic leader clings to a crumbling ideology.

Low hills and fertile valleys cover more than half the country, and 29 percent of the land is arable. In pre-Castro years, United States companies prospered from Cuban sugar production. Sugar, still the number-one export, was swapped for 13 million tons of Soviet oil yearly. Connoisseurs still rate the Cuban cigar, whose sale is illegal in the United States, as the most distinguished in the world. A nuclear-power program is being accelerated. There is free education and free health care. Nearly every one of the 10.5 million Cubans is literate, and the infant mortality rate has been lowered dramatically. More than 100,000 Cubans have emigrated to the United States. (Cuban troops intervened in Angolan and Ethiopian wars.) The capital of Havana (population: 2 million) is the largest city in the West Indies. In an unexpected about-face, the government is encouraging tourism, now the fastest-growing industry. For decades the United States prohibited its citizens from visiting Cuba.

DOMINICA, the last Caribbean island to be colonized and the region's most mountainous and wettest land, is a tiny, hurricane-assaulted, volcanic, tropical tract about a quarter the size of Rhode Island and the most northerly of the Windward Islands. It is a paradise of nature, densely rain-forested, which limits to 23 percent the land accessible to cultivation. There are 365 rivers, and there is an average of 300 inches of rain annually. The mountains are high enough to create wet, oceanic weather. The parliamentary democracy was a British possession for nearly two centuries, until 1978. Ninety-one percent of the population of 83,000 is African or mulatto, 80 percent is Roman Catholic, 80 percent is literate. The economy, which is poor, depends primarily on bananas and on tourism, pumice, citrus, and coconuts. In 1979, Hurricane David devastated Dominica, which had been christened by Columbus in 1493 for the Lord's Day.

DOMINICAN REPUBLIC, which was founded nearly five centuries ago and has been controlled by dictatorial regimes, is the largest and most populous democracy in the Caribbean. It occupies the eastern two thirds of the West Indies island of Hispaniola, where the Spanish by 1502 had begun to import Africans as slaves. (The western third is Haiti, from which the Dominican Republic became independent in 1844.)

The state, which is about the size of Vermont and New Hampshire combined, is bisected by a range of mountains. The geologic extremes are Pico Duarte, which is 10,417 feet high, and Lago Enriquilla, which is 131 feet below sea level. Sugar, cocoa, coffee, and tobacco are produced, and tourism is a fast-growing industry. Lack of irrigation is a major problem. Unemployment is high, and the rate of poverty is exceeded in the western hemisphere only by Haiti and Bolivia. Haitian children are kidnapped and forced to work in slavelike conditions in fields here. In periods of political instability, the Dominican Republic has been occupied by the United States military.

Seventy-three percent of the population of 7.1 million is mulatto, and 95 percent is Roman Catholic. Tourism is increasing, thanks to bargain prices and some of the world's most beautiful beaches. The capital, Santo Domingo, is the oldest city (1496) that was established by Europeans in the New World. Dominican Republic also has the hemisphere's oldest university (1538) and archbishopric (1547). Columbus established a home here, and many historians believe his bones are interred in the Santo Domingo Cathedral.

EL SALVADOR, which was discovered by the Spanish in 1523, is the smallest and the most densely populated republic in Central America. Of the seven Central American countries, only El Salvador lacks a Caribbean coastline. All but 10 percent of the state, which is the size of Massachusetts and has a population of 5.5 million (90 percent mestizo), is of volcanic origin.

For more than a decade, the United States provided military aid and a million dollars a day to the government's war against insurgents, who were bolstered by Cuba and neighboring Nicaragua. (The guerrillas deployed advanced weapons systems and inflicted billions of dollars in damage.) The Salvadoran military, for its part, was impervious to human-rights reforms; with impunity, government-sponsored death squads murdered at least 70,000 civilians. Raging class hatred and extreme social injustice fueled the civil war in the most unspeakably brutal theater of East-West conflict in the hemisphere.

El Salvador has a high Pacific coastal plain; it rises to a cooler plateau and a valley region conducive to grazing. The mountainous north has many volcanoes. The capital, San Salvador, has been destroyed by earthquakes nine times since 1700.

GRENADA is the smallest state in the western hemisphere—its 133 square miles are twice the area of Washington, D.C. It is a mountainous, stunningly beautiful eastern Caribbean island 90 miles northeast of Venezuela, and includes the islands of the southern Grenadines north of Little Martinique. Rainfall is heavy, the volcanic soils are rich, agriculture is the largest revenue earner. Grenada was discovered by Columbus in 1498, and it has been called the Isle of Spice.

In 1974, Grenada gained full independence from Great Britain amid the turmoil of a general strike. Five years later, the ideologically fanatic People's Revolutionary Government took command. That government toppled ally Prime Minister Maurice Bishop's popular, revolutionary Marxist-Leninist regime—the New Jewel Movement—prompting an invasion by the United States military and a token force from six Caribbean-area countries; their "rescue mission," it was claimed, aborted a Soviet-Cuban plan to subvert the Caribbean. The government today is democratic.

Eighty-two percent of the population of 84,000 is black, descendants from African slaves who worked indigo and sugar plantations and generated huge profits. Per capita income is $500. Unemployment tends to run high. Tourism helps to support the shaky economy; rum, cocoa, mace, and nutmeg are staples.

GUATEMALA is home to about a third of all the people of Central America and is the only country in North America with a predominantly Indian population. It has remarkably fertile volcanic soil, a cosmopolitan middle class, the largest manufacturing base on the isthmus—and poverty as grinding as any to be found throughout all of Latin America. So severe is the poverty that thousands of children are homeless and hungry and living in the streets, where they are often subjected to brutal treatment by the police. Guatemala has never lived up to its potential, because the state, which is the size of Ohio, has been governed by brutal, authoritarian dictatorships that use security forces as instruments of intimidation and terror. Since a CIA-backed coup in 1954, the security forces have been sponsored and trained by the United States. There have been a hundred thousand political murders. A guerrilla war with leftist rebels has been going on for decades. The violence has driven Guatemalans to emigrate to Mexico. A civilian president was elected, but the military exercises the real power, terrorizing with impunity. Guns are everywhere; they even appear on the national seal and coins.

The capital of Guatemala City lies in a level valley deeply eroded by ravines. The population of nearly 10 million is concentrated in the central highlands of the mountainous republic, which, west-east, is between the Pacific Ocean and the Gulf of Honduras and, north-south, is between Mexico, Belize, Honduras, and El Salvador. Guatemala City has the largest urban population (750,000) between Medellín, Colombia, and Mexico City. The northern rain forests and grasslands are sparsely populated and largely undeveloped. The southern plains are fertile. Much of the Pacific rain forest has been replaced by coffee, cotton, and sugar cane plantations. A program to plant 52 million trees to improve the ecology is under way. Sixteen percent of the land is arable. A wealthy few own half the farm acreage. The United Fruit Company was the largest property owner before much of its land was seized under an agrarian reform act. Billions of bananas are exported every year. The annual population growth is nearly 3 percent. Guatemala is a country of illiterates; nearly half the adults cannot read or write. It is a violent geologic region—there are more than 30 volcanoes. In 1976, there was a devastating earthquake. Guatemala was once the center of the Mayan empire. Claims on neighboring Belize were given up in 1985.

HAITI is the world's oldest black republic and the poorest country in the western hemisphere. It was born of a 13-year slave revolt, led by Toussaint L'Ouverture, which resulted in separation from France in 1804. The second independent country in the New World, Haiti is today a Third World state with a history of exploitation and terror. For much of this century, government policy has been one of repression, corruption, and human-rights abuses. (The United States occupied Haiti from 1915 to 1934.) The 30-year dictatorship from 1957 of the Duvaliers—father and then son—was secured by the brutality of paramilitary secret-police death squads, the Tonton Macoute, and by the army ("a kind of Mafia gang"). The president elected in 1990 had a mandate for radical change. A military coup in the fall of 1991 threw the country into chaos.

Haiti, which is about the size of Maryland, is the western third of the tropical Caribbean island of Hispaniola, in the West Indies. (The Dominican Republic occupies the eastern two thirds.) The land is as impoverished as the people. It is one-third semiarid, two-thirds mountainous. The influence of France persists. Most Haitians speak a patois. Blacks, descendants of slaves, make up about 95 percent of the population of 6.3 million. Mulattoes have traditionally formed the elite of government circles. The literacy rate is 23 percent. Unemployment estimates range up to 50 percent, and the annual rural income averages $60. People are the principal export, followed by coffee. Voodooism continues to be practiced in Haiti. Tourists have always been attracted to the vibrant and exotic land. Much art of a high caliber, employing vivid primary colors, is created here.

HONDURAS is a Central American democratic, constitutional republic that was once part of the Mayan empire. It borders on Nicaragua, Salvador, and Guatemala and has coastlines running 500 miles on the Caribbean Sea and 40 miles on the Pacific Ocean. Swamps and jungles form part of the so-called Mosquito Coast. Honduras is mainly mountainous and forested, with some fertile valleys; its chief export is bananas. Coal, lead, silver, cigars, and beans are also produced, but the economy remains underdeveloped.

Honduras was visited by Columbus in 1502. It won independence from Spain in 1821. Great Britain controlled the country until 1965, when a new constitution was framed. Ninety percent of the people are mestizo, and 97 percent are Roman Catholic. Honduras, which is about the size of Tennessee, has been a base for United States military efforts combating Communism in the region.

JAMAICA, an island in the West Indies slightly smaller than Connecticut, was discovered in 1494 by Christopher Columbus—"the fairest isle that eyes have beheld"—and settled in 1509 by Diego Columbus. Its immediate neighbors are Cuba and Haiti. Following attacks by the English, Jamaica was ceded by Spain to Great Britain in 1670. It was the largest British colony in the Caribbean. Independence came in 1962. A great many slaves were imported from Africa to work the sugar plantations. Seventy-six percent of the 2.5 million Jamaicans are of African descent. Sugar no longer drives the economy, which has experimented with socialism.

The "land of wood and water" is poor, largely mountainous, with deciduous forests covering nearly half the island; slightly less than a quarter of the land is arable. Jamaica is a major producer of bauxite, the reddish ore that is the principal source of aluminum. Jamaican rum is celebrated the world over. Marijuana is grown for trading underground in the United States. There are vast reservoirs of submerged peat awaiting exploitation. The burgeoning tourist business, stimulated by sparkling beaches, waterfalls, and unique flowering plants, is often set back by eruptions of social unrest. The literacy rate is 73 percent. Religious sects are prominently represented by the Rastafarian. In 1692, an earthquake dumped most of the large English town of Port Royal, the "buccaneer capital," into the sea and drowned thousands.

NICARAGUA, a tropical agricultural country the size of Iowa, and once the breadbasket of the region, has had as stormy a history as any Latin American state. In recent years, there have been assassinations, the treasury has been plundered, and political upheavals were influenced by the Soviet Union, Cuba, and the United States, which had occupied Nicaragua from 1926 to 1933. For four decades, the United States maintained the corrupt and repressive regime of dictator Anastasio Somoza Debayle, who was overthrown in July 1979. In the 1980s, the United States supported the contra rebels in their civil war with the Marxist Sandinista government.

The peaceful presidential victory in 1990 of Violeta Barrios de Chamorro removed the Sandinistas from power. She took a conciliatory "healing" approach to the opposition leadership—the military is still Sandinista-dominated—and therefore has been besieged on all sides. In 1991, the economy was in disarray, there were world-record inflationary rates, and the foreign debt was the largest per capita in the world. Military spending continued at high levels.

The Central American republic lies to the north and northwest of Honduras, to the west of the Caribbean, to the north of Costa Rica, and to the northeast of the Pacific Ocean. The highlands in the northwest are home to cattle. Some gold is mined. Ten percent of the land is arable, and bananas are the major crop. The 300-mile-long Mosquito Coast on the Caribbean is a sparsely populated area of swamps, pine forests, tropical jungles, and few roads. More than 9 out of 10 of the 3.6 million people are Roman Catholic; about 70 percent are mestizo. The capital is earthquake-prone Managua, celebrated in a popular song as "a wonderful place." Nicaragua, which gained its independence from Spain in 1821, has the longest, lowest geological gap in the hemisphere, which prompted engineers at the turn of the century to consider it as the western part of a canal between the oceans.

PANAMA, the narrow waistline of the western hemisphere, is a tropical isthmus, slightly larger than West Virginia, connecting in a low saddle the mountain chains of Central America and South America. It has become a strategic center and an international transfer point for trade and transportation since the opening of the United States–completed Panama Canal in 1914. It is also a smuggler's dream. The coasts are punctuated with coves, inlets, and bays, and there are more than 1,500 islands and a dense rain forest. The United States controls the Canal Zone, the "funnel of world commerce," five miles on either side of the 40.3-mile-long waterway, whose annual capacity is 25,550 ships.

Prior to construction of the canal, Panama was part of Colombia; the United States paid Colombia for the separation. In 1977, the United States agreed to turn over operation of the canal to the republic at the end of the century. Military and civilian clashes and a United States–imposed economic boycott preceded the 1989 U.S. invasion— Operation Just Cause—to capture one man, the dictator and alleged drug kingpin General Manuel Antonio Noriega, and to install a democratic government friendly to the United States. But since the imprisonment in Florida of General Noriega, the cocaine traffic through Panama has flourished at record levels.

S-curved, the pint-sized state stretches east and west for 400 miles and varies in width from 30 to 120 miles. The Pacific end of the canal, which gets 69 inches of rainfall a year, is 27 miles east of the Caribbean terminus, which averages 130 inches of rainfall annually. The Caribbean Sea is on the north, the Pacific Ocean is on the south; the waters abound in fish and shrimp. Trade winds help to keep the temperature even and bearable. Two curving mountain spines run the length of the isthmus. Volcán Baru is 11,401 feet high. Less than a quarter of the land is arable, and only half of that is of even marginal utility for intensive agriculture. Dense rain forests are in the east, and a thin forest is on the Atlantic coast between Costa Rica and Colombia. (A century and a half ago, the country was 92 percent forested; it is less than 40 percent today.)

Portobelo, the "beautiful port," was once the greatest trade crossroads in the New World. More than 100 international banks are currently based in Panama, and more than 12,000 ships fly the Panamanian flag. Most of the population of 2.4 million is mestizo and Roman Catholic. The language is Spanish. The leading export is bananas, which carpet much of the northwest with green. In Panama in 1513, Balboa gazed upon the Pacific Ocean—the first European to do so: the vista established the fact that the Americas were not part of Asia.

SAINT KITTS AND NEVIS, two tropical, scenic, small Leeward Islands set two miles apart east of Puerto Rico in the Caribbean Sea, gained independence from the United Kingdom in 1983. St. Kitts was settled in 1623 and is known as the Mother Colony of the West Indies. Ninety-five percent of the 40,000 who inhabit the two islands are black, and an extraordinarily high proportion (98 percent) is literate. The economic underpinnings are sugar, sea-island cotton, and tourism. (The reserves of phosphates were mined out.) Thirty-three percent of the people are Anglican, 29 percent are Methodist.

SAINT LUCIA, the mountainous, second-largest of the Windward Islands in the tropical eastern Caribbean Sea, is one fifth the size of Rhode Island. It gained independence from Great Britain in 1979 and enjoys relative political stability. Plans to bolster the agricultural economy have included a free-trade zone and an oil-transshipment terminal. Chief exports are bananas and cocoa. Ninety percent of the population of 150,000 is African–West Indian and Roman Catholic.

SAINT VINCENT AND THE GRENADINES are part of the hurricane-lashed Windward Islands chain in the tropical West Indies in the eastern Caribbean Sea. They are tiny (together, only twice the size of Washington, D.C.), and half the land is arable. Mountainous, volcanic, thickly wooded St. Vincent is the main island and a tourist attraction, but it is one of the poorest Caribbean states. The eruption of Mount Soufrière in 1979 was ruinous. The Grenadines, to the north, consist of hundreds of small islands. The population of 105,000 are mainly descended from African slaves. Forty-two percent are Anglican. Bananas represent 62 percent of the exports. Per capita income is $920, and the literacy rate is 96 percent.

TRINIDAD AND TOBAGO form a prosperous oil-exporting two-island republic about the size of Delaware, the only Caribbean state with substantial petroleum; they also refine and export Middle Eastern oil. The islands are 20 miles apart. Trinidad, separated from Venezuela by the seven-mile Gulf of Paria, was discovered by Columbus in 1498 but did not attract Spanish colonization, because there was no gold. Dutch, French, and English buccaneers raided the island, and in 1802 it was ceded by Spain to England. In 1962, Trinidad and Tobago became independent members of the British Commonwealth.

Tobago lies north of Trinidad and is a mountain ridge heavily forested with hardwoods. The racially diverse island is the largest of the Lesser Antilles chain and has the third-largest per capita income (around $7,000) in the western hemisphere. Tobago subsists mainly on agriculture and tourism. Trinidad's Pitch Lake yields an apparently inexhaustible supply of natural asphalt. Forty-one percent of the population of 1.2 million is East Indian, descendants of slaves brought by the British to work the sugar cane plantations. Many young Trinidadians have pursued their education in England, and some of them have had distinguished literary careers.

NETHERLANDS ANTILLES (a non-state), an integral, autonomous part of the Netherlands realm, comprises two groups of islands (310 square miles) between Puerto Rico and Venezuela in the West Indies in the Caribbean Sea. The group's colonial status was abolished in 1954, and Aruba was separated out in 1986. Tourism and refining crude oil from Venezuela are the principal industries, and corn, salt, and phosphates are the major resources. The prosperous, racially mixed population is 176,000. Curaçao is the largest of the islands, and, like Aruba and St. Martin, is a favorite of tourists. During the Second World War, oil refineries on Curaçao and Aruba were shelled by Nazi submarines.

NORTH AMERICA: the third-largest continent, and the one with the largest coastline, extends north from Central America. It is 4,000 miles wide in the far north, which is ice covered, but only 31 miles wide in the far south, which is thickly forested and sometimes steamy. Population density decreases south to north. Mexico City expects to have 25 million people presently; it is one of the world's most polluted cities. The United States has ten times the population of Canada, the largest country in North America and the second-largest country in the world. The highest point on the continent is Mount McKinley, Alaska, 20,320 feet; the lowest is Badwater, in Death Valley, California, 282 feet below sea level. The United States has high income and high consumption, devouring raw materials and energy at a very rapid rate; millions still yearn to live here.

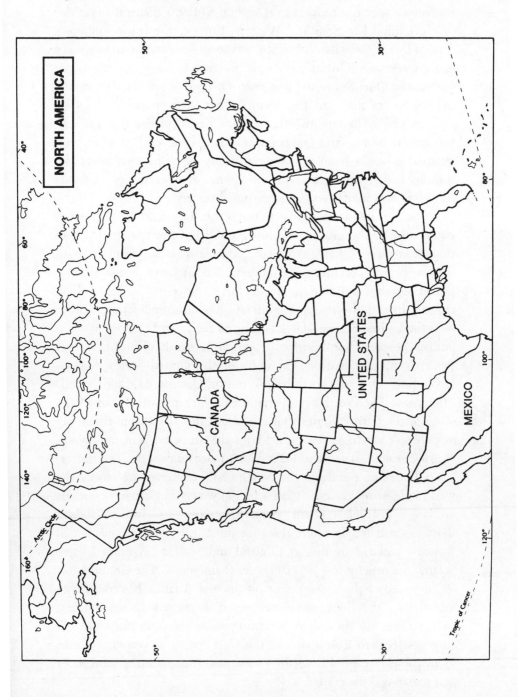

CANADA, scrubbed for eons by continental glaciations, is the second-largest country in the world, extending some 4,000 miles from the Atlantic Ocean to the Pacific Ocean. It is also one of the most prosperous: per capita income is around $13,500. With historic ties to the United Kingdom and Western Europe, the confederation—formed by the consensus of public servants—consists of 10 provinces and 2 territories. Ontario, the only province to border all five Great Lakes, and Quebec are the two most populous provinces, supporting the opinion of the western provinces that their concerns are of no interest to Parliament in Ottawa, back east. Quebec is a French-speaking province, and Quebecis have long believed they have been mistreated by the country's Anglo majority; they threaten to secede, wanting to be recognized as a francophone "distinct society." Canada has never had a civil war or widespread slavery.

Canada, like its neighbor to the south, is a great melting pot of nationalities. The city of Winnipeg publishes newspapers in more than 20 languages. Large groups of Ukrainians can be found throughout the Prairie provinces of Manitoba, Saskatchewan, and Alberta. Canada extends well into the Arctic, but 90 percent of its 26.4 million citizens live in the temperate climate near the undefended border with the United States, which is crossed each year by more than 100 million people. (Southernmost Canada is south of northernmost Pennsylvania and within 138 miles of the United States' Mason-Dixon Line. One goes north from Windsor, Ontario, to enter Detroit, Michigan.) The Northwest Territories is larger in area than all the states of the United States that are east of the Mississippi River, yet more people live in one east coast United States city, Winston-Salem, North Carolina, than in all 1.3 million square miles of the Territories.

Chunks of granite in the Northwest Territories are among the world's oldest rocks: more than 4 billion years. The Barren Grounds are low-level treeless plains, home to trappers and Inuits. Hudson Bay was squashed by a mile-thick ice sheet. Ellesmere Island, which is nearly as large in area as England and Scotland combined, is so hostile, the nearest tree is 2,100 miles to the south. The Grand Banks off the coast of the easternmost province of Atlantic-battered Newfoundland ("the Rock") long provided fishermen with dependable, rich catches, but the fish are thinning out in the cold, plankton-rich currents. British Columbia, on the west coast, is booming, thanks to infusions of capital from the Orient and the growth of Vancouver as a movie-making center.

Canada is a major wheat producer. In the west-central region is one of the world's largest oil reservoirs. One of the richest nickel deposits lies in the Canadian Shield, a mantle of mineral-rich Precambrian rock, which angles across northern Manitoba and Saskatchewan and nips the corner of Alberta. The most magnificent resource is a third of the world's supply of fresh water. (But the vast quantities in the barren north that melt each year spill uselessly into the Arctic and Pacific oceans.) "The Great Lone Land" suffers blistering summer heat and exceptionally harsh winter cold. In December 1991, Canada agreed to give Inuits political domain over a fifth of the country—a 750,000-square-mile territory extending north from Hudson Bay to the Arctic Ocean and Greenland: Nunavut—"our land."

Canada is the United States' largest trading partner. The two giants have a free-trade pact and have initiated the equivalent of a no-trade-barrier common market, which could be institutionalized by the turn of the century. Canada has a sense of its own nationalism quite separate from that of the United States. But meanwhile, many Canadian firms still relocate to the U.S. to take advantage of lower taxes and cheaper labor in the States; individual Canadians cross the border to stock up on cheaper merchandise. High unemployment rates have persisted countrywide for several years.

MEXICO is the land of Aztec, Mayan, and Iberian cultures. Its popular, pulsating, mountain-ringed capital city of Mexico City (the capital for six centuries, and home to at least 15 million residents) experienced in 1991 nearly four times the maximum acrid air-pollution limit; immediate closing of the city's largest government-operated oil refinery was ordered; children could not leave for school until the morning motor-vehicular rush hour was over and the air was not as fouled, and they had to be homeward bound well before the evening rush hour began. Breathing the air in the capital has been like smoking two packs of cigarettes a day. The city is trapped by unfavorable landscape and explosive population growth. For the past two decades, the government has actively encouraged a program of reducing the birth rate. Between 1973 and 1986, the average number of children per woman decreased from 6.3 to 3.8.

Mexico, a rising regional power, is Latin America's largest oil exporter and it has vast oil reserves. When oil prices collapsed in 1982, it became the developing world's most indebted country. During the current austerity program and fragile economic recovery, about half the population of 86.5 million languishes in poverty. About 44 million Mexicans are without sewers, about 21 million are without potable water. Japan, using Mexico's vast, low-cost labor supply to assemble products bound for United States consumers, has established a manufacturing beachhead in northern Mexico, which is becoming a booming industrial center, with nearly 1,000 assembly plants. Hard times have forced tens of thousands of peasants in the rural sections of the 31 states and the federal district (Mexico City) into the cultivation of marijuana and opium poppy.

Originally known as New Spain, the United Mexican States ranges from tropical jungles to desert plains. Most Mexicans live on the temperate central plateau, which is bounded by rugged mountains. Orizaba, at 18,700 feet, is the third-highest mountain in North America. There is a lack of sufficient rainfall. Immediately south of its "gringo" neighbors—California, Arizona, New Mexico, and Texas—this North American state has roots in the rich soil of an Indian past. It gave birth to great civilizations, such as the Olmec, Maya, Toltec, Mixtec, Zapotec, and Aztec. Their ruins are a formidable tourist attraction, as are many scenic wonders and splendid beach resorts. North America's largest rain forest covers more than half the state of Chiapas, which is about the size of Connecticut. Mexico is three times as large as Texas.

THE UNITED STATES, born of rebellion and the cult of independence, "the land of the free and the home of the brave," the arsenal of democracy and the world's richest country, unleashed the first nuclear bombs and landed 12 men on the Moon. Protected by its geographic isolation, the practically self-reliant superpower extends across the North American continent and from the 49th parallel in the north (the Canadian border) to the Gulf of Mexico in the south. And there are two of the 50 states that are not contiguous, not part of the Lower 48. Waterways, 30,000 kinds of soil, natural resources, and a temperate climate are the underpinnings of the prosperity; the great midwestern grain belt is blessed with deep layers of glacial till and wind-deposited silt.

One of the 50 states (the Hawaiian Islands, which were once on the bottom of the ocean) lies in the Pacific 2,000 miles west of

California, and another in the far northwest (Alaska) is separated by western Canada from the 48 contiguous states. Earth's most massive mountain is Mauna Loa on the Big Island of Hawaii. Thanks to two lengthy unimpeded shorelines, the Navy has immediate access to the world's sea lanes. Twenty million acres have been paved over with roads, but the United States is still the world's richest farm country. The United States–Canadian border is the longest undefended border anywhere: 5,527 miles. An electrified 1,000-mile-long fence—the "Tortilla Curtain"—has been planned for the border—La Línea—with Mexico to fend off illegal aliens.

Among the world's developed countries, the United States during the Reagan presidency went from leading credit extender to foremost debtor, and strangled on a trillion-dollar deficit. Foreign-aid spending accounts for only 1 percent of the federal budget; 5 percent is still spent on the military. To head off Soviet expansion, the United States established military alliances around the globe. Since the Monroe Doctrine (1823), Uncle Sam has been the policeman of Latin America. With only 5 percent of the world's population, the United States uses about 45 percent of its energy. It produces about 25 percent of the world's goods and services, and the most garbage (135 million tons annually). Per capita GNP is $21,737. It is the world's "breadbasket" and leading producer of corn and exporter of soybean products. Immense natural resources include large reserves of uranium. By the rule of natural size, the United States is "entitled" to around 18 percent of the world's wealth and power; in its heyday, it reportedly enjoyed an extraordinary 40 percent. Americans drive 135 million automobiles.

For a century, it has maintained a sometimes open-door policy toward the fearful, the oppressed, and the impoverished, and introduced human rights as a political point of reference. Millions, mostly from Eastern Europe, immigrated around the turn of the century and in the early decades of the 20th. In recent times, immigration has been the heaviest among Latin Americans and Asians, principally Filipinos, Koreans, and Chinese. About a million foreigners enter legally or illegally each year. There are upwards of 200,000 Arabs in and around Detroit, Michigan. Nearly one in every four Americans claims African, Asian, Hispanic, or Native American ancestry. By 1995, one third of American public-school pupils will be from minorities.

Eighty-five percent of the 250 million Americans are white,

nearly 12 percent are black. The life expectancy of blacks is the lowest of all groups. Twelve percent of American adults are functional illiterates; their skills do not go beyond the fourth grade. Once predominantly an agrarian society, only 24 percent of the people now live in rural areas. The average American moves his residence 18 times in his lifetime, and the average family watches 49 hours of television every week—2,548 hours a year! More than 33 million Americans were living in poverty in 1990, and the median household income was still in decline. Most Americans who have jobs say they are satisfied with them. The work week has increased to 46.8 hours, and leisure time has plummeted since the sixties, when middle-class Americans last saw tangible improvements in their incomes. Around 27 million Americans rely on food stamps to put food on the table. Every day, 7 percent of all Americans eat at McDonald's restaurants.

The United States has a frightful crime record; citizens have an unlimited access to firearms, and they use them. There were 24,000 murders in 1991 and Washington, D.C., was again the "murder capital." (The country is the leading arms exporter.) It is one of two developed countries (the Republic of South Africa is the other) that denies its citizens a national health service, and it lags behind many countries in infant mortality rates. Nearly a fourth of the 3.9 million American women who gave birth in 1990 were not married. Young people receive educations inferior to that of children in many other developed countries, and 5.5 million children are hungry because of their family's limited economic resources. Since the 1960s, American children have been more likely to be fat, to commit suicide, to be murdered, or to get low scores on standardized tests. The United States, which plants 6 million trees every day, has the highest greenhouse-gas emissions. New York City, the country's largest city, contains the world's largest penal colony. Life expectancy for women is 78.6 years; for men 71.6 years—an unusual discrepancy influenced by the high death rate of black males at an early age because of drug abuse and homicide. A self-governing territory is Puerto Rico, which shares citizenship with mainland residents; a large contingent of Puerto Ricans wants statehood. The United States Virgin Islands, a dependency, were purchased from Denmark during the First World War as a guard to the strategic entrance to the Caribbean.

INDEX

Only the most significant references and subentries are listed.

ABOUT THE AUTHORS

GEORGE J. DEMKO is a professor of geography and the director of the Nelson A. Rockefeller Center for the Social Sciences at Dartmouth College. He received his Ph.D. at Pennsylvania State University and has held faculty positions at Ohio State University and the University of Virginia. He has been a visiting professor at Moscow University and at Comenius University in Czechoslovakia, and has conducted research extensively in Eastern Europe and the former Soviet Union. He is a specialist on population and political geography and on spatial policies in centrally planned systems. He has written 10 books and more than 50 articles on issues ranging from spatial inequality in Eastern Europe to environmental problems and policies in the U.S.S.R. His most recent book, published in English and Russian, is *The Art and Science of Geography: A Soviet/American Perspective*. For the five years before his appointment to Dartmouth, Dr. Demko was Director of the United States Office of The Geographer, in The Department of State.

JEROME AGEL has written and produced 40 books, including collaborations with Carl Sagan, Marshall McLuhan, Stanley Kubrick, Herman Kahn, Isaac Asimov, and R. Buckminster Fuller. His new works include *Cleopatra's Nose. . . ., Dr. Grammar's Writes from Wrongs,* and (with Richard B. Bernstein) *Amending America: If We Love the Constitution So Much, Why Do We Keep Trying to Change It?*

EUGENE BOE has written and coauthored more than a score of books, including *Dr. Cott's Help for Your Learning-Disabled Child, Fasting: The Ultimate Diet, The Rutgers Guide to Lowering Your Cholesterol,* and (with Joan Rivers) *Having a Baby Can Be a Scream.* He wrote with Jerome Agel the nonfiction novels *Deliverance in Shanghai* and *22 Fires.*